Membranes, Metabolism, and Dry Organisms

A SPECTRUM OF DESICCATION

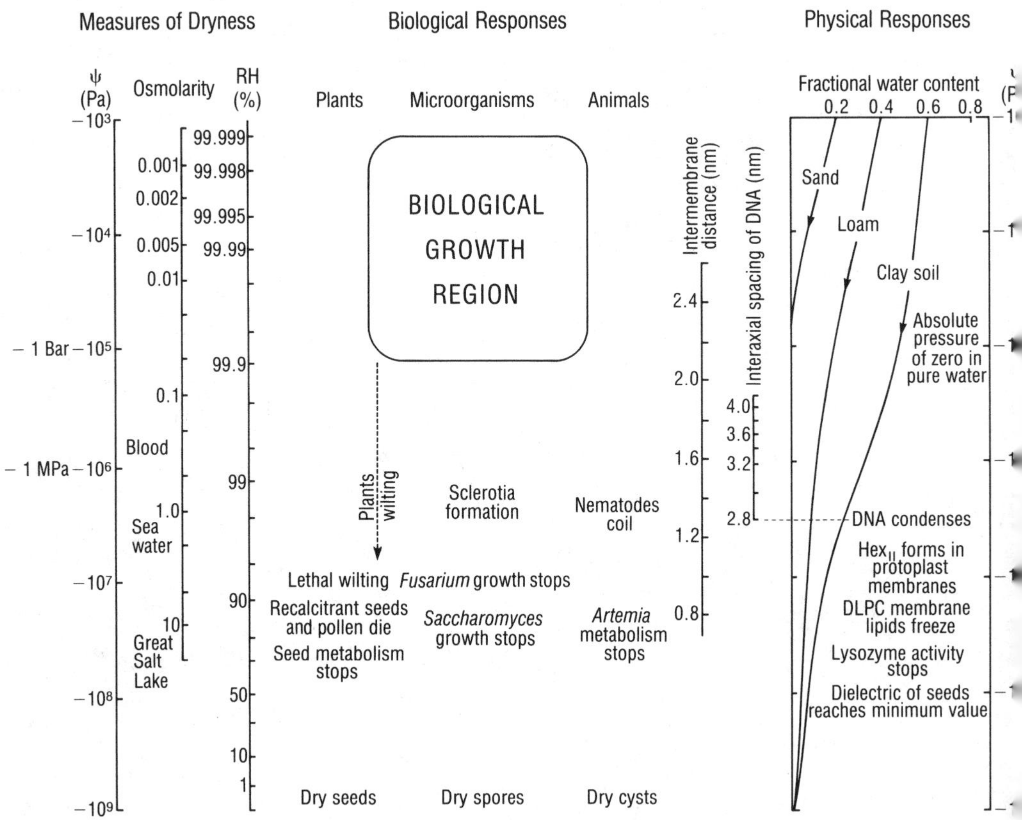

This diagram depicts representative physical, biochemical, and physiological changes associated with desiccation—the topics of this book. The three vertical scales on the left present the physical parameters of desiccation. The first is water potential (ψ), defined as the chemical potential of water divided by the molar volume at atmospheric pressure; it is given in Pascals. The next scale is osmolarity; the osmolarities of blood, sea water, and the water in the Great Salt Lake serve as reference points. The third scale is relative humidity (RH), or water activity (A_w); the values are entered on a log log scale.

In the middle and right-hand sections of the diagram, some biological and physical changes are entered at the appropriate values on the dryness scales. The biological changes shown include wilting of plants, death of recalcitrant pollen (discussed in Ch. 6) and of recalcitrant seeds (Ch. 20), and the termination of metabolism in orthodox seeds (Chs. 2 and 3). The values at which some representative microorganisms stop growing are depicted, as are the values at which nematodes become coiled (Ch. 9) and *Artemia* metabolism ceases (Ch. 10).

The physical changes illustrated include the decrease in distances between membrane layers, to the point at which membranes collapse or form a Hex_{II} configuration (Chs. 17 and 18), the folding of DNA molecules into liquid crystals (Ch. 18), the freezing of membrane lipids (Ch. 18), the cessation of the enzymic activity of lysozymes (Ch. 16), and the point below which the dielectric reaches a minimum in wheat seeds (Ch. 16). The changes in water potential of three types of soils during drying are also shown.

This diagram was designed by Joe Wolfe and A. Carl Leopold.

Membranes, Metabolism, and Dry Organisms

A. Carl Leopold, Editor

W. C. Crocker Scientist
Boyce Thompson Institute for Plant Research

Comstock Publishing Associates
A division of Cornell University Press
Ithaca and London

Library of Congress Cataloging-in-Publication Data

Membranes, metabolism, and dry organisms.

Bibliography: p.
Includes index.
1. Anhydrobiosis. 2. Membranes (Biology)
3. Metabolism. I. Leopold, A. Carl (Aldo Carl),
1919–
QH524.M46 1986 574.1 86-47646
ISBN 0-8014-1979-4 (alk. paper)

First published 1986 by Cornell University Press.

This book was typeset from author-prepared disks.

Printed in the United States of America

The paper in this book is acid-free and meets the guidelines for permanence and durability of the Committee on Production Guidelines for Book Longevity of the Council on Library Resources.

Contents

The Conference on Anhydrous Biology

The conference was held at Villa Serbelloni, Bellagio, Italy, October 21–25, 1985, under the auspices of The Rockefeller Foundation.

PLANNING COMMITTEE

A. Carl Leopold, *Conference Director*
James S. Clegg
John H. Crowe
Martin Caffrey

The session on methods of analysis of membranes and water was organized by Martin Caffrey.

List of Participants

J. DEREK BEWLEY, Professor of Biology, University of Guelph, Guelph, Ontario N1G 2W1, Canada

MICHAEL J. BURKE, Associate Dean and Director, Academic Programs, College of Agriculture, Oregon State University, Corvallis, Oregon 97331, USA

MARTIN CAFFREY, Planning Committee, Research Associate, Section of Biochemistry, Molecular and Cell Biology, and Cornell High Energy Synchrotron Source, Cornell University, Ithaca, New York 14853, USA

GIORGIO CARERI, Professor of Physics, Dipartimento di Fisica, Universita degli Studi di Roma, Roma, 1-00185, Italy

JAMES S. CLEGG, Planning Committee, Director, Bodega Bay Marine Laboratory, Bodega Bay, California 94923, USA

JOHN H. CROWE, Planning Committee, Professor of Zoology, University of California, Davis, California 95616, USA

LOIS M. CROWE, Senior Research Associate, Department of Zoology, University of California, Davis, California 95616, USA

EDITH EBERT, CIBA-Geigy, Postfach 4002, Basel, Switzerland

DIANA W. FRECKMAN, Associate Research Nematologist, Department of Nematology, University of California, Riverside, California 92521, USA

BRUCE P. GABER, Deputy Head, Bio/Molecular Engineering Branch, Naval Research Laboratories, Washington, D.C. 20375, USA

GRAHAME W. GOULD, Colworth Laboratory, Unilever Research, Bedford MK44 1LQ, United Kingdom

HELMUT HAUSER, Professor of Biochemistry, Eidgenössische Technische Hochschule, Zürich, CH-8092, Switzerland

FOLKERT A. HOEKSTRA, Plant Physiology Department, Agricultural University, Wageningen, 6703 BD, The Netherlands

A. CARL LEOPOLD, Conference Director, W. C. Crocker Scientist, Boyce Thompson Institute for Plant Research, Ithaca, New York 14853, USA

BRYAN D. MCKERSIE, Associate Professor of Crop Science, Ontario Agricultural College, University of Guelph, Guelph, Ontario N1G 2W1, Canada

ANITA D. PANEK, Instituto de Quimica, Departamento de Bioquimica, Bloco A, sala 547, 5° andar, Ilha do Fundao, Cidade Universitaria, 21941, Rio de Janeiro, RJ, Brazil

ADRIAN PARSEGIAN, Physical Sciences Laboratory, Division of Computer Research and Technology, National Institutes of Health, Bethesda, Maryland 20205, USA

JOHN A. RUPLEY, Professor of Biochemistry, University of Arizona, Tucson, Arizona 85721, USA

PHILLIP C. STANWOOD, Research Agronomist, National Seed Storage Laboratory, Colorado State University, Fort Collins, Colorado 80523, USA

GEORGE STRAUSS, Professor of Chemistry, Rutgers University, New Brunswick, New Jersey 08903, USA

MICHAEL TOMBS, Colworth Laboratory, Unilever Research, Bedford MK44 1LQ, United Kingdom

CHRISTINA W. VERTUCCI, Section of Plant Biology, Cornell University, Ithaca, New York 14853, USA

JOE WOLFE, School of Physics, University of New South Wales, Kensington NSW 2033, Australia

Additional Contributors

E. BERNARDES, Instituto de Quimica, Departamento de Bioquimica, Bloco A, sala 547, 5° andar, Ilha do Fundao, Cicade Universitaria, 21941, Rio de Janeiro, RJ, Brazil

INDIRA CHANDRASEKHAR, Naval Research Laboratories, Washington, D.C. 20375, USA

MICHAEL F. DOWGERT, Director of Research, Glie Farms, 1600 Bathgate, Bronx, New York 10457, USA

ANDREA GIANSANTI, The Rockefeller University, New York, New York 10021, USA

ALLISON R. KERMODE, Department of Botany, University of Guelph, Guelph, Ontario N1G 2W1, Canada

RONALD M. LYNCH, Laboratory for Kidney and Electrolyte Metabolism, National Institutes of Health, Bethesda, Maryland 20892, USA

BERNARD MAIER, School of Physics, University of New South Wales, Kensington, NSW 2033, Australia

ANA C. PANEK, Instituto de Quimica, Departamento de Bioquimica, Bloca A, sala 547, 5° andar, Ilha do Fundao, Cicade Universitaria, 21941, Rio de Janeiro, RJ, Brazil

NAGARAJAN PATTABIRAMAN, Computer Graphics Laboratory, University of California, San Francisco, California 94132, USA

DONALD RAU, National Institutes of Health, Bethesda, Maryland 20205, USA

TISSA SENARATNA, Department of Crop Science, Ontario Agricultural College, University of Guelph, Guelph, Ontario N1G 2W1, Canada

LINDA SIEMANKOWSKI, Department of Biochemistry, University of Arizona, Tucson, Arizona 85721, USA

PETER L. STEPONKUS, Professor of Agronomy, Cornell University, Ithaca, New York 14853, USA

JOSHUA ZIMMERBERG, Physical Sciences Laboratory, Division of Computer Research and Technology, National Institutes of Health, Bethesda, Maryland 20205, USA

Acknowledgments

Grateful acknowledgment is made to the following sources of support for the Conference on Anhydrous Biology:

The Rockefeller Foundation
The National Science Foundation
Monsanto Corporation
ARCO Plant Cell Research Institute
United States Department of Agriculture
CIBA-Geigy Incorporated
Boyce Thompson Institute for Plant Research

Preparation of this volume was aided by a grant from the United States Department of Energy.

A. CARL LEOPOLD

Ithaca, New York

Membranes, Metabolism, and Dry Organisms

Introduction

A. Carl Leopold

Until recently the question of how organisms can tolerate desiccation has been almost a cryptic one, potentially interesting but undeciphered. Its antecedent historically was the question of *whether* organisms could ever tolerate desiccation—or, in a more formidable sense, whether organisms could "enjoy the advantage of real resurrection after death" (Spallanzani, 1776). In the eyes of early naturalists, desiccation was tantamount to death.

In 1702 when van Leeuwenhoek described having observed the springing to life of dry objects (rotifers) when they were provided with water, he was rewarded with little attention. When Needham (1745) made similar observations of desiccated eelworms taken from dry wheat seeds, and noted that water restored the animals to moving, wriggling organisms, he was contradicted by the illustrious Spallanzani, who asserted that reviviscence was not possible. The wheat grains were no doubt every bit as dry as the eelworms, although the grains' condition was clearly overlooked—poignantly so, it seems to me. The ability of dry seeds to survive has been common knowledge since before historic time.

Even though the early history of anhydrous biology is a fascinating story (Keilin, 1959), I will mention only one more turning point. In 1878 Claud Bernard pieced together the wider picture of anhydrous biology by noting that the ability to withstand desiccation was a characteristic widespread among many lower animals, eelworms, tardigrades, and rotifers and among plant seeds generally. With awareness that this remarkable (and previously considered unbelievable) feature of some organisms occurred frequently, the issue of anhydrous biology emerged as an important problem in biological science.

Realizing that desiccation tolerance occurs over an enormous range of taxa, including numerous invertebrates, bacteria, fungi, yeasts, lichens, seeds, and pollen, one may reasonably ask what advantages such an adaptation might confer. There are three: (1) desiccation provides a means of avoiding periods of environmental stress, (2) desiccation permits a suspension of metabolic activity in a time of stress, and (3) desiccation provides a state of tolerance to temperature extremes ranging as low as −270°C and as high as 370°C. In view of the anhydrous adaptation, Keilin wrote, "life may be a discontinuous process."

The Desiccated State

If desiccation is abruptly imposed on organisms, tolerance is not expressed. It appears to be a general feature of desiccation tolerance that the drying process must be slow. One might well infer that the tolerance depends upon the synthesis of some protective materials during the drying process.

The drying process imposes enormous changes in sheer volume upon the organism; seeds commonly lose two-thirds or even three-fourths of their volume as they dry. Similar shrinkage occurs in nematodes, *Artemia* cysts, and yeasts. Such drastic changes in architecture often involve the deformation of the structural parts of the organism to accordionlike folds, or the organism may become spirallike or looplike in form. As the shrinkage occurs, the plasmalemma in many instances is observed to become looped or folded upon itself.

The desiccated organism may exhibit little or no respiratory activity, depending upon the extent of dehydration. As the metabolic activity is suppressed, the dry organism becomes vulnerable to the accumulation of aberrant stresses or stochastic reactions, for it is incapable of repairing damages during the quiescent stage. The genetic information within the organism may be altered by the dry experience; the ribosomes may become detached from the endoplasmic reticulum, and the new messenger transcribed after rehydration may be substantively different from the messenger RNA previously present. Many organisms which show desiccation tolerance exhibit considerable stress response as they become rehydrated. This list of features of the desiccation experience includes some of the most costly hazards associated with the anhydrous type of adaptation.

The Pathway of Research in Desiccation

Research on desiccation tolerance was mobilized by the publication of two books: *Anhydrobiosis*, edited by J. Crowe and J. Clegg (1973), which brought together a range of papers relating to desiccation tolerance, and *Dry Biological Systems*, again edited by J. Crowe and J. Clegg (1978), the proceedings of a conference on anhydrous biology. The research that was subsequently stimulated has raised several highly interesting and relevant questions and has provided some impressive answers.

In view of the dependence of membrane structure upon hydrophilic and hydrophobic interactions, it seemed reasonable to question whether the membranes remained intact in desiccated organisms (Simon, 1974). The early arguments against the retention of membrane integrity rested heavily on experiments with membranes from desiccation-intolerant biological materials; while early experiments supported the idea that membranes lost integrity, it ultimately became clear that in desiccation tolerant materials the membranes retain a good bilayer configuration and are not structurally disrupted.

Experiments involving the physical state of water in desiccated systems have utilized an enormous range of experimental techniques, for example measuring proton spin with NMR, dielectric properties, heat capacities, and nuclear scattering and using differential thermal analysis and calorimetry and differential scanning calorimetry. The results of the diverse approaches have been substantially reconciled very recently with the description of the states of water in protein model systems under various states of hydration (Rupley et al., 1983). This work has permitted changes in physical or metabolic characteristics at various levels of hydration to be interpreted as reflecting ranges of water binding to surface groupings of various levels of hydrophilicity.

A central question about desiccation tolerance has concerned the nature of the tolerance mechanism. By far the majority of organisms are destroyed by desiccation, and the tolerance of this stress as a phenomenon has recently excited the attention of researchers. Early progress in this connection was made when it was observed that several compounds were synthesized in tolerant organisms during the drying phase. Such compounds as glycerol, glycogen, and trehalose were observed to accumulate in various tolerant organisms; there followed the landmark suggestion by Webb (1965) and Crowe (1971) that polyhydroxy compounds might protect membranes during desiccation by substituting for the hydrogen bonded water as the water was withdrawn. This theory led promptly to the finding that trehalose did indeed protect membranes, and a central key to the mechanism of tolerance in numerous animal and microbiological organisms was at hand. Evidence regarding the mechanism of such protection came to light when physicochemical techniques were used (Crowe et al., 1984).

A further question concerning the tolerance of desiccation relates to deteriorative reactions that occur in the dry state, leading to loss of viability.

Our understanding of the ways in which desiccated organisms are protected from destruction by reactions in the dry state is in an early state of development.

In this volume, contributions to each of these pathways of research are reported. The wide range of experimental and theoretical approaches, from biological to physicochemical, covers an exciting spectrum, from natural history to reductionist experimentation, and blends physiological, chemical, and physical methods in the pursuit of an understanding of anhydrous biology.

Prospects

Our increasing understanding of anhydrous biology holds promises, not only as it may illuminate events in natural history, but also as it may find technological applications. For example, knowledge of the mechanism of desiccation tolerance should lead to improved technologies in seed storage, gene banks, and the preservation of dry foods and pharmaceutical products. New information regarding the biochemistry of tolerance should open possibilities for genetic engineering.

The rapid development of this subject in biology and biochemistry is adding to our comprehension of the anhydrous state, a form of life that fails each of the usual tests for living systems: anhydrous organisms lack the ability to metabolize, to respond to the environment, to grow, or to reproduce. They truly represent, then, a discontinuity in life processes.

REFERENCES

Bernard, C. 1878. Leçons sur les phénomènes de la vie communs aux animaux et vegetaux. Paris. (Cited by Keilin, 1959.)

Clegg, J. 1967. Metabolic studies of cryptobiosis in encysted embryos of *Artemia salinov*. Comp. Biochem. Physiol. 20: 801-809.

Crowe, J.H. 1971. Anhydrobiosis: an unsolved problem. Am. Naturalist 105: 563-574.

Crowe, J.H. and Clegg, J.S. 1973. Anhydrobiosis. Dowden, Hutchinson and Ross, Stroudsburg, PA.

Crowe, J.H. and Clegg, J.S. 1978. Dry Biological Systems. Academic Press, New York.

Crowe, J.H., Crowe, L.M. and Chapman, D. 1984. Preservation of membranes in anhydrobiotic organisms: the role of trehalose. Science 223: 701-703.

Crowe, J.H. and Higgins, R.P. 1967. The revival of *Macrobiotus areolatus* (Tardigrada) from the cryptobiotic state. Am. Microscop. Soc. Trans. 86: 286-294.

Keilin, D. 1959. The problem of anabiosis or latent life: history and current concept. Proc. Royal Soc. London B150: 149-191.

Leeuwenhoek, A. van. 1702. On certain animalcules found in the sediment in gutters of the roofs of houses. *In* The Select Works of Antony van Leeuwenhoek, letter 144. (Transl. by Samuel Hoole, 1798. Ref. vol. 2, pp. 207-213. Cited by Keilin, 1959.)

Needham, J.T. 1745. New microscopical discoveries. (Of eels in blighted wheat, pp. 85-89.) London. (Cited by Keilin, 1959.)

Rupley, J.A., Gratton, E. and Careri, G. 1983. Water and globular proteins. Trend. Biochem. Res. 8: 18-22.

Simon, E.W. 1974. Phospholipids and plant membrane permeability. New Phytol. 73: 377-420.

Spallanzani, M. 1776. Opuscules de physique animale et vegetale. (Transl. from Italian by J. Senebier, 1787. Ref. vol. 2, pp. 203-285. Cited by Keilin, 1959.)

Webb, S.J. 1965. Bound Water in Biological Integrity. Thomas, Springfield: 187 pp. (Cited by Crowe and Higgins, 1967.)

1 Physical Attributes of Desiccated Seeds

A. Carl Leopold
Christina W. Vertucci

While the desiccated state provides several advantages for biological organisms, the physiological mechanisms of tolerance to desiccation remained completely beyond understanding until Webb (1965) and Crowe (1971) suggested that water was replaced by some polyol and Crowe et al. (1983) established experimentally that trehalose could serve as such a replacement. Even as the trehalose explanation of desiccation tolerance is emerging, those of us who deal with desiccation-tolerant seeds and with pollen and angiospermous resurrection plants still lack an explanation, for to the best of our knowledge trehalose is either absent or very nearly absent from these plant structures. As a first generalization, then, we suggest that there must be at least two separate mechanisms for desiccation tolerance, one based on trehalose and one not based on trehalose.

The trehalose-based system is associated with the capabilities of the glucose disaccharide to protect the intactness of membranes in the dry state (Chapters 11, 12, and 13). A first question for the plant physiologists, then, might be whether membranes remain intact in the dry state. While some early evidence suggested that the membranes of seeds were not intact in the dry state (Simon, 1974), and early electron micrographs suggested that membrane integrity was lost (e.g. Webster and Leopold, 1977), evidence has gradually accumulated to indicate that plant membranes do indeed remain intact in the desiccation-tolerant condition (e.g. Opik, 1981; Priestley and de Kruijff, 1982).

Before considering possible alternatives to the trehalose protective system, we may constructively examine the status of water and water binding in relation to the moisture content of seeds.

Evidence of Water Binding in Seeds

A convenient method for seeking evidence about water binding in seeds comes from studies of moisture isotherms. When biological materials are equilibrated at various relative humidities, the amounts of water adsorbed at each level of water activity describe a moisture isotherm (Gregg and Sing, 1982; Iglesias and Chirife, 1982). The sigmoid shape of the isotherm for soybean cotyledon tissue (Figure 1-1) reflects three levels of tissue/water interactions. Calculations of the binding affinity of water as the enthalpy of sorption using van't Hoff analysis provide more explicit evidence for three sectors of water binding (Figure 1-2): a relatively high enthalpy of binding for water at water contents below 8%, a relatively weaker enthalpy in a middle region of water contents (8-24%), and a low affinity at higher water contents (Vertucci and Leopold, 1984).

Three sectors of water binding have been described as they exist for various biological materials such as proteins (see review of Rupley et al., 1983); the decreasing enthalpies of binding at successively higher water contents have been ascribed to the binding of water to molecular sites of decreasing hydrophilicity (Chapter 15).

The use of moisture isotherms for measuring water binding has an inherent limitation in that one cannot accurately identify the upper limit of water binding. At 100% relative humidity, equilibrium cannot be determined. We have two lines of evidence from which we may define the upper limits of water binding in soybeans. Both of these involve the determination of the lower limits of freezable water. Using both differential thermal analysis apparatus and differential scanning calorimetry (DSC), we have been unable

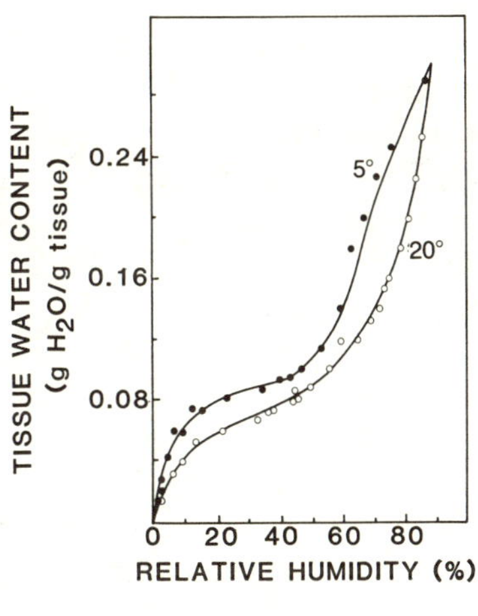

Figure 1-1. Moisture isotherms for pellets of soybean seed tissue at two different temperatures. (From Vertucci and Leopold, 1984, courtesy of *Plant Physiology*.)

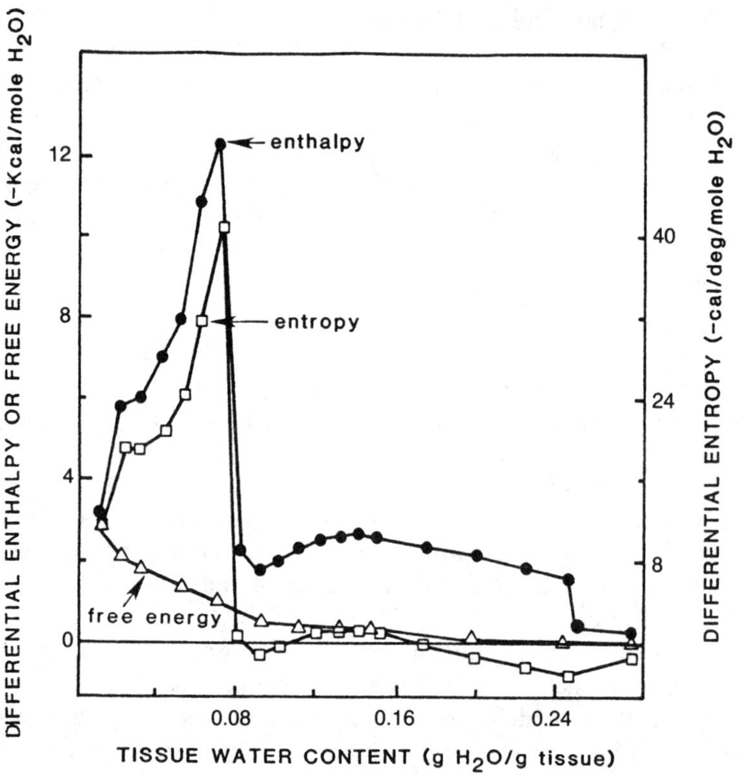

Figure 1-2. Differential energies and entropies of water sorption at different moisture contents. Calculated from data in Figure 1-1. (From Vertucci and Leopold, 1984, courtesy of *Plant Physiology*.)

to detect melting exotherms of ice even after temperatures below −65°C for soybean tissues at water contents below about 35% (dry weight basis). A similar lack of melting exotherms is found using DSC below 38% water content. The lack of melting exotherms is consistent with the finding that freezing damage to the seed occurs only at water contents above 35% (Figure 1-3). From these lines of evidence, we suggest that the upper limit of water binding in soybean seed is between 32 and 38% water content.

In an earlier study (Seewaldt et al., 1981), we reported proton NMR data for soybean seed tissue which indicated that free water was first detectable at water contents of about 12%; we are presently in the process of further studies in efforts to clarify the nature of the NMR signal at water contents below those we consider to be the limits of bound water. While the NMR data indicate an increased mobility of protons beginning at 12% water, this

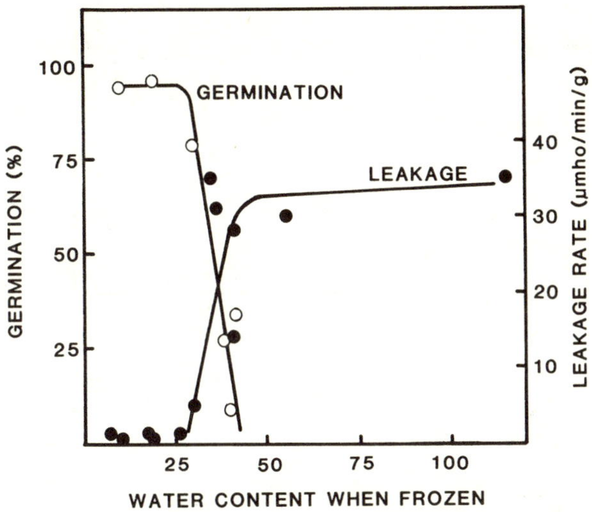

Figure 1-3. Effects of freezing (−65°C) of soybean seeds at various moisture contents, showing the loss of germinability and the increase in solute leakage at the higher moisture contents.

signal need not represent free water. The data of Careri (Chapter 16) are analogous to the NMR data and again are measurements of proton mobility in the stricter sense.

From the current evidence, it is our opinion that water in soybean seeds is in bound forms at all moisture contents up to about 32-38% water content. At this level of hydration the seed is very soft and spongy. Chapter 2 will compare biological reactions which can occur in the various regions of water binding.

Physical Characteristics Associated with Seed Water Content

The ability of desiccated systems to withstand high temperatures is a consequence of the increased resistance of proteins to denaturation upon the withdrawal of water. The tolerance of proteins to high temperatures may reflect the immobilization of protein structures as the hydrogen bondings with water are lost (Jaenicke, 1981). To illustrate the effects of desiccation on protein denaturation in soybean seeds, Figure 1-4a shows the effects of brief high temperatures on the loss of a proteinaceous trypsin inhibitor. In this case, a temperature of 90°C is completely ineffective in removing the trypsin inhibi-

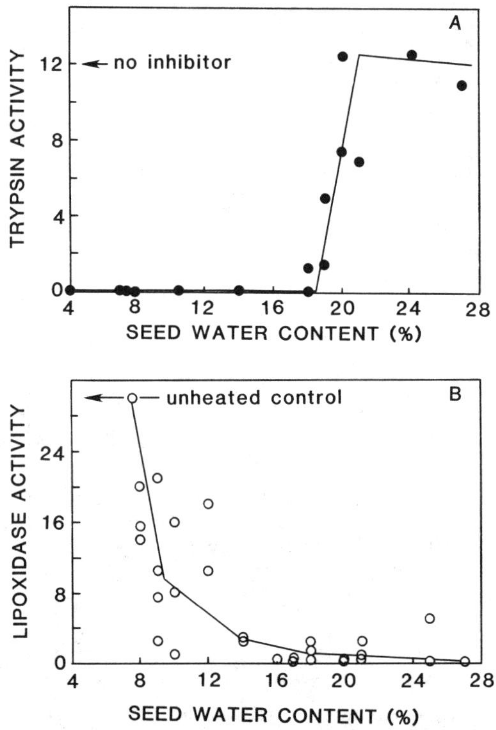

Figure 1-4. Effects of heating (90°C for 10 min) on a trypsin inhibitor (a) and lipoxygenase activity (b) in soybean seeds. When seeds were at higher moisture contents, the trypsin inhibitor was destroyed and the lipoxygenase activity was lost.

tor at seed moisture contents below about 18% (dry weight basis). If the seeds contained 20% water or more at the time of the temperature stress, this protein was completely inactivated. Lipoxygenase is known to be sensitive to high-temperature destruction, and similar experiments were carried out on the activities of this enzyme in soybean seeds after the same high-temperature challenge. The data in Figure 1-4b illustrate that this enzyme is destroyed by the high temperature at markedly lower moisture levels; a water content of about 10% was sufficient for the complete loss of this enzyme from the tissue.

The survival of soybean seeds at high temperature is also very sensitive to the water content. Seed viability was depressed by high temperatures which were markedly below those necessary to bring about protein denaturation. As in the protein denaturation experiments, damage was avoided during heat treatments if the moisture content was below 20%, and markedly lower

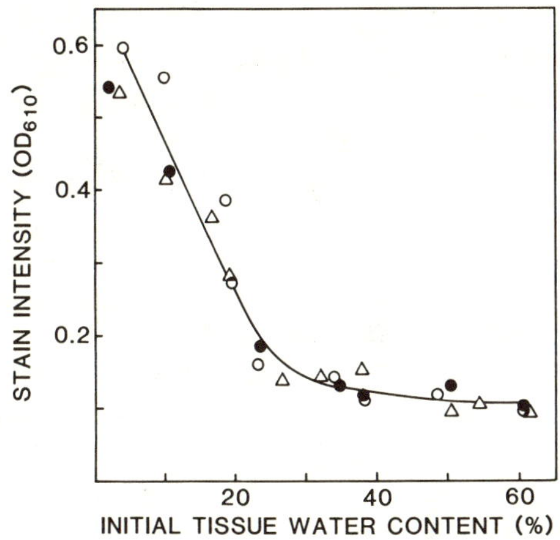

Figure 1-5. Evidence for imbibitional damage to soybean seeds after imbibition in freestanding water. The seeds were stained with Evans blue, and then extracted in aqueous TCA and assayed as OD_{610}. When the seeds were imbibed from low initial moisture contents, there were substantial increases in staining intensity, indicating cellular damage. The three symbols represent three separate experiments.

temperatures were needed to erase germinability than were necessary to destroy the proteins mentioned above; in this case, a 75°C experience was sufficient to destroy germinability of soybeans at any moisture content above about 20%. Thus, the desiccated state imparts tolerance both to low temperatures and to high-temperature stresses. Sakai and Noshiro (1975) have reported that seeds of several species survive temperatures of liquid nitrogen (−196°C) if their water contents are below 17-24%. The characteristic of stability at low temperatures when levels of hydration are low is an important part of seed storage technology (Chapter 20).

As desiccated tissues are rehydrated, damage can occur in connection with the imbibition event (Hobbs and Obendorf, 1972; Powell and Matthews, 1978; Duke and Kakefuda, 1981). In soybean seeds the damage is exacerbated by low initial water content, increasing greatly if the initial water content is less than 14% (Vertucci and Leopold, 1984). Figure 1-5 describes the quantitative relationship between the extent of damage and the initial water content, showing the extent of tissue damage after imbibition as estimated by the extent of staining with Evans blue. In this case, tissue damage was evident after rapid imbibition from any moisture contents below about 24%. Earlier experiments in which water viscosity and wetting abilities have been varied have led us to suggest that a major component of imbibitional damage is associated with the wetting step upon initial contact with water and is not closely associated with the water's rates of flow into or through the tissue (Vertucci and Leopold, 1983).

Having, then, a special interest in the initial wetting component of imbibition, we have measured the temperature exotherms, or the heats of wetting, of soybean tissue as related to the initial water content. We have obtained maximal heats of wetting for seed tissue in the range of 1% water content, and the magnitude of the exotherm decreased logarithmically as water contents increased (Figure 1-6). No exotherms were detectable at initial tissue water contents of 20% or more. The maximal value extrapolated from the log curve is estimated to be about 11.5 cal/g for dry tissue; this figure is comparable to the values that have been reported for the heats of wetting of dry cellulose or starch (Wahba and Nashed, 1950) and may be interpreted as changes in the heat capacity of the tissue components during wetting. Similar heats of wetting for wheat seeds have been reported by Maximov (1955).

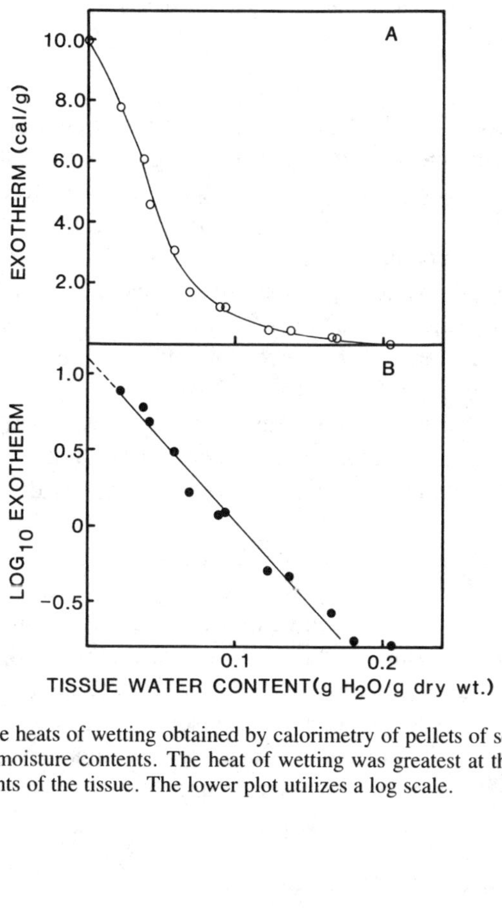

Figure 1-6. The heats of wetting obtained by calorimetry of pellets of soybean seeds at various initial moisture contents. The heat of wetting was greatest at the lowest initial moisture contents of the tissue. The lower plot utilizes a log scale.

With this brief survey, then, we have illustrated some characteristics of desiccated tissue which are quantitatively related to the extent of desiccation; these include changes in the affinity of water binding, in the resistance to high temperature damage, in resistance to freezing damage, in imposition of imbibitional damage, and in the heat of wetting. For each of these properties, soybean seeds show the greatest changes in the first and second sectors of water binding.

Contrasts between Desiccation Tolerance and Intolerance

In pursuing an understanding of the mechanism of desiccation tolerance in higher plants, it is reasonable to compare characteristics of tolerant and intolerant materials. It is possible to cause soybean seed axes to lose desiccation tolerance using a pretreatment of 24 hr imbibition, followed by redrying in the fashion described by McKersie and Stinson (1980 and Chapter 5). In this manner, then, we have obtained comparable tissue (soybean radicals)

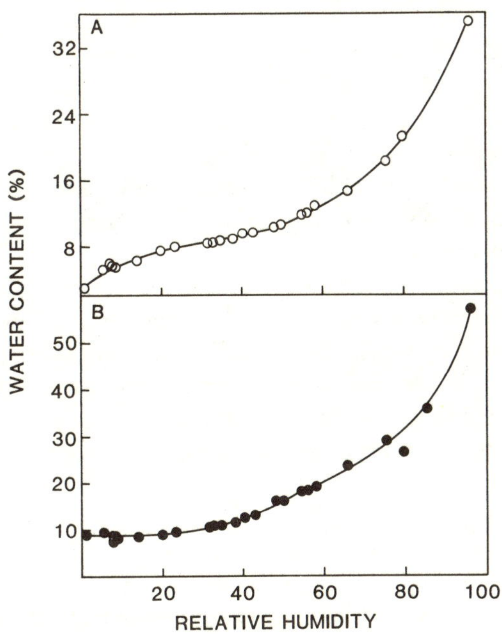

Figure 1-7. Moisture isotherms for soybean axis tissue at 15°C, comparing the isotherm for ungerminated axes (a) with axes of seeds that had been germinated for 24 hr (b).

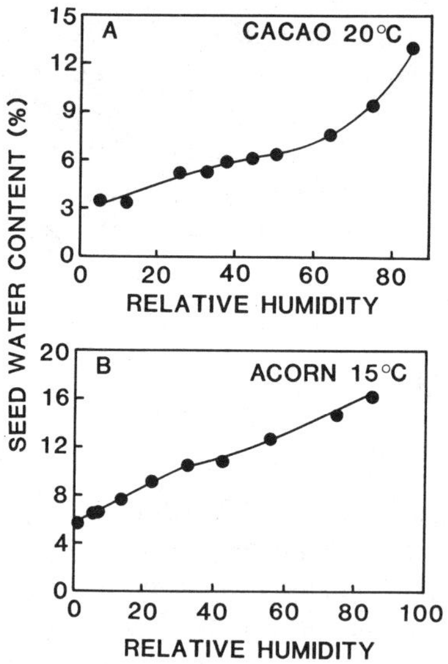

Figure 1-8. Moisture isotherms for tissues from two recalcitrant seeds, that is, seeds which are not tolerant to desiccation. Tissues of cacao seeds (a) and acorns of red oak (b).

which show desiccation tolerance and intolerance. Comparing the moisture isotherms for these two conditions, we find that with the loss of desiccation tolerance there is an associated change from a double sigmoid (Figure 1-7a) to a simple rising curve (Figure 1-7b).

The configuration of moisture isotherms has also been determined for several recalcitrant seeds—that is, seeds which do not tolerate desiccation. In contrast to the double-sigmoid isotherm shown in Figure 1-1, the recalcitrant seeds of oak and cacao show a simple rising moisture isotherm with increasing water activity (Figure 1-8a, b).

As an interpretation of these two types of isotherms, Labuza (1984) has suggested that materials which enter into crystalline forms upon drying may display the simple rising moisture isotherms. Our comparative data for tolerant and intolerant tissues lead us to suggest, on the basis of Labuza's interpretation, that the desiccation tolerant state may be one in which some tissue components are able to resist crystallization during drying.

A Hypothesis Concerning Desiccation Tolerance

With the data which we have accumulated, we can make a tentative suggestion about a possible mechanism of desiccation tolerance. Using the hypothesis that desiccation tolerance involves some mechanism by which certain polyols replace water on hydrophilic end groups of membrane phospholipids (Crowe, 1971; also Chapters 11 and 12), we propose that polyols or sugars may provide this protective effect only if they do not become crystallized. This statement is intuitively reasonable if the hydroxyl groups on the polyol or sugar are essential for the replacement of water on hydrophilic sites. The feature of desiccation tolerance, under these assumptions, would be that (1) some polyol or sugar must be present to protect the hydrophilic membrane components during drying and (2) the sugar component must resist crystallization during loss of water.

Studies of the crystallization of sucrose have been made by Smythe (1967), who placed pieces of crystalline sucrose into syrups of either sucrose or sucrose with other sugar addenda and measured the growth of the sucrose crystals gravimetrically. His experiments clearly show that certain sugars suppress the growth of sucrose crystals; the most effective inhibitor was raffinose, and the next most effective inhibitor was stachyose (Figure 1-9). Smythe offered the theoretical explanation that the oligosaccharide raffinose competitively inhibited sucrose crystallization because of its similarity to sucrose in the fructose-glucose moieties; its lack of fit onto the surface of the growing crystal was due to the additional galactose moiety. Trehalose had no effect on sucrose crystallization.

Some experiments by Makower and Dye (1956) may be closer analogues to the situation in seed tissues; they followed the crystallization of amorphous sucrose under controlled levels of humidity, using the changes in weight of the sucrose sample as an indication of crystallization. At humidities from 4 to 34%, the sucrose samples adsorbed water from the surrounding atmosphere; at humidities of 24% or more, there was an initial absorption of water, and then a subsequent decrease of sample weight which was interpreted as crystallization and the attendant exclusion of water. X-ray diffraction was used to verify the occurrence of crystallization.

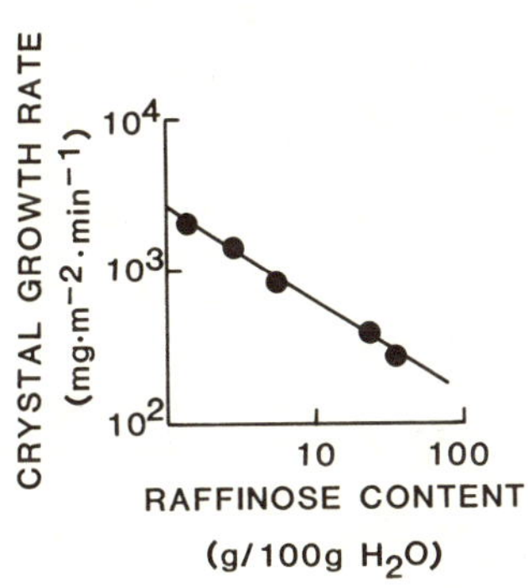

Figure 1-9. The effects of various concentrations of raffinose on the growth rate of sucrose crystals. The raffinose was incorporated into a sucrose syrup. (Data from Smythe, 1967.)

In Chapters 11 and 12 the Crowes have compared the effectiveness of various sugars in preserving sarcoplasmic reticulum membrane stability during desiccation, and they report that trehalose is most effective, but substantial effects were also obtained for glucose, fructose, sucrose, and raffinose. While trehalose stands out as the best single sugar, it is possible that a combination of sucrose and raffinose (the most abundant sugars in soybean seeds) would be effective, with raffinose serving to inhibit the crystallization of sucrose.

Our tentative hypothesis, then, is that desiccation tolerance mediated by trehalose may depend in part on the noncrystallizing characteristic of this sugar. In tissues in which trehalose is absent, as in the seeds of angiosperms, sucrose may fill the role of membrane protectant provided that there is sufficient oligosaccharide (raffinose and/or stacchyose) to inhibit the crystallization of the sucrose and hence to prevent the loss of its protective potential.

In conclusion, it appears reasonable to presume that membranes are sensitive sites of desiccation damage and that tolerance to desiccation involves the provision of some protection to membranes. In the case of spores, tardigrades, and soil nematodes, trehalose appears to be the agent providing membrane protection; according to our interpretation an important characteristic of trehalose in this function would be its resistance to crystallization. In the case of angiosperm materials which show desiccation tolerance, as in seeds, pollen, and some resurrection plants, trehalose does not appear to be present in appreciable amounts, and sucrose plus raffinose and/or stachyose may possibly provide the protective effect for membranes. In our interpretation, the mixture of sugars may serve to defer crystallization, thus maintaining the effectiveness of sucrose in replacing water in the hydrophilic sites of the cellular membranes.

REFERENCES

Clegg, J. 1967. Metabolic studies of cryptobiosis in encysted embryos of *Artemia salinov*. Comp. Biochem. Physiol. 20: 801-809.

Clegg, J.S., P. Seitz, W. Seitz and C.F. Hazelwood. 1982. Cellular responses to extreme water loss: the water-replacement hypothesis. Cryobiology 19: 306-316.

Crowe, J.H. 1971. Anhydrobiosis: an unsolved problem. Am. Naturalist 105: 563-574.

Crowe, J.H., L.M. Crowe and D. Chapman. 1984. Preservation of membranes in anhydrobiotic organisms: the role of trehalose. Science 223: 701-703.

Crowe, J.H., L.M. Crowe and S.A. Jackson. 1983. Preservation of structural and functional activity in lyophilized sarcoplasmic reticulum. Arch. Biochem. Biophys. 220: 477-484.

Crowe, J.H. and R.P. Higgins. 1967. The revival of *Macrobiotus areolatus* (Tardigrada) from the cryptobiotic state. Am. Microscop. Soc. Trans. 86: 286-294.

Crowe, L.M., R. Mouradian, J.H. Crowe, S.A. Jackson and C. Womersly. 1984. Effects of carbohydrates on membrane stability at low water activities. Biochim. Biophys. Acta 769: 141-150.

Crowe, J.H., M.A. Whittam, D. Chapman and L.M. Crowe. 1984. Interactions of phospholipid monolayers with carbohydrates. Biochim. Biophys. Acta 769: 151-159.

Darbyshire, B. 1974. The function of the carbohydrate units of three fungal enzymes in their resistance to dehydration. Plant Physiol. 54: 717-721.

Duke, S.H. and G. Kakefuda. 1981. Role of the testa in preventing cellular rupture during imbibition of legume seeds. Plant Physiol. 67: 449-456.

Gregg, S.J. and K.S.W. Sing. 1982. Adsorption, Surface Area and Porosity. Academic Press, New York: 303 pp.

Hobbs, P.R. and R.L. Obendorf. 1972. Interaction of initial seed moisture and imbibitional temperature on germination and productivity of soybean. Crop Sci. 12: 664-667.

Iglesias, H.A. and J. Chirife. 1982. Handbook of Food Isotherms. Academic Press, New York: 347 pp.

Jaenicke, R. 1981. Enzymes under extremes of physical conditions. Annu. Rev. Biophys. Bioengr. 10: 1-67.

Labuza, T.P. 1984. Moisture Sorption. Amer. Assoc. Cereal Chemists, St. Paul: 150 pp.

McKersie, B.D. and R.H. Stinson. 1980. Effect of dehydration on leakage and membrane structure in *Lotus* seeds. Plant Physiol. 66: 316-320.

Madin, K.A.C., J.H. Crowe and S.H. Loomis. 1978. Metabolic transitions in a nematode during anhydrobiosis. *In* Dry Biological Systems (J.H. Crowe and J.S. Clegg, eds.). Academic Press, New York: pp. 155-174.

Makower, B. and W.B. Dye. 1956. Equilibrium moisture content and crystallization of amorphous sucrose and glucose. J. Agric. Food Chem. 4: 72.

Maximov, G.A. 1955. Heat and moisture exchange in plant seeds during imbibition. Fiziol. Rastenii 2: 81-83.

Opik, H. 1980. The ultrastructure of coleoptile cells in dry rice grains. New Phytol. 85: 521-529.

Palmer, K.J., W.B. Dye and D. Black. 1956. X-ray diffractometer and microscopic investigation of crystallization of amorphous sucrose. J. Agric. Food Chem. 4: 77-81.

Powell, A.A. and S. Matthews. 1978. The damaging effect of water on dry pea embryos during imbibition. J. Exp. Bot. 29: 1215-1229.

Priestley, D.A. and B. de Kruijff. 1982. Phospholipid motional characteristics in a dry biological system. Plant Physiol. 70: 1075-1078.

Rockland, L.B. 1969. Water activity and storage stability. Food Technol. 23: 11-19.

Rupley, J.A., E. Gratton and G. Careri. 1983. Water and globular proteins. Trends Biol. Sci. 1983: 18-22.

Sakai, A. and M. Noshiro. 1975. Factors in survival of crop seeds cooled to the temperature of liquid nitrogen. *In* Crop Genetic Resources for Today and Tomorrow (O.H. Frankel and J.G. Hawkes, eds.). Cambridge Univ. Press, Cambridge: pp. 317-326.

Schneider, M.J.T. and A.S. Schneider. 1972. Water in biological membranes. J. Membrane Biol. 9: 127-140.
Seewaldt, V., D.A. Priestley, A.C. Leopold, G.W. Feigenson and F. Goodsaid-Zalduondo. 1981. Membrane organization in soybean seeds during hydration. Planta 152: 19-23.
Simon, E.W. 1974. Phospholipids and plant membrane permeability. New Phytol. 73: 377-420.
Smythe, B.M. 1967. Sucrose crystal growth. Aust. J. Chem. 20: 1097-1114.
Vertucci, C.W. and A.C. Leopold. 1983. Dynamics of imbibition by soybean embryos. Plant Physiol. 72: 190-193.
Vertucci, C.W. and A.C. Leopold. 1984. Bound water in soybean seed and its relation to respiration and imbibitional leakage. Plant Physiol. 75: 114-117.
Wahba, M. and S. Nashed. 1950. Change in the heat of wetting of cellulose in water. Nature 166: 998.
Webb, S.J. 1965. Bound Water in Biological Integrity. Thomas, Springfield, Ill.: 187 pp. (Cited by Crowe and Higgins, 1967.)
Webster, B.W. and A.C. Leopold. 1977. Ultra-structure of dry and imbibed cotyledons of soybean. Am. J. Bot. 64: 1286-1293.

2 Physiological Activities Associated with Hydration Level in Seeds

Christina W. Vertucci
A. Carl Leopold

The essential element needed to convert a seed from a dry quiescent object into a germinating plant is water. Chapter 1 describes the initial stage of water uptake as a wetting process in which there are three levels of water binding marked by successively lower binding strength because of the decreasing hydrophilicity of binding sites. Thermodynamic data calculated through van't Hoff analysis illustrate these three water binding levels for soybean, pea, and apple cotyledons (Figure 2-1). In this chapter we will try to reveal changes in the organization and function of cell components with hydration. We will first discuss some changes in physiological activity with hydration and then present evidence that conformational changes occur which may be contributing to the observed physiological changes.

There have been a few studies relating physiological activity to hydration level (Acker, 1969; Careri et al., 1980; Clegg, 1978). Most of them involved model systems of enzymes in which activity was not observed below a minimum moisture content (about 8%), and further additions of water caused much greater activity (Acker, 1969; Careri et al., 1980). A classic study of physiological activity in an anhydrobiotic organism was done in Clegg's lab using *Artemia* cysts. Clegg (1978) separated the ''resting'' state of the cysts into three levels of metabolic activity which correspond to hydration: an ametabolic state where no enzyme-mediated reactions were observable, a state with intermediate metabolism in which simple reactions could occur, and finally a fully active metabolic state in which mitochondrial electron transport was observed.

In our experiments, we have studied the changes in physiological activity as a function of hydration level in seeds in an attempt to answer two ques-

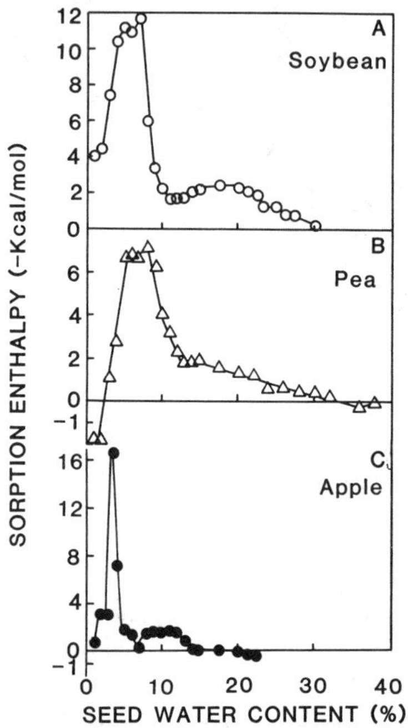

Figure 2-1. Differential enthalpies of water sorption at different moisture contents for (a) soybean, (b) pea, and (c) apple embryos. Values are calculated for sorption isotherms at 15 and 25°C (soybean and pea) and 5 and 15°C (apple).

tions: (1) at what levels of hydration do seeds begin to exercise the physiological activities associated with the imbibed state and (2) what does water do to a dry cell such that metabolic activity can be reinitiated. The physiological activities on which we will focus are the damages incurred during hydration, oxygen uptake and respiration, light-activated electron transport and light energy transfer, and dormancy breaking. In most cases, moisture content was manipulated by equilibrating seeds in closed jars over saturated salt solutions (Vertucci and Leopold, 1984). This procedure allowed us to obtain tissues with rather precise levels of hydration. Activity level was then measured by means that required no further additions of water. From these experiments, we have observed several discrete levels of physiological activity which correspond to the three levels of water binding.

Physiological Events Related to Water Content

Imbibitional Damage

The early stages of water uptake are a crucial time for the seed; sensitivities to rapid imbibition, chilling, and anoxia may affect seedling establishment and later yields (Obendorf and Hobbs, 1970). Illustrative examples of the damaging effects associated with chilling temperatures (5°C for 10 min), rapid imbibition (seeds submersed in 25°C water for 5 min), heat stress, and denaturants (55°C or 33% MeOH for 5 min) on imbibing soybean seeds of various initial moisture contents are described in Figure 2-2a-d, respectively. In this experiment, protein leakage rate after the seeds were fully imbibed (measured as A_{260}/min/g) is used as an index of damage incurred from the early stress treatment. All the data are normalized to a control.

Between water contents of 11 and 15%, the sensitivity of soybeans to rapid imbibition and to chilling is reduced (Figure 2-2a and b) but the sensitivity to denaturants is increased. Thus, at initial moisture contents below 11%, the imbibing cotyledons are sensitive to those stresses typical of a dry system hydrating. At moisture contents above 15%, cotyledons are sensitive only to stresses of a fully hydrated system. It follows that, for the case of stresses during imbibition, there may be a transition between dry and wet systems at 11-15% water content where the level of hydration is within binding region II (Vertucci and Leopold, 1984).

Respiration

The several studies investigating the reinitiation of mitochondrial oxygen consumption as a seed imbibes have concluded that, in the dry state, mito-

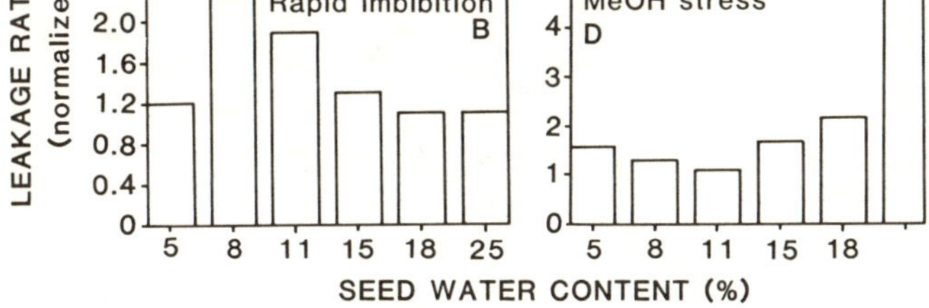

Figure 2-2. Leakage from fully imbibed soybean embryos following imbibitional stresses applied at the indicated initial moisture contents. Seed treatments were (a) imbibed at 5°C for 10 min, (b) submerged in 25°C water for 25 min, (c) imbibed at 55°C for 5 min, and (d) immersed in 33% MeOH for 5 min. Solute leakage was measured as A_{260}/min/g dry weight of tissue. The leakage rates from stressed seeds are normalized to leakage rate values from unstressed seeds at the same initial moisture content.

chondria from seeds are "incomplete" and cytochrome *c* is not retained within the membrane (Wilson and Bonner, 1971; Nawa and Asahi, 1973). We have attempted to study in detail the effects of water on respiratory activity in seeds.

In order to define the water contents at which mitochondrial activity was possible, O_2 uptake rates from several seed species were measured in a Gilson differential respirometer (Vertucci and Leopold, 1984). The tissues were equilibrated to specific moisture contents in jars with saturated salt solutions. For the apple seeds, equilibration was conducted in the cold. Manometric measurements were then carried out at 25°C over a 3-day period.

The data in Figure 2-3 show a similar relationship between water content and O_2 uptake rate among seeds of different species. At the lowest levels of hydration, O_2 consumption was very low. Then, at moisture contents above 14% for apple, 20% for corn, 24% for soybean, and 26% for pea, dramatic increases in the rate of O_2 uptake occurred. These water contents are higher than those observed for the transition in stress sensitivity (Figure 2-2) and in fact coincide reasonably well with the transition of water binding from level II to level III (Figure 2-1). We suggest that mitochondrial respiration begins at the onset of level III. In the case of soybean and corn, O_2 uptake reaches a steady level at about 35 and 40% water, respectively. For soybean this point coincides with water content at which all water binding is complete (Chapter 1).

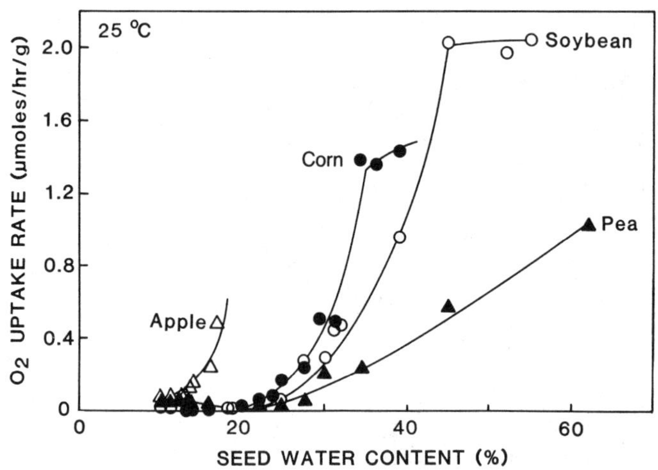

Figure 2-3. Oxygen uptake rates as a function of water content in apple, corn, soybean, and pea seeds. Measurements, expressed as μmole/O_2/hr/g, were taken in a Gilson respirometer at 25°C.

Since respiration is the oxidation of carbon-containing compounds, we expect to see CO_2 evolution occurring in amounts equivalent to the O_2 consumption. Measurements of CO_2 evolution in dry soybean seeds using infrared gas analysis show that a trend exists for CO_2 evolution similar to that observed for O_2 uptake in soybean (Figure 2-4). The rate of CO_2 evolution is low at moisture contents of 7-25%, but above 25% there is a dramatic increase in CO_2 evolution. At 35% water, CO_2 evolution peaks and levels off.

The respiratory quotient (RQ) is a measure of how much CO_2 is evolved per O_2 consumed. For mitochondrial respiration, the RQ should be close to 1.0. In soybeans, when moisture contents were below 25%, the RQ calculated from the data in Figures 2-3 and 2-4 was <0.5. When the moisture content was 10%, it was even lower, being approximately 0.2. These low values suggest that the oxidative activities occurring at low moisture contents were dominated by oxidations other than mitochondrial metabolism. At hydration levels above 25%, the RQ increased to 1, and so we are confident that we are observing mitochondrial electron transport when the hydration level is in sector III.

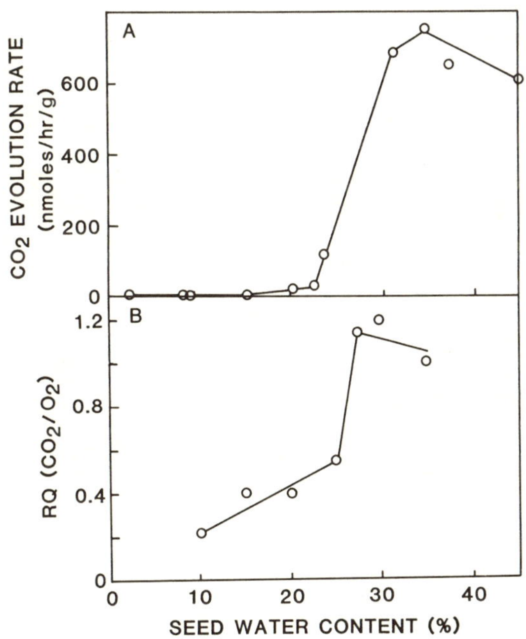

Figure 2-4. CO_2 evolution and respiratory quotient as a function of water content of soybean seeds. (a) CO_2 evolution measured by infrared gas analysis, (b) respiratory quotient calculated from the data in Figures 2-3 and 2-4.

Photosynthesis

We have demonstrated that mitochondrial electron transport is operational when the seed hydration is in level III. Photosynthetic electron transport can also be measured taking advantage of the chlorophyllous feature of dry pea cotyledons. Light-induced electron transport was studied in vivo using fluorescence induction kinetics, a method that measures the activity of photosystem II (Vertucci et al., 1985).

When dry pea cotyledons were illuminated with red light, there was an initial rise in fluorescence followed by a slow decay. The area under the decay curve, called fluorescence yield (Fy), was used as a measure of electron transport. The Fy changes with hydration level (Figure 2-5): when tissue water content was 1-4%, Fy decreased with increasing water content to a minimal value; this condition was maintained until the moisture content exceeded 12%. With water contents of 12-18%, Fy increased and was maximal at moisture contents greater than 20%. That light-induced electron transport was in fact being measured at water contents greater than 12% was substantiated by evidence that DCMU, an inhibitor of photosynthetic electron transport, progressively depressed Fy at moisture contents above 12%

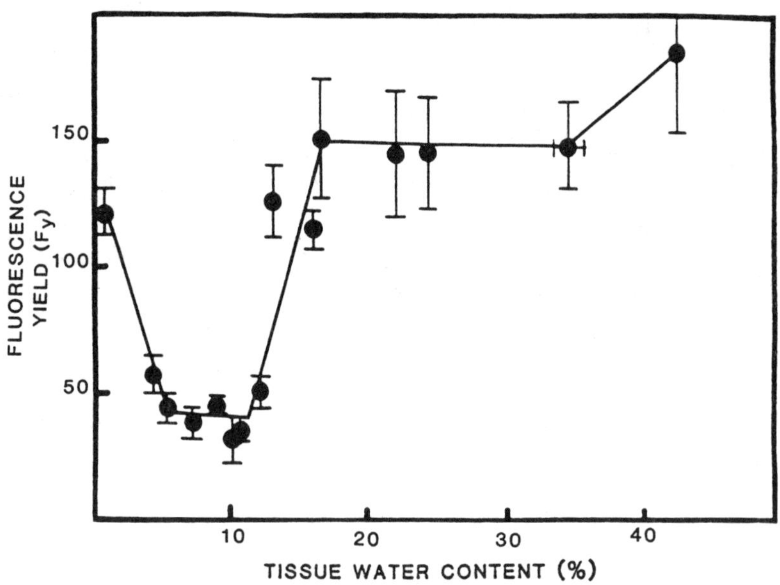

Figure 2-5. Changes in fluorescence yield from dark-adapted pea cotyledons of different moisture contents. Fluorescence yield was calculated from the area under the fluorescence induction decay curve. Error bars represent the standard error calculated from five replicates. (From Vertucci et al., 1985, courtesy of *Plant Physiology*.)

(Table 2-1). We conclude that electron transport within the chloroplast was possible when the water content was 12% and became fully operative when the water content was 20%. Therefore, photosynthetic electron transport can occur at lower water contents than respiratory electron transport.

Another aspect of photosynthesis that is affected by hydration level is the ability of pigments to transfer absorbed light energy (Vertucci et al., 1985). Presumably this transfer results when the water content is sufficient to induce clustering of pigments, effectively increasing the efficiency of energy transfer. The excitation of chlorophyll *a* fluorescence was used to demonstrate this phenomenon. With the emission wavelength set at 728 nm (a region that is strongly emitted by chlorophyll *a*), pea cotyledons of varying moisture contents were illuminated with light at wavelengths between 400 and 700 nm, and the excitation spectrum was recorded (Figure 2-6). At higher levels of hydration (18 or 35%), there was an increased contribution to the fluorescence spectrum between 450 and 490 nm. These are wavelengths at which carotenoids and chlorophyll *b* absorb light. The results suggest that those pigments may be transferring their excited state to chlorophyll *a* only at the higher levels of hydration. The contribution of these subsidiary pigments to chlorophyll *a* fluorescence is restricted at 7% water content (water binding level I); it is facilitated at 18% (water binding level II); and at 35% water content (level III) the effect is maximal.

Table 2-1
Effects of DCMU on fluorescence yield in pea cotyledons of different water contents

Water content (%)	Preimbibition treatment	Fy
1.8	H_2O	19.3±3.9
	DCMU	17.1±2.9
9.7	H_2O	7.7±1.4
	DCMU	7.5±7.5
18.1	H_2O	33.1±2.7
	DCMU	20.2±1.0
28.2	H_2O	52.9±3.9
	DCMU	38.1±3.3
50.5	H_2O	159.7±32.8
	DCMU	52.1±3.0

Note: DCMU blocks photosynthetic electron transport.
Source: Vertucci et al., 1985.

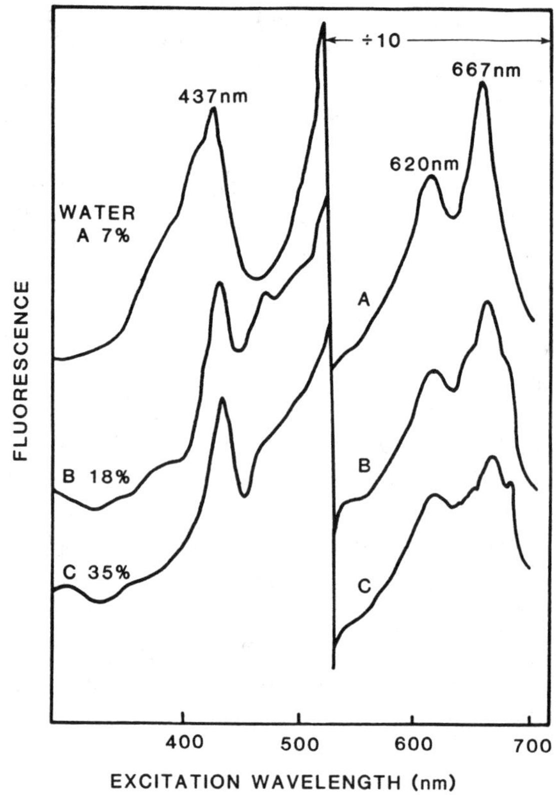

Figure 2-6. Fluorescence excitation spectra of chlorophyll in pea cotyledons at (a) 7%, (b) 18%, and (c) 35% water. The emission wavelength was set at 728 nm. Signal gain was reduced by 10 at wavelengths above 540 nm. (From Vertucci et al., 1985, courtesy of *Plant Physiology*.)

Dormancy Breaking

Dormancy breaking is yet another physiological activity that can be studied as a function of water content. A seed is said to be dormant if it fails to germinate in warm, moist conditions or if it produces an abnormal seedling. Though such a phenomenon is common, it is poorly understood.

We have used apple seeds as an example of seeds which require stratification (cool, moist conditions for several weeks) for dormancy breaking. If apple seeds are subjected to warm, moist conditions instead of stratification, a secondary dormancy sets in. The purpose of our experiment was to determine the moisture content at which apple seeds were able to respond to the cold by breaking dormancy.

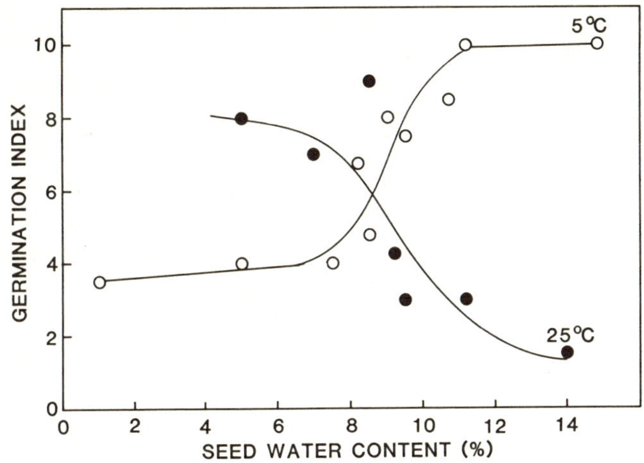

Figure 2-7. The effectiveness of temperature on dormancy breaking of apple seeds "stratified" at different moisture contents. The germination index used ascribes a 1.0 to seeds which germinate normally, 0.5 to seeds which germinate but are abnormal, and 0.0 to ungerminated seeds.

In order to stratify apple seeds at different moisture contents, we placed seeds in jars over saturated salt solutions and stored them at 5°C or 25°C for 10 weeks. The seeds were then germinated and scored: 0, no germination; 0.5, germinated but abnormal; and 1.0, germinated and normal (Williams, 1984).

When seeds were stratified at 5°C with moisture contents greater than 8%, there was an increase in the degree of germinated, normal seedlings (Figure 2-7). Dormancy was completely broken when the seeds were stored in the cold at 11% moisture or greater. A reverse trend was observed for seeds stored at 25°C, suggesting that, at water contents greater than 8%, apple seeds can sense temperature and respond by either breaking dormancy or going into secondary dormancy. Here again, three distinct levels of response relate to the hydration level: below 8% water, there was no temperature response; between 8 and 11% the response was proportional to hydration; and above 11% the response was complete. Recall that respiration was observed at water contents above 14% in apple seeds.

Summary

The above-mentioned experiments describe repeating correlations of physiological activity with the three levels of water binding. Seed tissues with water contents in level I are quiescent, with little or no observed physiological activity. They are particularly sensitive to imbibitional stresses. Tissues with water in level III show behavior similar to that of wetted tissue. Tissues with water in level II are in transition from the dry state to the wet state (Table 2-2). What is happening within a cell to cause such large changes in cellular activity over a rather narrow range of water content?

Membranes from animal tissues (Crowe and Crowe, 1982; Finean, 1969; Luzatti and Husson, 1962) have been found to undergo severe structural changes with desiccation. Simon (1974) suggested that similar changes occurred in plant membranes as they dried. However, the few studies investigating changes in seed and pollen membranes have shown no evidence for dramatic changes in their structure upon desiccation (McKersie and Stinson,

Table 2-2
Physiological activities associated with the three regions of water binding in soybean, pea, and apple seeds

		Regions of water binding	
Activities:	I	II	III
		SOYBEAN	
Water content:	1-9%	11-22%	24-32%
O_2 uptake:	Not detected	Very low, with low RQ	Mitochondrial e^- transport
Imbibition damage:	Imbibitional chilling	Imbibitional damage MeOH damage	Heat damage MeOH damage
		PEA	
Water content:	4-11%	12-21%	24- %
O_2 uptake:	Not detected	Very low	Mitochondrial e^- transport
Photosynthetic electron transport:	Low	Chloroplast e^- transport starts	Chloroplast e^- transport maximal
Light energy transfer:	No	Pigment energy transfer starts	Pigment energy transfer maximal
		APPLE	
Water content:	1-6%	8-12%	14- %
O_2 uptake:	Not detected	Low	Mitochondrial e^- transport
Stratification:	No temp. response	Stratification proceeds	Stratification maximal

1980; Seewaldt et al., 1981; Priestley and de Kruijff, 1982; Thomson and Platt-Aloia, 1982). Protein conformation studies have led to similar conclusions: structure was conserved as a protein was dried or rehydrated (Schneider and Schneider, 1972; Careri et al., 1980). Nonetheless, because dramatic changes in physiological activity occur at discrete hydration levels, we suggest that changes in the organization and structure of cell constituents occur with hydration.

Conformational Events Related to Water Content

Seed Swelling

There is an enormous increase in seed volume with imbibition (Leopold, 1983). This swelling is most likely due to the unfolding of proteins (Mayer and Poljakoff-Mayber, 1963), and contrary to earlier findings (Schneider and Schneider, 1972; Careri et al., 1980), the swelling feature strongly suggests hydration-induced conformational changes in these macromolecules. In an effort to define seed polymer swelling with hydration, the volume of imbibing soybean seeds was measured by water displacement. The volume of the seed polymers alone was estimated by subtracting the volume of water in the imbibing seed from the total seed volume, assuming that the specific volume of water at all water contents was 1 ml/g. There is evidence for small changes

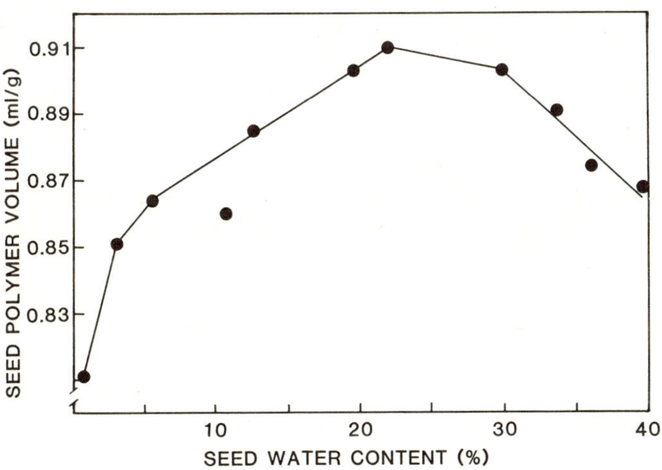

Figure 2-8. Changes in soybean seed polymer volume with water content. Seed polymer volume is calculated by subtracting the volume of water from the total seed volume at each given seed moisture content.

in specific volume of water in dry organisms (Clegg, 1984), but we have chosen the simpler assumption in order to facilitate our estimates.

The data in Figure 2-8 show a very large change in estimated soybean seed polymer volume as the water content increased from 4 to 8%. Between 8 and 21%, polymer swelling was less dramatic but still rather large. As moisture content exceeded 21%, the estimated polymer volume did not show further increases. These data suggest that the volumes of the polymers change with water content, with the greatest increases occurring at water contents below 20%. The inflection points in polymer swelling in soybean cotyledons seem to coincide reasonably well with the boundaries between the sectors of water binding.

Fluorescence Spectral Changes

Other evidence for changes in the organization of cell constituents with hydration is more subtle. We have already discussed how changes in the arrangement of subsidiary pigments around chlorophyll *a* with hydration could cause increased efficiency of light energy transfer in pea seeds. The chlorophyll fluorescence excitation spectra (Figure 2-6) suggest that other hydration-induced changes occur in the environment of chlorophyll *a* as well.

The excitation peak at 667 nm is most pronounced at low water contents. A similar absorbance peak is observed at this wavelength when chlorophyll a is dissolved in lipophilic solvents. Perhaps, then, excitation at 667 nm reflects a nonpolar environment in the dry seed. At moisture contents of 12-20%, there is an increased contribution of a peak at 683 nm, analogous to the fluorescence peak in hydrated leaves which occurs at 680 nm. This observation suggests that an increase in hydration and the associated increased polarity of the environment induces some conformational change in chlorophyll, which in turn affects its absorbance wavelength. The trend begins when water is sufficient to satisfy level I binding and is completed when there is sufficient water to satisfy level II binding.

Melting Transitions

Another method we have used to examine the effect of water on the organization of cell constituents involves studies of storage lipid transitions in soybean using differential scanning calorimetry (DSC). In these experiments, we observed the melting transitions of triglycerides in ground soybean cotyledons at various moisture contents. Figure 2-9 illustrates the two types of thermograms observed: a single peaked curve with several shoulders was obtained when the moisture content ranged from 3 to 21%, and a double

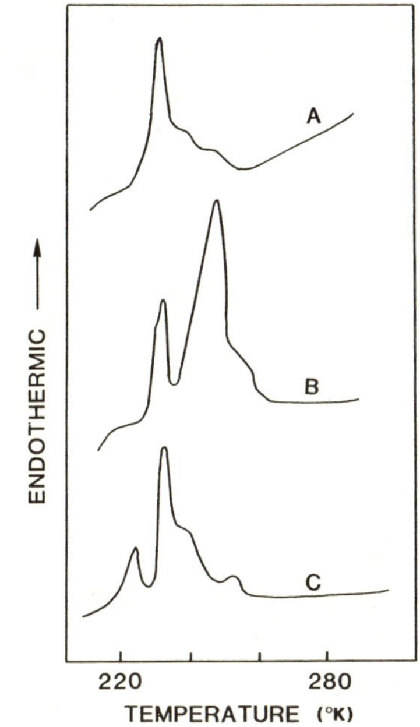

Figure 2-9. Sample thermograms of differential scanning calorimetry heating runs from (a) ground soybean pellets with water contents <21%, (b) soybean pellets with water contents between 22 and 38%, and (c) lipids extracted from soybean with chloroform and methanol.

peaked curve when the moisture content ranged from 22 to 38%. The temperature and the enthalpy of the transition increased as the water content increased (data not shown). Thermograms from extracted lipids were also double peaked (Figure 2-9c) and resembled those from the wettest soybean pellets. The endotherms of the extracted lipids suspended in water were similar to those of the dry extracted lipid except for a very large water melting endotherm at −15°C (data not shown).

Since extracted triglycerides would not be expected to interact with water, it was puzzling that they should be affected by moisture content in vivo. We do have structural evidence that in dry tissues the lipid droplets are small and that these droplets coalesce into larger bodies upon hydration (Webster and Leopold, 1977; Chabot and Leopold, 1985). The melting transitions of the small lipid droplets may be affected by the phospholipid membrane and protein matrix which surround the lipid. The interaction between the stored lipids and the surrounding membrane and protein matrix may become less important as the size of the lipid droplets increases. Thus, in dry seed pellets where the droplets are small, the low transition enthalpies occurring as broad peaks at relatively low temperatures in the DSC give the appearance of impurities. As the moisture content of the seed increases, the lipid droplets coalesce because of hydrophobic interactions. The melting transitions of the lipids within these larger liposomes resemble purer substances with higher transition temperatures and enthalpies and narrower endothermic peaks. Since the effect becomes prominent at water content of about 21%, we suggest that hydrophobic interactions become substantial at this point.

Conclusions

In this chapter we have examined changes in several types of physiological activities in dry seeds, and we have assembled various indications of changes in macromolecular structure with various levels of hydration.

Our data suggest that there is no electron transport in the first hydration level. Photosynthetic electron transport was observed in the second, and both photosynthetic and mitochondrial electron transport were observed in the third water binding level. A shift between states of dormancy and nondormancy in apple seeds occurred if their moisture contents were in or above binding level II. Presumably these findings reflect changes in the metabolic capabilities of the seeds with hydration level. Evidence of major changes in the structure of soybean seed components as water content increased was demonstrated by changes in estimates of seed polymer swelling. Most active polymer swelling occurred in the I and II hydration sectors. In pea seeds there was a tendency toward an increasing polar environment as water contents increased from level I to level II, as evidenced by shifts in the environment of chlorophyll *a*. The changes in triglyceride melting characteristics in soybeans may reflect a coalescing of nonpolar moieties as water contents increase from hydration level II to III; we suggest the aggregation of hydrophobic components in this hydration region.

In the literature as well as in our own studies, the three levels of water binding in seeds appear to have particular biophysical significance. It is in the second level of water binding that ''liquid water'' is first observed by proton NMR (Seewaldt et al., 1981) and that proton mobility is possible (Chapter 16). It is in the third level of water binding that changes in phospholipid bilayer structure occur, as have been previously reported (Luzatti and Husson, 1962; Finean, 1969; Seewaldt et al., 1981). Because the three regions sustain different types of physiological activity, as well as showing distinctive biophysical features, we suggest that they represent central steps in the shift from an anhydrobiotic to a hydrated state.

REFERENCES

Acker, L.W. 1969. Water activity and enzyme activity. Food Technol. 23: 27-40.
Careri, G., Gratton, E., Yang, P.H. and Rupley, J.A. 1980. Correlation of IR

spectroscopic, heat capacity, diamagnetic susceptibility and enzyme activity measurements on lysozyme powder. Nature 284: 572-573.

Chabot, J.F. and Leopold, A.C. 1982. Ultrastructural changes of membranes with hydration in soybean seeds. Am. J. Bot. 69: 623-633.

Clegg, J.S. 1978. Interrelationships between water and cellular metabolism in *Artemia* cysts. J. Cell Physiol. 94: 123-138.

Clegg, J.S. 1984. Interrelationships between water and cellular metabolism in *Artemia* cysts. XI. Density measurements. Cell Biophys. 6: 153-169.

Crowe, J.H. and Crowe, L.M. 1982. Induction of anhydrobiosis: membrane changes during drying. Cryobiology 19: 317-328.

Finean, J.B. 1969. Biophysical contributions to membrane structure. Quart. Rev. Biophys. 2:1-46.

Leopold, A.C. 1983. Volumetric components of seed imbibition. Plant Physiol. 73: 677-680.

Luzzati, V. and Husson, F. 1962. The structure of the liquid- crystalline phases of the lipid-water systems. J. Cell Biol. 12: 207-219.

McKersie, B.D. and Stinson, R.H. 1980. Effect of dehydration on leakage and membrane structure in *Lotus* seeds. Plant Physiol. 66: 316-320.

Mayer, A.M. and Poljakoff-Mayber, A. 1963. Factors affecting germination. *In* The Germination of Seeds. Pergamon Press, New York.

Nawa, Y. and Asahi, T. 1973. Relationship between the water content of pea cotyledons and mitochondrial development during the early stage of germination. Plant and Cell Physiol. 14: 607-610.

Obendorf, R.L. and Hobbs, P.R. 1970. Effect of seed moisture on temperature sensitivity during imbibition of soybean. Crop Sci. 10: 563-566.

Priestley, D.A. and de Kruijff, B. 1982. Phospholipid motional characteristics in a dry biological system. Plant Physiol. 70: 1075-1078.

Schneider, M.J.T. and Schneider, A.S. 1972. Water in biological membranes: Adsorption isotherms and circular dichroism as a function of hydration. J. Membrane Biol. 9: 127-140.

Seewaldt, V., Priestley, D.A., Leopold, A.C., Feigenson, G.W. and Goodsaid-Zalduondo, F. 1981. Membrane organization in soybean seeds during hydration. Planta 152: 19-23.

Simon, E.W. 1974. Phospholipids and plant membrane permeability. New Phytol. 73: 377-420.

Thomson, W.W. and Platt-Aloia, K. 1982. Ultrastructure and membrane permeability in cowpea seeds. Plant Cell Environ. 5: 367-373.

Vertucci, C.W., Ellenson, J.L. and Leopold, A.C. 1985. Chlorophyll fluorescence characteristics associated with hydration level in pea cotyledons. Plant Physiol. 79: 248-252.

Vertucci, C.W. and Leopold, A.C. 1984. Bound water in soybean seed and its relation to respiration and imbibitional damage. Plant Physiol. 75: 114-117.

Webster, B.D. and Leopold, A.C. 1977. The ultrastructure of dry and imbibed cotyledons of soybean. Am. J. Bot. 64: 1286-1293.

Williams, K.M. 1984. ATPase activity and Water Uptake in Apple Seeds during Stratification. Ph.D. thesis, Cornell University.

Wilson, S.B. and Bonner, W.D. 1971. Studies of electron transport in dry and imbibed peanut embryos. Plant Physiol. 48: 340-344.

3 A Study of Metabolism in Dry Seeds of *Avena fatua* L. Evaluated by Incubation with Ethanol-1-^{14}C

Ronald M. Lynch
James S. Clegg

Most phyla contain representative organisms that can lose very large fractions of their intracellular water in a reversible way (Crowe and Clegg, 1978; Chapters 10-11). This capability is of interest in an evolutionary sense and provides some interesting questions at all levels of biological organization. One of these concerns the metabolic status of the organisms as they undergo water loss: does the system reach a water content at which metabolism comes to a *reversible standstill*, or does the system continue to express a metabolism, albeit ever so slow, down to the point where virtually no water exists? That question is of some importance when we evaluate the biological significance of the reversibly dried state. It has been proposed that any biological system whose water content has been reduced to, say, 0.1 g H_2O/g dry weight (or g/g for short) cannot carry out a metabolism, given a reasonable definition of that term (Clegg, 1978). The basis for that view comes chiefly from studies on the physical properties of *Artemia* cysts as a function of water content. However, a number of published papers claim that metabolism is, indeed, occurring at this water content or even lower (Chen, 1972, 1978; Edwards, 1976; Wilson et al., 1979). We have reexamined one of these studies and report our results here.

Chen (1972, 1978) obtained evidence for the existence of metabolism in the dry state by exposing seeds of wild oat (*Avena fatua*) to radioactive ethanol for 3 days at room temperature and subsequently analyzing them for

We thank S. S. C. Chen for supplying us with the seeds used in this study and Vivien Roe for manuscript preparation. The research was supported by grants from the U.S. National Science Foundation (PCM 76-24037, 79-25609, 82-17311).

distribution of radioactivity in various extracts. The seeds contained 10-11% water (≈0.1 g/g). He reported that significant incorporation of ^{14}C-ethanol occurred, the label being present in sucrose, glucose, fructose, a broad spectrum of free amino acids, and protein (peptide-bonded amino acids); somewhat lower incorporation into nonamino organic acids was also reported (Chen, 1972). He concluded that the seeds were metabolizing and suggested that this metabolism could be associated with afterripening. Finding this observation to be at variance with our views on the matter, we corresponded with Dr. Chen, who very kindly sent us samples of the seed population he used in his work. The history of these seeds is given by Chen (1978).

All experiments to be described here were performed during the interval September-April 1979. Isotopically labeled ethanol-1-^{14}C (specific activity 6.9 mCi/mmole, New England Nuclear Corporation) was obtained in 100 μCi quantities provided in breakseal tubes. Two separate tubes were used for these studies, their upper chambers providing the seed exposure chamber as described by Chen (1972 and 1978). In effect, we utilized all of his incubation techniques in our experiments, and further details can be found in his papers. Three experimental designs were followed.

Experiment 1

Air-dried seeds were dehulled, surface sterilized in 100% 2-propanol for 10 min, and dried under a germicidal lamp for 24 hr. Fifty-five percent of a 22-seed sample germinated when placed in water for 5 hr. Of 20 seeds, 5 were used to determine the initial moisture content (24 hr at 104°C), and another 5 seeds were used to determine final moisture content. The seeds were exposed to the labeled ethanol for 3 days in the upper chamber of the first breakseal tube at room temperature (22-24°C).

The moisture content of these seeds prior to incubation was 9.78%; after 3 days' incubation, the content rose to 13.22%. That this weight increase is chiefly due to water and not to the uptake of ethanol is indicated by the following: the total mass of 10 seeds prior to ethanol exposure was 109.49 mg, and after exposure these same seeds weighed 112.69 mg. The *total* mass of ^{14}C-ethanol in the tube was only 0.9 mg (Nuclear-Chicago specifications), which is about 28% of the 3.20 mg net weight gain of the seeds. We suggest that the gain in weight is due largely to water uptake from the atmosphere in the breakseal tube during incubation. Thus, we consider the water content of our seeds to be 10-13% during incubation with ^{14}C-ethanol.

After the 10 seeds had been exposed to ^{14}C-ethanol, they were thoroughly homogenized in a Ten-Broeck tissue homogenizer with 2 ml of 80%

ethanol and the homogenate centrifuged for 15 min at 12,000 × g. The pellet was washed twice with 2 ml of 80% ethanol and the supernatants combined. Aliquots of ethanol-soluble fractions and ethanol-insoluble fractions were plated on stainless steel planchets, dried at 60-65°C, and assayed for radioactivity using a Nuclear-Chicago thin-window Geiger counter (<4 cpm background). Because the samples are thoroughly dried, no *free* labeled ethanol will appear in any of our data, which therefore represent other forms of the isotope. After counting in duplicate, 0.5 ml of 0.05 N HCl was added to one of the duplicate planchets and 0.5 ml of 0.5 N NH_4OH was added to the other. These were redried in the same fashion and recounted. Addition of distilled water and redrying showed that no significant change in radioactivity levels resulted from the solvation and recounting process itself.

The results are summarized in Table 3-1. Substantial nonvolatile radioactivity was associated with the seeds, about 92% of this being ethanol soluble, a result also observed by Chen (1972, 1978). Note that treatment with acid liberated almost all of the homogenate radioactivity. The effect of base was much less dramatic. Further analysis of the ethanol-soluble fraction was performed by ascending chromatography using Whatman #1 paper and a neutral solvent system [n-butanol (5): 95% ethanol (4): acetone (3): water (2)]. Developed chromatograms of the extracts were cut into 1.5 cm strips and each strip eluted with distilled water directly onto planchets, which were then dried and assayed as usual. After counting, the planchets were treated with 0.5 cc of 0.05 N HCl, dried, and recounted (Figure 3-1).

The open bars in Figure 3-1 represent ^{14}C present in the serially eluted strips. We noted, as did Chen (1972, 1978), that the migration of most of the

Table 3-1
Radioactivity associated with different fractions of wild oat seeds after incubation with gaseous 1-^{14}C-ethanol for 3 days

Fraction	CPM/10 seeds[a]	Decrease (acid/base) (%)
Homogenate	72,200	—
+HCl	2,950	96
+NH_4OH	52,000	38
Ethanol soluble	66,300	(92% of homogenate)
Ethanol insoluble	2,750	(4% of homogenate)

Note: All data represent nonvolatile radioactivity; thus, all free ^{14}C-ethanol associated with the samples would be driven off during sample drying.

[a]The dry weight of these 10 seeds was 97.35 mg, corrected for residual water content.

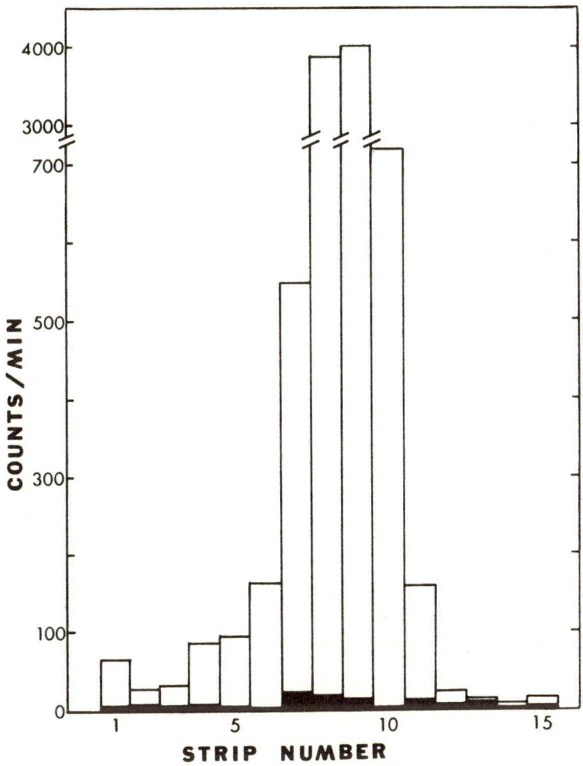

Figure 3-1. The distribution of water soluble radioactivity on chromatograms of the ethanol-soluble fraction of wild oat seeds incubated with ethanol-1-^{14}C (open bars). Strip 1 is the origin and strip 15 is the solvent front. Authentic samples of sucrose, glucose, and fructose migrate to strips 7 through 10. Black bars are the results obtained when the planchets are redried in the presence of 0.05 N HCl.

radioactivity coincided with that of sucrose, glucose, and fructose (strips 7-10). However, when we added dilute acid (0.05 N HCl) to these same planchets and then dried and recounted them, only 2% of the applied radioactivity remained (black bars). Clearly, this result would not have occurred had the radioactivity been present in the covalently bound carbons of these sugars.

The findings suggested that all or most of the radioactivity present might result from adsorption to seed components rather than from metabolic processing of the labeled ethanol by the seeds prior to their extraction. Experiment 2 was designed to test this possibility further.

Experiment 2

Fifteen seeds were dehulled, were surface sterilized as above, and were divided into three groups of 5. One group served as a control (viable), another was heated at 104°C for 16 hr (heat killed), and the remaining group was homogenized with 80% ethanol and the homogenate redried. All three groups were allowed to equilibrate with laboratory humidity (about 65%) and were then exposed to ethanol-1-^{14}C in a second breakseal tube, as previously described. After the 3-day incubation period, the three samples were separately homogenized, extracted, and assayed for radioactivity as before. In this case, additional treatment of the ethanol-insoluble fraction was carried out to purify total protein: after suspension in 0.5 N NaOH and incubation at 45°C for 3 min, the preparation was centrifuged and the supernatant treated with two volumes of 10% trichloroacetic acid (TCA). The resulting insoluble fraction was collected by centrifugation, resuspended in 5% TCA and heated for 90 min at 95°C to hydrolyze nucleic acids. The resulting protein was collected by centrifugation, washed twice with 5% TCA, and dissolved with 77% formic acid for radioactivity determinations.

Table 3-2 shows that all preparations contained appreciable amounts of ^{14}C. The fact that the homogenized preparation exhibited over three times more radioactivity than did viable seed is consistent with the interpretation of nonmetabolic binding. The homogenized preparation can be expected to exhibit an increase in surface area and exposed binding sites compared with intact seed. Likewise, heat-killed seed may exhibit less isotope because of decreased binding potential as well as because of a possible decrease in seed permeability to the labeled ethanol, although we cannot be sure.

Figure 3-2 illustrates the results from chromatographic analysis of the ethanol-soluble fractions. In this case, unlike that shown in Figure 3-1, the strips were placed directly on planchets and were not eluted with water. The bars represent homogenates (black bars), extract of heat-killed seeds (center

Table 3-2
Incubation of viable, heat-killed, and homogenized seeds of *Avena fatua* with ethanol-1-^{14}C for 3 days

	CPM/mg dry weight of preparation		
Fraction	Viable	Heat killed	Homogenized
Homogenate	207	106	733
Ethanol soluble	166	41	541
Ethanol insoluble	41	54	178

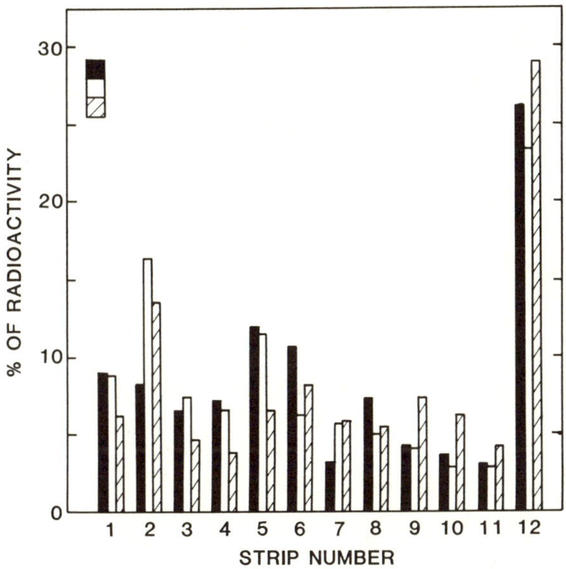

Figure 3-2. The distribution of total radioactivity on chromatograms of the ethanol-soluble fraction from preparations of wild oat seeds incubated with ethanol-1-^{14}C. Strip 1 is the origin and strip 12 is the solvent front. The central bar for each strip represents extracts of heat-killed seeds; that to the left represents seed homogenates and that to the right the distribution of radioactivity from extracts of viable seeds.

clear bars), and extract of viable seeds (right hatched bars). The distribution of radioactivity along the paper for viable seed extracts differed significantly from that obtained for viable seed in Experiment 1. The difference is presumably due to the method of assay (water elution versus total radioactivity). The relative distributions of radioactivity on the chromatograms of all three seed preparations are essentially the same. In our judgment such a result would be impossible if viable seeds were metabolizing the radioactive ethanol. On thermodynamic grounds, we reject the possibility that seeds pretreated for 16 hr at 104°C could carry on any significant metabolism.

Table 3-3 summarizes our results from further study of the ethanol-*insoluble* fraction of the three preparations. In this case we purified the total protein as described above. It is evident that virtually all of the isotope was removed by the purification procedure. The same result would not have occurred if the radioactivity were present in peptide-bound amino acids. Furthermore, all seed preparations exhibited nearly the same small percentage of radioactivity in their protein fraction, in spite of the fact that these were markedly different preparations. In summary, we feel that the outcome

Table 3-3
Analysis of radioactivity in the ethanol-insoluble fraction and purified protein from different preparations of *Avena fatua* seeds incubated with ethanol-1-^{14}C

	CPM/mg dry weight seeds		
Preparation	Ethanol-insoluble fraction	Extracted protein	(%)
Viable	33	1.1	3.3
Heat killed	55	2.1	3.8
Homogenized	102	3.2	3.1

of Experiment 2 strongly argues against significant metabolic incorporation of ethanol-1-^{14}C into viable dry oat seed.

Experiment 3

In the final study, three purified macromolecules in powdered form were exposed to labeled ethanol to evaluate the nonmetabolic binding of radioactivity from the ethanol sample. Samples of yeast RNA (Miles Laboratories), bovine hemoglobin (Sigma Chemical Corporation), and oyster glycogen (Mann Research Laboratories) were weighed and then exposed to ethanol-1-^{14}C in the upper chamber of the second breakseal tube for the 3-day incubation period in the same fashion as in the previous experiments. After exposure, the samples were reweighed to evaluate weight change and were separately suspended in 2.0 ml of water. Aliquots (0.2 ml) were counted as usual, then were treated with 0.5 ml of 0.05 N HCl before being dried and recounted.

We reasoned that, if the presence of radioactivity in viable seeds did result from nonmetabolic adsorption onto seed components, then it should be

Table 3-4
Radioactivity associated with powdered macromolecules after incubation with ethanol-1-^{14}C

Macromolecule	CPM/mg dry weight	Radioactivity removed by acid treatment (%)
Hemoglobin	1879	27
RNA	36	49
Glycogen	65	74

Note: See text.

possible to observe such adsorption using purified compounds. Table 3-4 summarizes the outcome of such a study, which did indeed reveal substantial binding of radioactivity, particularly by hemoglobin. Also, varying amounts of adsorbed isotope were labile under mild acid treatment. Furthermore, the kinetics of binding have been followed for hemoglobin. The result is that of an adsorption isotherm, with saturation occurring in between 1-2 days of exposure (data not shown).

Discussion and Conclusions

In our opinion the results of these three experiments cast considerable doubt on the conclusion that air dry seeds of wild oat, containing about 11% moisture, *metabolize* exogenous ethanol to an appreciable extent (Chen, 1972, 1978). Rather, we propose that the earlier results and ours can be adequately explained on the basis of nonenzymatic binding of radioactivity to seed components. It is very difficult to understand how ethanol itself could be held with sufficient energy to seed components such that it would be retained on the planchets after complete drying of the fractions at 60-65°C. We reject the possibility that ethanol interactions could occur only by hydrophobic interactions or by weak polar bonding (hydrogen bonds, ion-dipole interactions, etc.). These could not withstand our treatments before counting. Moreover, the fractionation itself was carried out with 80% ethanol, and the complete exchange of ^{12}C-ethanol for the isotopic form might be expected to occur under such conditions. In view of all these observations, we consider it most likely that one or more radioactive volatile impurities are present in the commercial ethanol preparation and that these are involved in binding.

We had neither the equipment nor the time to evaluate the kinds of volatile contaminants that might be present in these samples. However, a discussion with the technical experts at New England Nuclear Corporation (in August 1979) leads us to view acetaldehyde and its "trimer" (para-acetaldehyde: 2,4,6-trimethyl-1,3-5-trioxane) as prime suspects. The latter is apparently a notorious binder of a wide variety of organic compounds and forms spontaneously from acetaldehyde under "dry" conditions, precisely those of our experiments.

Finally, we evaluate the fraction of the available ethanol that is associated with the seeds after 3 days of incubation. Calculating from the value for the acid stable ^{14}C in Table 3-1, we estimate that about 0.004% of the available radioactivity was associated with 10 seeds after 3 days of exposure. Chen (1978) also obtained incorporation percentages of the same order as ours or less. In view of these considerations, it would seem very risky to assume that the association of ^{14}C with the seeds had a metabolic origin.

While we believe that the use of ^{14}C-ethanol has *not* shown that these seeds carry out a metabolism (given a reasonable definition), we also emphasize that the same studies do not exclude its presence. As discussed at length elsewhere (Clegg, 1973, 1978; Chapter 10), it seems unlikely that any biochemical experiment of this sort can achieve that goal with confidence. Nevertheless, we maintain our position that cells containing 0.1 g/g or less of water have lost the physical properties upon which metabolism operates; therefore, they cannot metabolize. This position is similar to that taken in Chapter 2.

One final point should be made regarding the foregoing. We accept the possibility that isolated enzymatic reactions may occur in cells at or below water contents of 0.1 g/g and that these may be reflected in the chemical modification of some isotopically labeled precursor presented to the cells. However, that possibility should not, in our judgment, be taken as evidence for the existence of a "metabolism" if by that term we mean the operation of *integrated pathways* of enzyme-catalyzed reactions that are regulated in rate and direction and that contribute to the maintenance of the cells in which they occur. We propose that one must carefully distinguish between metabolism and adventitious chemistry and physics. The latter, indeed, eventually kills the dried cell rather than being of value to it (cf. Chapter 5): the conditions that suppress such chemical reactions are precisely those that contribute to the survival time (Clegg, 1978).

REFERENCES

Chen, S.S.C. 1972. Metabolic activities of dormant seeds during dry storage. Naturwissenschaften 59: 123.

Chen, S.S.C. 1978. Application of a volatile tracer to detect metabolism in dry oat seeds. *In* Dry Biological Systems (J.H. Crowe and J.S. Clegg, eds.), pp. 175-184. Academic Press, New York.

Clegg, J.S. 1978. Hydration dependent metabolic transitions and the state of water in *Artemia* cysts. *In* Dry Biological Systems (J.H. Crowe and J.S. Clegg, eds.), pp. 117-154. Academic Press, New York.

Crowe, J.H. and Clegg, J.S. (eds.). 1973. Anhydrobiosis. Dowden, Hutchinson and Ross, Stroudsburg, PA.

Crowe, J.H. and Clegg, J.S. (eds.). 1978. Dry Biological Systems. Academic Press, New York.

Edwards, M. 1976. Metabolism as a function of water potential in air-dry seeds of charlock. Plant Physiol. 58: 237-239.

Wilson, A.J., Vickers, M. and Mann, L.R.B. 1979. Metabolism in dry pollen: a novel technique for studying anhydrobiosis. Naturwissenschaften 66: 53-54.

4 Alteration of Genetically Regulated Syntheses in Seeds by Desiccation

Allison R. Kermode
J. Derek Bewley

During its development, the plant seed passes through various contiguous phases. The first phase, histodifferentiation, or initial morphogenesis (Long et al., 1981), commences with pollination and, as a consequence of rapid cell division, leads into the formation of the embryonic axis and of tissues that eventually accumulate reserve materials (Adams and Rinne, 1980). There follows the maturation, or seed expansion phase, during which there is a marked increase in dry weight; cell expansion occurs as reserve deposition takes place. The third and final phase—maturation drying, or desiccation—is characterized by a general reduction in metabolism as water is lost from the seed.

Upon imbibition of this mature seed, there is a reactivation of existing metabolic systems, supplemented by synthesis of new components, which leads to renewed cell expansion and subsequent cell division. Mobilization of stored reserves is an essential component of this latter event, and hence quantitative and qualitative changes in catabolic enzymes, particularly in the storage organs, are commonly observed events.

Even a casual consideration of seed development and germination/growth reveals that the major metabolic events associated with the former are distinctly different from those involved in the latter. During development the metabolism of many cells in the seed is geared towards the massive synthesis of reserve materials such as protein, lipid, and carbohydrate; following ger-

Allison R. Kermode is the recipient of a Natural Sciences and Engineering Research Council of Canada postgraduate scholarship and a Ralph Steinhauer Award of Distinction. This chapter reports on work supported by NSERC Grant A6352.

mination the reverse process, namely catabolism of reserves, is effected. In the vast majority of seeds, a desiccation period is naturally interpolated between development and germination, and it is logical to surmise that this loss of water plays some role in the "switch" in cellular activities from an exclusively developmental programme to an exclusively germination/growth-oriented programme.

For several reasons, it is not feasible to use the mature dry seed to study the role of desiccation in terminating development and eliciting germination. For example, it is not clear whether the metabolic changes elicited during the maturation drying period are due ultimately to water loss from seed tissues or are a preprogrammed part of development which would occur even if the seed was not allowed to dry out (Bewley, 1979, 1982). Hence, several researchers have taken the approach of subjecting the developing seed to *premature* desiccation (i.e. they have taken the seed from the mother plant at some time before its normal development is completed and have determined the effect of an artificially imposed drying period). The fundamental questions that can be addressed using this approach are two. If a seed is dried prematurely, upon subsequent rehydration is its metabolism still geared to resume those developmental events which were incomplete at the time of desiccation? Or does the prematurely imposed drying period cause the cessation of developmental events and lead, upon subsequent rehydration, to a metabolic programme associated exclusively with germination and growth? The following account attempts to answer these questions.

Normal Maturation and Seed Drying

During the latter part of the seed expansion (or maturation) phase of development, most seeds lose water as reserves are deposited primarily within storage tissues. In the castor bean, for example, the decline in whole seed water content and percentage water content from 20 to 50 days after pollination (DAP) (the period of dry weight gain) is likely to be due to the displacement of water by protein, phytin, and lipid reserves deposited within the endosperm (Greenwood and Bewley, 1982; Kermode and Bewley, in press *a*). Then, as the seed undergoes maturation drying and approaches quiescence, there is a period of fresh weight loss accompanied by a rapid decline in whole seed water content (Figure 4-1).

Little is known about the mechanism and route of water loss from seeds. Some studies suggest the existence of a passive mechanism whereby water is lost primarily by evaporation from the surface of surrounding seed structures (Nechiporenko and Rybalova, 1983; Lee and Atkey, 1984; Goncharova et

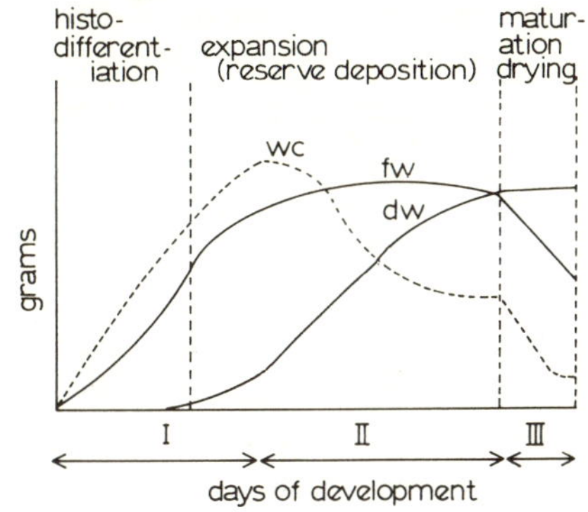

Figure 4-1. A general scheme of changes in whole seed fresh weight (fw), dry weight (dw), and water content (wc) during the histodifferentiation, expansion, and maturation drying phases of seed development. Three major periods are noted: I. rapid fresh weight gain, II. rapid dry weight gain, III. fresh weight loss. For data collected during castor bean seed development, see Greenwood and Bewley (1982) and Kermode and Bewley (in press *a*).

al., 1985). Another suggestion is that water moves from the seed to the mother plant by a metabolically active process (i.e. the plant actually "pumps" the water from the seed, Meredith and Jenkins, 1975).

In the castor bean, the desiccation period is most likely initiated by the severing of the vascular supply to the seed (funiculus detachment occurs between 40 and 50 DAP) and senescence of the capsule (Greenwood and Bewley, 1982). This event would suggest that relocation of water from the seed to the mother plant is not the means by which water loss occurs. Similarly, pectic substances in the lumina of xylem elements of the rachis of wheat and barley (laid down during the final stages of grain maturation) may lead to the progressive dehydration of the ear by cutting off its water supply (Cochrane, 1985).

The Acquisition of Desiccation Tolerance

That drying is important for the acquisition of germinability has been well documented for cereals (see the review in Mitchell et al., 1978). As early as 1852, Duchartre noted the beneficial effect of drying upon the subsequent germination of rye, wheat, and barley grains. Harlan and Pope (1922) showed that barley will germinate at 5 days postanthesis if dried in situ on the straw; similar observations were made that immature wheat grains will germinate after drying (Wellington, 1956; King, 1976). In fact, not only do air-dried grains of wheat germinate at an earlier stage of development than nondried grains, but at later stages they may also germinate at a faster rate than their nondried counterparts (Mitchell et al., 1978). These observations are not unique to cereals, since immature seeds of some legumes also will not germinate on water when removed from the mother plant in the fully hydrated state but will germinate after drying (Adams and Rinne, 1981; Long et al., 1981; Dasgupta et al., 1982).

Seeds appear not to tolerate drying at all stages during their development, however. Indeed, they may undergo a transition from a desiccation-intolerant state to a desiccation-tolerant state at a particular time in the course of their development. Seeds of *Phaseolus vulgaris* (French bean), for example, will not germinate if removed fresh from the pod and placed in water at stages of development prior to 40 days. Seeds at 26-32 days of development can be induced to germinate to increasing extents if they are first dried over silica gel, whereas those dried at 22 days fail to germinate when rehydrated and eventually deterioriate (Dasgupta et al., 1982). Subsequent studies have confirmed that at 22 days of development the seeds do not recover their full cellular and metabolic integrity following the premature drying treatment. At later times of development, however, the seeds acquire a tolerance of desiccation and are induced to germinate by this treatment. A similar situation exists for the castor bean seed. Here germinability is not achieved until after some 50-55 days of development, whereas premature drying will promote the germination of seeds as young as 25 days (Table 4-1; Kermode and Bewley, in press *a*). At earlier stages of development, drying not only fails to induce germination but also causes the death of the seed. For both *P. vulgaris* and castor bean, the seeds acquire a tolerance of desiccation and an ability to be induced to germinate by this treatment about 25 days after development commences.

In relation to drying and germinability, it is noteworthy that seeds of *P. vulgaris* and castor bean when removed in the fresh state from the pod will germinate only after achievement of maximum dry weight, which is subsequent to the beginning of water loss in situ (Dasgupta et al., 1982; Greenwood and Bewley, 1982). Full germination is not achieved until the seeds of these species have nearly completed the natural desiccation phase of their development, however (Table 4-1) (Kermode, Gifford, and Bewley, in press).

The transition to a desiccation-tolerant state approximately midway through development is also characteristic of other seeds, e.g. *Pisum sativum*, *Glycine max*, *Agrostemma githago* (Rogerson and Matthews, 1977; Adams and Rinne, 1981; de Klerk, 1984). Tolerance of desiccation is gained over only a few days of development (e.g. between 20 and 25 days in castor bean); it is achieved well prior to the completion of major developmental events such as reserve deposition and the commencement of normal maturation drying (Kermode, Dasgupta, Misra, and Bewley, in press). For example, in castor bean, at 25 days of development (the earliest desiccation-

Table 4-1
Germination of dried or fresh intact seeds of castor bean harvested at various times after pollination

DAP	Fresh or dried	Germination	Germination 7 days after imbibition (%)	Percentage showing normal postgermination growth[a]
20	F	−	0	0
	D	−	0	0
25	F	−	0	0
	D	+	75	0-33
30	F	−	0	0
	D	+	100	25-75
35	F	−	0	0
	D	+	100	90-100
40	F	−	6	0
	D	+	100	90-100
45	F	−	0	0
	D	+	100	100
50	F	+	70	40
	D	+	100	100
55	F	+	77	22-30
	D	+	100	100
60[b]	F	+	100	100

[a]The percentage of developing seeds achieving the normal radicle lengths of mature germinated seeds 48-120 hr after imbibition.
[b]Mature dry seed.

tolerant stage), the seed has reached only 20% of the maximum dry weight attained during development (Kermode and Bewley, in press *a*). At this time, the endosperm has accumulated 17% of its total protein reserves and 13% and 3% of the amount of lipid and phytin, respectively, present in the mature dry seed (Greenwood et al., 1984). Even at 30 days of development, the stage at which full germination can occur after drying, the seed has achieved only about 40% of maximum dry weight. Thus, the capacity to germinate, which can be expressed following premature drying, does not require the completion of major developmental events related to storage deposition. Of course, we do not know whether a seed can acquire germinability during the intolerant state of development because the only means known to express that ability—premature drying—results in seed death.

Capacity for Normal Growth

In wheat (Symons et al., 1983), corn (Sprague, 1936), and cotton (Kulieva et al., 1985), the capacity for normal germination and growth has been related to the water content of the seed (either at the time of harvest or after premature drying). In corn a decrease in water content (of freshly harvested developing seeds or immature seeds dried for various periods of time) is correlated with an increase in the percentage of seeds germinating and a decrease in the variability in time required to complete germination (Sprague, 1936). A similar correlation has been observed in cotton (Kulieva et al., 1985).

The capacity of prematurely harvested castor bean seeds to achieve normal postgerminative growth also undergoes a change during development (Kermode and Bewley, in press *a*). For example, although approximately 75% of seeds at 50 to 55 days after pollination germinate when removed from the capsule and hydrated in the fresh state, only about one-third of these seeds achieve the normal radicle lengths exhibited by germinated mature (60 DAP) seeds imbibed for the same lengths of time (Table 4-1). In contrast, when these developing seeds (50 to 55 DAP) are prematurely dried and then imbibed, they exhibit their full potential for normal radicle growth.

In wheat, premature drying of intact grains increases the percentage and rate of germination. Furthermore, whereas removal of the outer pericarp at various stages during development permits qualitatively different types of growth (i.e. callus, coleoptile, or normal), depending upon the total grain moisture content, drying of intact grains promotes normal germination only (Symons et al., 1983).

When developing castor bean seeds are prematurely dried and then imbibed, 90 to 100% of those at 35 DAP and older achieve normal radicle growth (Table 4-1). There is a lower capacity for normal growth following desiccation/rehydration at stages of development earlier than 35 DAP. The relationship between developmental stage and the capacity for normal postgerminative growth following rehydration may be related to interruption of reserve deposition by the drying period. The full capacity for normal postgerminative growth in castor bean is acquired later during development (35 DAP) than is full germinability (30 DAP) and may depend to a greater extent upon processes related to reserve deposition (Kermode and Bewley, in press *a*).

The Rate of Desiccation

The rate at which drying occurs during the tolerant phase of development is critical for the subsequent expression of germinability. For example, while gradual rates of water loss result in the germination of castor bean seeds as young as 25 DAP (Table 4-1), rapid drying over silica gel is fatal to seeds younger than 55 DAP (Kermode and Bewley, in press *a*).

Gradual water loss may allow protective changes to occur and hence may increase the seed's resistance to disruption by dehydration. Rapid drying predisposes seeds to imbibitional injury, as indicated by increased rates of solute leakage (Adams et al., 1983), a symptom of cellular membrane disruption. Likewise, slowing the rate of hydration may prevent a loss of germinability; for example, germinability of soybean seeds (following an accelerated aging treatment) is increased from 10 to 90% by controlling the rate of imbibition (Tilden and West, 1985).

The capacity to withstand rapid or slow desiccation during the tolerant phase of development varies with the species. An intolerance of rapid desiccation exists during the first 42 days of soybean seed development, although these seeds can tolerate slow drying at stages as early as 26 DAP (Adams et al., 1983). *P. vulgaris* seeds, on the other hand, can survive and germinate following rapid drying over silica gel at relatively early stages during development (Long et al., 1981; Dasgupta et al., 1982). Thus, when it is stated that a seed acquires a tolerance of desiccation at a particular stage during its development, it is necessary to define the rate of water loss to which it is subjected.

The Transition to Desiccation Tolerance

Premature desiccation during the early developmental stages of *P. vulgaris* (up to 22 days of development) drastically reduces the metabolic and cellular integrity of the axis upon subsequent rehydration. Particularly evident is a loss in the capacity to recover polyribosome levels and to resume protein synthesis (Dasgupta et al., 1982). Ultrastructural observations indicate considerable damage to cellular organelles (including protein bodies and mitochondria) and to the nuclear membrane. In contrast, such severe perturbations do not occur following desiccation/rehydration at a tolerant stage (e.g. at 32 days of development). Drying may inflict limited damage to cells during the desiccation-tolerant stage of development, but the capacity to reverse such changes (or to effect "repair") upon subsequent rehydration is probably an integral feature of desiccation tolerance (Bewley, 1979).

During germination, seeds initially remain tolerant of desiccation, but at some stage after axis elongation there is a loss of this tolerance. Germinating soybean seeds are tolerant of drying during the early stages, up to 6 hr after commencing imbibition, but they become increasingly intolerant after this time (see Chapters 1 and 5). Thus, desiccation at 36 hr after the start of imbibition leads to a complete loss of seed viability (Senaratna and McKersie, 1983). The membranes appear to be a major site of damage in seeds during the desiccation-intolerant stage of germination, as indicated by ultrastructural studies (Crevecoeur et al., 1976) and by the observed increases in solute and electrolyte leakage upon subsequent rehydration (Senaratna and McKersie, 1983). There is some evidence that drying during the intolerant stage induces a deesterification of membrane phospholipids, with a resultant increase in the free fatty acid content of the membranes. Increased membrane fluidity (and hence increased permeability) and altered phase properties of the lipid bilayer ensue (Senaratna et al., 1984). A free radical-mediated mechanism of injury during desiccation has been proposed (Senaratna et al., 1985 and Chapter 5).

What changes, then, occur within the seed during that brief period of development when it changes from being intolerant of desiccation to being tolerant? Membranes are widely suspected of playing a key role in this transition (Rogerson and Matthews, 1977; Dasgupta et al., 1982), as most of the contributions to this volume indicate.

Carbohydrates such as trehalose (a non-reducing disaccharide of glucose) are effective in preserving the structural and functional integrity of membranes in vitro at low water contents (Chapters 10-13). Perhaps the acquisition of desiccation tolerance during seed development is accompanied by an increased amount of, or heightened capacity to synthesize, this or similar carbohydrates in response to desiccation (Adams et al., 1982 and Chapter 1).

The acquisition of desiccation tolerance during the maturation or seed expansion stage of development occurs well prior to the completion of reserve deposition. However, it is possible that the attainment of a critical level of reserves is required before the seed can withstand desiccation. Cells that are highly vacuolated (and hence contain little reserve material) may undergo severe mechanical disruption such as tearing or shearing of membranes (or other cellular components), leading to irreversible changes in their internal morphology. The presence of a critical level of cellular reserves would limit such changes. The quantity of reserves may also merit consideration in relation to the loss of tolerance during seed germination (Chapter 5). The major breakdown of reserves within the cotyledons is a postgerminative event; however, catabolism of reserves within the axes occurs relatively early (i.e. during germination). The decline of reserves below a critical level within the axes may contribute to a loss of desiccation tolerance within this germinating tissue.

Germination in the Absence of a Drying Period

It is possible for some developing embryos to germinate without first being dried. Seeds of some species (e.g. mangrove and certain viviparous mutants of cereals) will not tolerate drying (Chapter 20). Immature embryos of rape, French bean, cotton, soybean, rice, and wheat will germinate when excised from the developing seed and placed in tissue (embryo) culture conditions (Ihle and Dure, 1972; Quebedeux et al., 1976; Crouch and Sussex, 1981; Long et al., 1981; Triplett and Quatrano, 1982; Ackerson, 1984a; Obendorf and Wettlaufer, 1984; Stinissen et al., 1984). The isolated embryos may germinate immediately (e.g. cotton; Ihle and Dure, 1972) or may continue to develop and accumulate storage proteins for several days in culture as they would in the intact seed before becoming briefly quiescent and then germinating (e.g. French bean; Long et al., 1981). Presumably the environment of the developing embryo on the mother plant is normally such that it encourages development and prevents germination; removal from this environment causes a hiatus such that development ceases. The controlling factor(s) provided by the mother plant (or by maternal tissues surrounding the embryo) are not known, although addition of the plant hormone abscisic acid (at physiological concentrations) to developing embryos of several species in culture prevents their germination and encourages the continuation of developmental processes. In several cases its addition to the culture medium is a prerequisite for normal embryogeny to proceed (Ammirato, 1977; Karssen et al., 1983). In such cultures, abscisic acid (ABA) enhanced the following processes:

- the accumulation of stored reserves, particularly storage proteins, in embryos of rape, soybean, wheat, and French bean (Sussex and Dale, 1979; Crouch and Sussex, 1981; Triplett and Quatrano, 1982) but apparently not in cereal endosperms (Quatrano, personal communication);
- the development of enzyme activities in late embryogenesis (e.g. malate synthetase in cotton; Choinsky and Trelease, 1978; Choinsky et al., 1981);
- the production of mRNA and protein fractions characteristic of late embryogenesis in cotton (Dure et al., 1981).

These late proteins are not synthesized when embryos precociously germinate in the absence of ABA.

Hence, for embryos in culture, ABA appears to suppress germination and prolong developmental events. It has been surmised that a steady supply of ABA (or some other translocatable factor that ABA wholly or partially mimics) from the mother plant to the embryo maintains its metabolism in a developmental (i.e. largely anabolic) mode. Reduction in ABA content (or some other regulatory agent) could release seeds and allow germination. Desiccation is the normal terminal event in development of most seeds, and this may itself lead to a loss of control by the mother plant, perhaps by reducing the flow of germination-suppressing substances to the embryo (e.g. ABA) or by altering the sensitivity of the embryo to suppressive or promotive substances. For the time being, this must remain a topic for imaginative speculation. In mangrove (Sussex, 1975) and other viviparous seeds (e.g. viviparous maize mutants; McDaniel et al., 1977; Smith et al., 1978; Robichaud et al., 1980), a low sensitivity to or a low level of abscisic acid may explain why these seeds pass from a developmental to a germinative mode without an intervening quiescent period.

Desiccation Causes a Switch in Metabolic Activity

A major metabolic event in the development of the castor bean endosperm is the deposition of lipids within lipid bodies and the deposition of two types of protein (insoluble crystalloid protein and soluble lectin/albumin proteins) within protein bodies (Tully and Beevers, 1976; Youle and Huang, 1981). Following germination, these reserves are mobilized to provide nutrients for the growing seedling. Numerous enzymes for the hydrolysis and conversion of the major reserves are produced de novo. The quantitative changes in castor bean endosperm lipid and protein reserves during development and during subsequent germination and growth are illustrated in Figure 4-2. To determine whether drying plays a role in causing this major change in metabolism, the types of proteins synthesized in vivo during rehydration have been examined (Kermode and Bewley, in press *b*; Kermode, Dasgupta, Misra, and Bewley, in press; Kermode, Gifford, and Bewley, in press).

With one-dimensional (SDS) polyacrylamide gel electrophoresis, it has been demonstrated that the pattern of soluble protein synthesis in endosperms from developing castor bean seeds (20-50 DAP) is distinctly different from that carried out by endosperms isolated from mature-imbibed seeds (i.e. during germination and growth; Kermode and Bewley, in press *b*). Storage proteins (lectin and albumin) accumulate throughout a major part of seed development (20-50 DAP). Following imbibition of the mature seed, however, this type of protein synthesis ceases within 12 hr and a pattern of synthesis distinctly different from that occurring during development commences. Thereafter, especially 48 hr and later, there occurs the synthesis of proteins which can be identified as being unique to a germination/growth programme, including those soluble proteins (enzymes) directly involved in the breakdown and utilization of protein and lipid reserves (e.g. leucine β-naphthylamidase and isocitrate lyase, respectively). At this time (i.e. 48 hr and later) the major breakdown of endosperm storage proteins commences (Gifford et al., 1983; Kermode and Bewley, in press *b*). Thus, it appears that

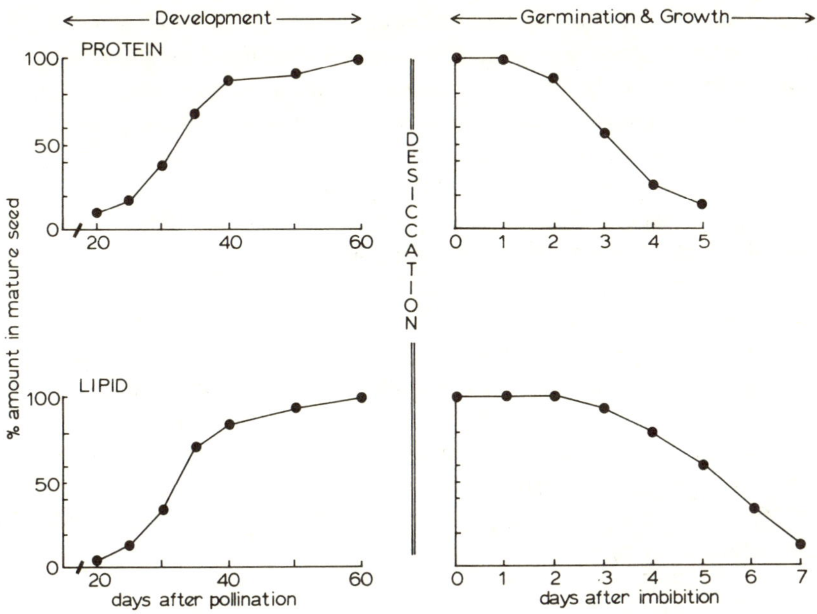

Figure 4-2. Quantitative changes in protein and lipid within the castor bean endosperm during development (from 20 to 60 days after pollination or DAP) and during germination and growth (from 1 to 7 days after imbibition or DAI). Values expressed as percentages in mature seed. Data from Muto and Beevers (1974); Greenwood et al. (1984); Kermode, Dasgupta, Misra, and Bewley (in press).

very early after the mature seed has imbibed, the metabolism of the endosperm is redirected towards the events of germination and growth. Soluble protein syntheses associated with development may occur during very early imbibition (at 5 hr) because of the presence of developmental mRNAs which remain stable through desiccation and rehydration. These mRNAs are rapidly lost (between 5 and 12 hr after imbibition) and are replaced by mRNAs involved in the production of proteins associated with germination and growth.

Following premature desiccation at 30 or 40 DAP, the pattern of protein synthesis upon subsequent rehydration is virtually identical to that following imbibition of the mature dry seed (Kermode and Bewley, in press *b*). In contrast, hydration of these developing (30 or 40 DAP) seeds immediately following detachment from the mother plant results in a continuation of the developmental pattern of protein synthesis. A change in direction of protein synthesis has been elicited by premature drying.

A similar dramatic change in the direction of protein synthesis in response to premature drying has also been demonstrated within axes of *Phaseolus vulgaris* seeds. Here again, the switch from development to germination is mirrored in the pattern of protein synthesis upon rehydration in that some proteins unique to development (including the storage protein, phaseolin) are no longer synthesized, while others recognizably associated with germination are produced (Dasgupta and Bewley, 1982). Thus, at a desiccation-tolerant stage of development, premature drying results in the suppression of the developmental programme and a switching on of the germination/growth programme (Misra et al., 1985; Kermode and Bewley, in press *b*; Kermode, Dasgupta, Misra, and Bewley, in press; Kermode, Gifford, and Bewley, in press). At 22 days of development of *P. vulgaris* axes, during the desiccation- intolerant stage, drying and rehydration do not lead to germination (Dasgupta et al., 1982). Moreover, the full capacity for protein synthesis is not recovered after drying, although some proteins are made, including both development and germination-related proteins (Dasgupta and Bewley, 1982). At this early stage of development the genome cannot respond effectively to a drying treatment.

Fate of the Protein-Synthesizing Complex during Maturation Drying

The protein synthesis associated with seed development ceases as desiccation proceeds during the final stages of maturation. It is pertinent to ask, therefore, whether some or all of the components involved in protein synthesis during development are destroyed as a consequence of desiccation or whether they are conserved in the dry seed and reutilized upon subsequent hydration.

In general, polyribosome levels decline during desiccation of seed storage organs, cereal embryos, or embryonic axes; polyribosomes are absent from these mature dry tissues (reviewed in Bewley, 1979, 1982; Bewley and Black, 1985). The absence of polyribosomes from dry seeds is thus a general phenomenon. Ribosomes and tRNA are found in abundance throughout maturation drying in potentially active states. A similar conservation of other components involved in protein synthesis (e.g. initiation and elongation factors, aminoacyl tRNA synthetases) has been noted. Their preservation throughout maturation drying is important, since it would allow a rapid resumption of cellular activity on exposure of the seed to water and would thus permit the completion of processes essential for germination (Bewley, 1982).

Studies of mRNA populations during seed development/germination have shown that there are changes in the frequency distribution of mRNAs during development of the seed and that a large proportion of embryonic mRNAs persist and are stored in the mature dry seed. These persistent mRNAs fall into two general classes.

Residual mRNAs. These are mRNAs synthesized and expressed during seed development which are not destroyed during late maturation and desiccation. A subset of messenger RNAs comprising principally the storage protein mRNAs declines during late maturation (Dure et al., 1981, 1983; Galau and Dure, 1981; Goldberg, Hoschek, Ditta, and Breidenbach, 1981; Goldberg, Hoschek, Tam, Ditta, and Breidenbach, 1981). However, in some dry seed (e.g. French bean, kidney bean, mung bean, soybean, wheat, and rye) there are residual mRNAs coding for developmental (storage) proteins (Mori et al., 1978; Bollini and Chrispeels, 1980; Carlier et al., 1980; Peumans et al., 1982; Misra et al., 1985; Misra and Bewley, 1985, in press). These residual messages may be translated during the early stages of germination but are simultaneously degraded and appear not to be essential to the germination process per se (Carlier et al., 1980; Peumans et al., 1982).

Stored or conserved mRNAs. A complex set of mRNAs synthesized during seed development is stored in the dry seed to be available for immediate translation upon hydration (reviewed in Payne, 1976). In contrast to the residual mRNAs, these "stored" messages code for proteins presumed to play an integral role in early germination. These messages have been subcategorized into those coding for (a) enzymes essential for intermediary metabolism, and (b) proteins essential for the successful completion of germination per se, culminating in the elongation of the radicle (Bewley and Black, 1985).

Compartmentation in the nucleus may be a mechanism for protecting messages (Hammett and Katterman, 1975; Payne et al., 1977). Following imbibition, stored messages may be released from the nucleus to the protein-synthesizing sites within the cytoplasm. On the other hand, the accumulation of mRNAs in cytoplasmic informosomes (messenger ribonucleoproteins, mRNPs) during development of wheat and rye embryos has been advocated as a protective mechanism (Ajtkhozhin et al., 1976; Peumans and Carlier, 1977). Unprotected messages (e.g. storage protein mRNAs) are presumably hydrolyzed during late maturation and desiccation or during early germination. It is not known whether the changes in the protein-synthesizing complex are due ultimately to water loss from the seed tissues or whether they are a preprogrammed part of development which would occur even if the seed was prevented from drying out (Bewley, 1982). The decline in storage protein mRNAs during late maturation, for example, may simply take place as a consequence of the attainment of a specific physiological age and may not occur as a direct result of drying.

Protein synthesis begins within minutes of hydration of the seed, and some of the conserved mRNA is utilized to direct this (Bewley and Black, 1985). However, within a few hours there is synthesis of new mRNAs, and as the conserved ones are degraded, protein synthesis leading to germination becomes increasingly independent of them. Some of the newly synthesized mRNAs may code for the same proteins as conserved messages, whereas others are for distinctly different products, perhaps proteins that are essential for the commencement of radicle elongation. Such proteins remain to be identified (Bewley and Black, 1985).

Changes in mRNA Populations in Response to Premature Drying

The amount and composition of the mRNA complement reflects changes in translational and transcriptional control during development and germination. Changes in the extent of storage protein synthesis during seed development appear to be related to the amount of appropriate mRNAs available (Goldberg, Hoschek, Ditta, and Breidenbach, 1981; Goldberg, Hoschek, Tam, Ditta, and Breidenbach, 1981). This relationship is indicative of a regulatory process effected primarily at the level of transcription (i.e. mRNA synthesis) and mRNA stability. The transition from development to germination could be effected primarily at a transcription (mRNA synthesis) level (i.e. premature drying could result in a loss of mRNAs for developmentally-related proteins as well as a permanent suppression of their production and an induction of those mRNAs for germination proteins). Or alternatively, the switch could be at the level of translation (i.e. the mRNAs for both development and germination are present within the cells of the germinating seed, with only the latter being selectively translated).

To examine these possibilities, the effect of premature desiccation upon the status of the mRNAs for developmental and germination proteins in axes of *P. vulgaris* seeds has been analyzed (Misra et al., 1985; Misra and Bewley, 1985). These studies have revealed that the types of mRNA present in undesiccated developing 32-day-old *P. vulgaris* axes and in those prematurely dried at this time of development are very similar. In contrast, the mRNA fraction in 12-hr rehydrated axes following premature drying (at 32 days of development) codes for a different protein set. For example, the mRNAs for several major developmental proteins (including the storage protein phaseolin) are no longer present at 12 hr from the start of rehydration following premature drying even though they were present in the dried axes. Also, mRNAs for germination proteins become available for utilization in the rehydrated axes. Thus, it appears that drying alone does not effect the "switch" in message population; rather this event is elicited by the subsequent rehydration phase after a short lag. The rehydration phase appears to be crucial to the loss of any long-lived developmental messages (i.e. those that remain stable throughout desiccation).

These studies of developing *P. vulgaris* axes strongly suggest that there is a control by drying at the level of translatable mRNAs. Messages for storage proteins which are present in the prematurely dried axes (at 32 DAP) are labile (or unprotected) and are degraded rapidly during the early stages of germination. These messages cannot be replaced, because their genes have been "switched off" directly or indirectly as a result of the drying treatment. The rapid rise in new mRNAs, which are for germination-related proteins, suggests that a "switch on" of their genes is effected upon rehydration (Misra et al., 1985; Misra and Bewley, 1985). It is noteworthy, that in both normally and prematurely dried axes of *P. vulgaris*, the same temporal sequence of changes to the mRNA population occurs during subsequent germination.

Desiccation of *P. vulgaris* axes at 22 DAP (the intolerant stage) does not lead to any change in their complement of translatable messages during the early stages of germination. The mRNAs present upon rehydration code largely for developmental proteins; germination messages are not produced. Thus, when the seed is still desiccation intolerant, drying does not elicit "germination" gene activity, nor does it lead to a decline in developmental messages following rehydration (Misra and Bewley, 1985).

Cessation of Developmental Protein Syntheses

Transcriptional Control. It appears, then, that desiccation plays a role in the declining capacity for storage protein synthesis during late maturation and subsequent germination by effecting qualitative and quantitative changes at the mRNA level. A decline in the synthesis of storage protein mRNAs during drying could be due to a diminished capacity to initiate their transcription and/or a reduction in the rate of transcription. With regard to the speed of transcription, several levels of control by drying may operate, including a reduction (in vivo) in RNA polymerase II activity and a decreased availability and/or mobility of the required substrates (e.g. the nucleoside triphosphates). Drying (or osmotic/ionic changes caused by drying) may invoke modifications to the genome itself. A condensation of specific regions of the chromatin may render the DNA coding for developmental proteins inaccessible to RNA polymerase II and may hence prevent the initiation of transcription of their mRNAs. These genes would presumably remain suppressed (i.e. "transcriptionally silent") following subsequent hydration.

In regulating transcription, then, desiccation may have a direct role by effecting changes in the genome itself. However, drying could also act indirectly, for example, through changes in the levels of hormones (or their receptors) involved in the expression of storage protein genes. Abscisic acid has been implicated in the regulation of storage protein synthesis; whether this hormone acts specifically to increase the production of storage protein mRNAs during the maturation or seed expansion phase, when maximum synthesis occurs, remains to be determined. Some evidence lends support to this contention (Bray and Beachy, 1984; Finklestein et al., 1985; Crouch et al., in press), and it has been suggested that ABA plays some role in maintaining the appropriate areas of the chromatin in a transcriptionally active state (R. Quatrano, personal communication). Hence, the maturation drying period, by effecting changes in the level of, or sensitivity to, abscisic acid, may partially or wholly remove the effects of this "positive modulator" of developmental gene expression.

Immature embryos of *Brassica napus* will germinate precociously when excised from the seed and placed in culture (in the absence of ABA) without a requirement for desiccation (Crouch and Sussex, 1981). However, recent studies have shown that, although these embryos germinate in culture, they do not turn off storage protein genes as in the case of normal germination. Also, storage protein gene expression remains responsive to ABA in the precociously germinating embryo, unlike mature tissue which lacks this response (Crouch et al., in press). Thus, ABA alone may not be sufficient to regulate the activity of these genes. In fact, desiccation may be a prerequisite for turning off storage protein genes.

Regulation at the Translational and Posttranslational Levels. Utilization of messages by the translational machinery depends upon the successful completion of the initiation step, the binding of several components—met $tRNA_i$, ribosomal subunits, initiation factors, nucleoside triphosphates, mRNA—to form a functional complex. Hence, regulation of initiation could have a profound effect upon protein synthesis (Browder, 1984). It has been suggested (on the basis of studies of pollen) that changes in the concentration of ions within the cell are responsible for the cessation of protein synthesis during the final stages of maturation drying (Bashe and Mascarenhas, 1984). Specifically, the rising concentration of K^+ (and possibly other ions) during desiccation becomes increasingly inhibitory to the initiation of translation, possibly by disrupting the binding of individual components of the initiation complex. The "breakdown" of polyribosomes (i.e. release of ribosomes from the mRNA) and formation of single ribosomes is thus favored, causing a termination of the seed's developmental synthetic activities. With hydration

Table 4-2
Possible roles of desiccation in the cessation of developmental protein syntheses

Metabolic event	Specific process affected	Expected cellular response to desiccation
Developmental protein synthesis		
1. Transcriptional level: mRNA synthesis	Decreased rate of transcription of storage protein mRNAs.	Reduced availability of substrates (ATP, GTP, CTP, UTP); lower RNA polymerase II activity in vivo; substrate mobility decreased.
	Decreased capacity to initiate transcription.	Hormonal or osmotic changes may lead to a modified configuration or compaction of chromatin, rendering the DNA inaccessible to RNA polymerase II and hence prevent the initiation of transcription.
2. Posttranscriptional level: mRNA longevity	Decline in storage protein mRNA stability; increased turnover rates.	''Long-lived'' developmental mRNAs for storage protein destroyed during rehydration due to their presence in the cytoplasm in an unprotected form. Conserved mRNAs required for germinative events stored in nucleus or as mRNP particles in cytoplasm.
3. Translational level: polyribosome activity	Initiation of translation prevented.	Ionic changes disrupt binding of components of initiation complex. Polyribosome ''breakdown'' during maturation drying; individual ribosomes predominate.
4. Co- and posttranslational levels: ER/golgi activity	Reduced capacity to modify nascent developmental polypeptides co- or posttranslationally.	Reduced activity of signal peptidase and glycosylation enzymes; lowered substrate (carbohydrate) mobility.
		Modified configuration of ER/golgi body membranes, leading to a loss in function. Diminished association of polyribosomes with the ER during late maturation.

of the seed during germination, the K^+ ions reach a concentration that is favourable or optimal for a reinitiation of protein synthesis. Provided that messages have different K^+-concentration optima for binding with the initiation complex, the changing concentration of ions (during early germination) may provide a mechanism by which cells of the imbibing seed preferentially "select" certain mRNAs for translation (Bashe and Mascarenhas, 1984).

Storage protein synthesis appears to occur on polysomes in association with ER. In addition to other levels of control, it is possible that there is a reduction in the capacity of the ER membranes to modify nascent storage proteins after their translation. During maturation drying in lima beans there is a diminished association of the protein-synthesizing complex with the ER just prior to the "breakdown" of polyribosomes and increase in free (single) ribosomes (Klein and Pollock, 1968).

The various ways in which drying may terminate developmental protein syntheses are summarized in Table 4-2.

Desiccation Promotes the Synthesis of Postgerminative Enzymes

α-Amylase production by the aleurone layer of cereals cannot occur in response to gibberellic acid (GA_3) unless the grain has undergone drying (Bilderback, 1971; Evans et al., 1975). Sensitivity to GA_3 is achieved only if the aleurone layer of developing wheat grains is dried to a water content of less than 25%. A mechanism related to membrane changes and perhaps hormone receptors in the membrane has been invoked to explain the nature of the requirement for drying to such a critical water content (Norman et al., 1982).

In dicot seeds, Adams et al. (1983) demonstrated that immature soybean seeds cannot produce malate synthetase or isocitrate lyase activities unless the seeds are first subjected to slow drying. Such is also the case in the castor bean endosperm. Desiccation, prematurely imposed after 40 days of development, causes an increase in activity of isocitrate lyase (an enzyme of the glyoxylate cycle involved in gluconeogenesis of lipid reserves) and LeuNAase (a proteolytic enzyme) to levels comparable to those produced following germination of the mature dry seed (Kermode and Bewley, in press *b*; Figure 4-3). In the absence of a drying period at 40 days, seeds removed from the capsule and testa but maintained in the hydrated state are incapable of increasing the activity of the two enzymes (Figure 4-3). Thus, desiccation is not only important in redirecting the metabolism from a developmental to a

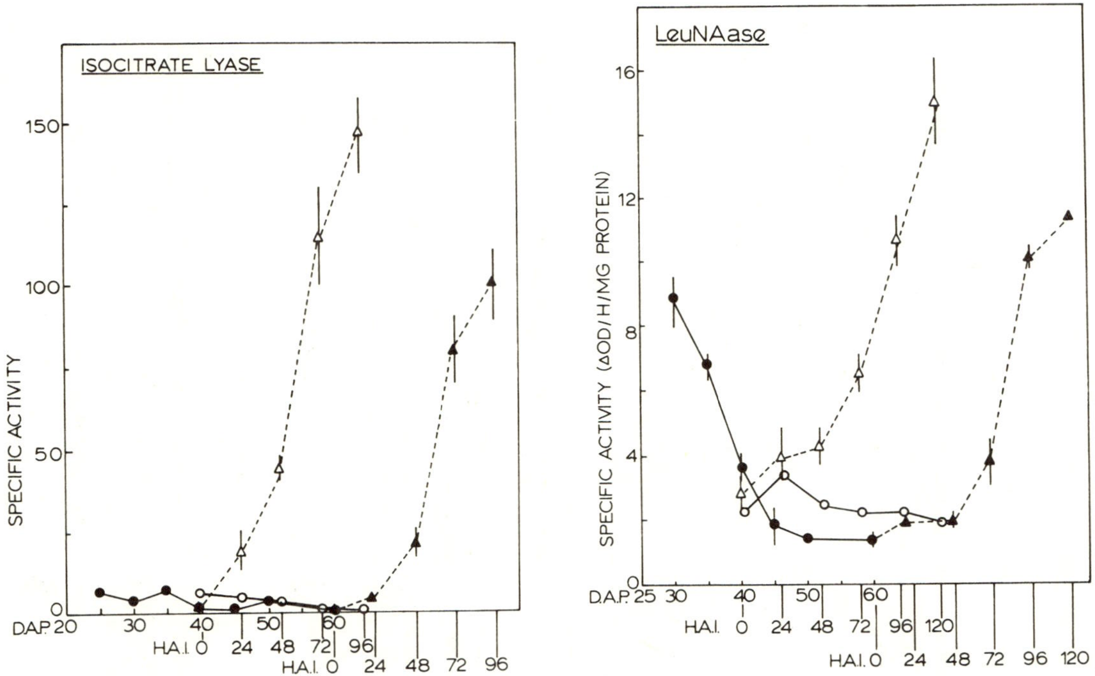

Figure 4-3. Changes in isocitrate lyase (ICL) and leucine β-naphthylamidase (LeuNAase) specific activity in extracts of castor bean endosperms during development from 25 to 50 DAP (●), following imbibition of mature seeds (▲), following rehydration of seeds dried at 40 DAP (△), and in hydrated (undried) 40 DAP seeds (○). Each point is an average of 2-5 determinations ± SD. Extraction procedures for enzymes were as given in Tully and Beevers (1978) and Gifford et al. (1983); see these references for LeuNAase determination. ICL was determined according to Cooper and Beevers (1969); calculations assumed a molar extinction coefficient of 1.7 x 10^4 $M^{-1}cm^{-1}$. HAI is hours after imbibition.

germinative mode but is also a prerequisite for the production of enzymes essential to the postgerminative phase of seedling development.

Summary and Speculations

It is apparent that desiccation, whether natural or imposed prematurely, may play an important role in switching seeds from a developmental mode to one that is essential for germination. Our work on castor bean (as well as some other studies) shows that premature drying not only redirects metabolism from a developmental to a germination programme but does so permanently, thus effecting an irreversible "switch." This switch cannot be brought about at all stages of development, for the seed must first develop a competence to withstand the drying treatment, a process which presumably involves various metabolic and/or structural changes. At this time we can only speculate that this competence is achieved simultaneously by the genome.

The plant hormone, abscisic acid, may play a role in regulating the switch from development to germination. It has been postulated, for example, that the level of ABA synthesized within the embryo itself or in the surrounding maternal tissue is important, with high amounts of ABA inhibiting germination but permitting developmental or maturation events to proceed (Long et al., 1981; Crouch and Sussex, 1981). Drying may affect the hormonal balance of the seed: ABA breakdown may be enhanced by the natural drying process in cereals (King, 1976), and soybeans can be induced to germinate precociously by a slow-drying treatment, which depletes the pool of endogenous ABA (Ackerson, 1984b). Rather than have an effect upon the hormone level itself, drying may decrease the sensitivity of the embryo to ABA (Finkelstein et al., 1985) because of a change in receptor levels or conformation.

Membrane changes have been invoked to explain the sensitization of the cereal aleurone layer to gibberellin after drying. It has been postulated (Norman et al., 1982) that the membrane undertakes compositional adjustments to its lipids in response to desiccation in order to maintain a functional state; there follows a subsequent permanent change in conformation of membrane components. Such changes, which might not be possible during the desiccation-intolerant stage of development, could lead to changes in sensitivity to hormones such as GA_3. The "membrane hypothesis" may be superimposed on the "ABA hypothesis" if the receptor(s) to this hormone are membrane-associated. At the present time the research accomplished is far too little to favour or discount these hypotheses. A summary of the possible levels of control by drying in effecting the switch from development to germination is proffered in Figure 4-4.

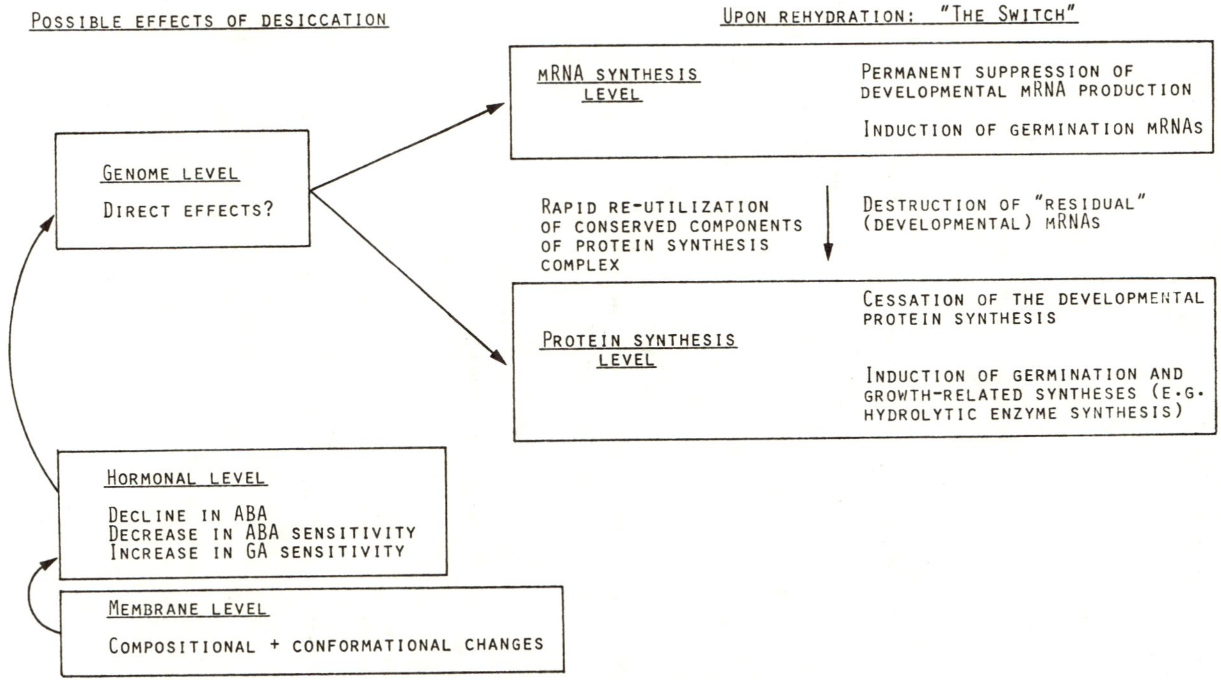

Figure 4-4. Possible levels of control of the "switch" elicited by premature or natural drying of developing seeds.

REFERENCES

Ackerson, R.C. 1984a. Regulation of soybean embryogenesis by abscisic acid. J. Exp. Bot. 35: 403-413.

Ackerson, R.C. 1984b. Abscisic acid and precocious germination in soybeans. J. Exp. Bot. 35: 414-421.

Adams, C.A. and Rinne, R.W. 1980. Moisture content as a controlling factor in seed development and germination. Int. Rev. Cytol. 68: 1-8.

Adams, C.A. and Rinne, R.W. 1981. Seed maturation in soybeans (*Glycine max* L. Merr) is independent of seed mass and of the parent plant, yet is necessary for production of viable seeds. J. Exp. Bot. 32: 615-620.

Adams, C.A., Fjerstad, M.C. and Rinne, R.W. 1983. Characteristics of soybean seed maturation: Necessity for slow dehydration. Crop Sci. 23: 265-267.

Ajtkhozhin, M.A., Doschanov, K.H.J. and Akhanov, A.J. 1976. Informosomes as a stored form of mRNA in wheat embryos. FEBS Lett. 66: 124-126.

Ammirato, P.V. 1977. Hormonal control of somatic embryo development from cultured cells of caraway. Interactions of abscisic acid, zeatin, and gibberellic acid. Plant Physiol. 59: 579-586.

Bashe, D. and Mascarenhas, J.P. 1984. Changes in potassium ion concentrations during pollen dehydration and germination in relation to protein synthesis. Plant Sci. Lett. 35: 55-60.

Bewley, J.D. 1979. Physiological aspects of desiccation-tolerance. Annu. Rev. Plant Physiol. 30: 195-238.

Bewley, J.D. 1982. Protein and nucleic acid synthesis during seed germination and early seedling growth. *In* Encycl. Plant Physiol. New Series, Nucleic Acids and Proteins in Plants I (D. Boulter and B. Parthier, eds.). Springer-Verlag, Berlin, Heidelberg. Vol. 14A. Pp. 559-591.

Bewley, J.D. and Black, M. 1985. Seeds. *In* Physiology of Development and Germination. Plenum Press, New York and London.

Bilderback, D.E. 1971. Amylases in developing barley seeds. Plant Physiol. 48: 331-334.

Bollini, R. and Chrispeels, M.J. 1980. Germinating kidney beans synthesize vicilin and other seed proteins utilizing pre-existing mRNA. J. Cell Biol. 75: 332a.

Bray, E.A. and Beachy, R.N. 1984. Effect of abscisic acid on the accumulation of β-conglycinin in soybean cotyledons. Plant Physiol. 75 (Suppl.): 196.

Browder, L.W. 1984. Developmental Biology. 2nd ed. Saunders, Philadelphia.

Carlier, A.R., Manickam, A. and Peumans, W.J. 1980. Characterization of a maturation-specific mRNA in dry mung bean embryonic axes. Planta 149: 227-233.

Choinsky, J.S. and Trelease, R.N. 1978. Control of enzyme activities in cotton cotyledons during maturation and germination. Plant Physiol. 62: 142-145.

Choinsky, J.S., Trelease, R.N. and Doman, D.C. 1981. Control of enzyme activities in cotton cotyledons during maturation and germination. III. In-vitro embryo development in the presence of abscisic acid. Planta 152: 428-435.

Cochrane, M.P. 1985. Assimilate uptake and water loss in maturing barley grains. J. Exp. Bot. 36: 770-782.

Cooper, T.G. and Beevers, H. 1969. Mitochondria and glyoxysomes from castor bean endosperm. Enzyme constituents and catalytic capacity. J. Biol. Chem. 244: 3507-3513.

Crevecoeur, M., Deltour, R. and Bronchart, R. 1976. Cytological study of water stress during germination of *Zea mays*. Planta 132: 31-41.

Crouch, M.L. and Sussex, I.M. 1981. Development and storage-protein synthesis in *Brassica napus* L. embryos in vivo and in vitro. Planta 153: 64-74.

Crouch, M.L., Tanberge, K., Simon, A., Finklestein, R., Scofield, S. and Solberg, L. In press. Storage protein mRNA levels can be regulated by abscisic acid in *Brassica* embryos. *In* Molecular Form and Function of the Plant Genome (L. van Vloten-Doting, G.S.P. Groot and T.C. Hall, eds.). NATO Advanced Studies Institute Proceedings. Plenum Press, New York and London. Pp. 555-566.

Dasgupta, J. and Bewley, J.D. 1982. Desiccation of axes of *Phaseolus vulgaris* during development causes a switch from a developmental pattern of protein synthesis to a germination pattern. Plant Physiol. 70: 1224-1227.

Dasgupta, J., Bewley, J.D. and Yeung, E.C. 1982. Desiccation-tolerant and desiccation-intolerant stages during development and germination of *Phaseolus vulgaris* seeds. J. Exp. Bot. 33: 1045-1057.

deKlerk, G.J. 1984. Changes in protein synthesis during the development of *Agrostemma githago* L. seeds. Physiol. Plant. 62: 607-614.

Duchartre, M.P. 1852. Note sur la germination des cereales recoltées avant leur maturité. Comptes rendus hebdomadaires des Séances de l'Académie des Sciences, Paris 35: 940-942.

Dure, L.S. III, Greenway, S.C. and Galau, G.A. 1981. Developmental biochemistry of cottonseed embryogenesis and germination: Changing mRNA populations as shown by in vitro and in vivo protein synthesis. Biochemistry 20: 4162-4168.

Dure, L.S. III, Pyle, J.B., Chlan, C.A., Baker, J.C. and Galau, G.A. 1983. Developmental biochemistry of cottonseed embryogenesis and germination. XVII. Developmental expression of genes for the principal storage proteins. Plant Mol. Biol. 2: 199-206.

Evans, M., Black, M. and Chapman, J.M. 1975. Induction of hormone sensitivity by dehydration is the one positive role for drying in cereal seed. Nature 258: 144-145.

Finkelstein, R.R., Tenbarge, K.M., Shumway, J.E. and Crouch, M.L. 1985. Role of ABA in maturation of rapeseed embryos. Plant Physiol. 78: 630-636.

Galau, G.A. and Dure, L.S. III. 1981. Developmental biochemistry of cottonseed embryogenesis and germination: Changing mRNA populations as shown by reciprocal heterologous cDNA-mRNA hybridization. Biochemistry 20: 4169-4178.

Gifford, D.J., Greenwood, J.S. and Bewley, J.D. 1983. Vacuolation and storage protein breakdown in the castor bean endosperm following imbibition. J. Exp. Bot. 34: 1433-1443.

Goldberg, R.B., Hoschek, G., Ditta, G.S. and Breidenbach, R.W. 1981. Developmental regulation of cloned superabundant embryo mRNAs in soybean. Dev. Biol. 83: 218-231.

Goldberg, R.B., Hoschek, G., Tam, S.H., Ditta, G.S. and Breidenbach, R.W. 1981. Abundance, diversity and regulation of mRNA sequence sets in soybean embryogenesis. Dev. Biol. 83: 210-217.

Goncharova, E.A., Udovenko, G.V., Nechiporenko, G.A. and Zholkevich, V.N. 1985. Water exchange of vegetable marrow fruits as studied with the aid of a tritium label. Sov. Plant Physiol. 31: 657-661.

Greenwood, J.S. and Bewley, J.D. 1982. Seed development in *Ricinus communis* (castor bean). I. Descriptive morphology. Can. J. Bot. 60: 1751-1760.

Greenwood, J.S., Gifford, D.J. and Bewley, J.D. 1984. Seed development in *Ricinus communis* cv. Hale (castor bean). II. Accumulation of phytic acid in the developing endosperm and embryo in relation to the deposition of lipid, protein and phosphorus. Can. J. Bot. 62: 255-261.

Hammett, J.R. and Katterman, F.R. 1975. Storage and metabolism of poly (adenylic acid)-mRNA in germinating cotton seeds. Biochem. 14: 4375-4379.

Harlan, H.V. and Pope, M.N. 1922. The germination of barley seeds harvested at different stages of growth. J. Hered. 13: 72-75.

Ihle, J.N. and Dure, L.S. III. 1972. The developmental biochemistry of cottonseed embryogenesis and germination. III. Regulation of the biosynthesis of enzymes utilized in germination. J. Biol. Chem. 247: 5048-5055.

Karssen, C.M., Brinkhorst-vanderSwan, D.L.C., Breekland, A.E. and Koornneeff, M. 1983. Induction of dormancy during seed development by endogenous abscisic acid: Studies on abscisic acid deficient genotypes of *Arabidopsis thaliana* (L.) Heynh. Planta 157: 158-165.

Kermode, A.R. and Bewley, J.D. In press *a*. The role of maturation drying in the transition from seed development to germination. I. Acquisition of desiccation-tolerance and germinability during development of *Ricinus communis* L. seeds. J. Exp. Bot.

Kermode, A.R. and Bewley, J.D. In press *b*. The role of maturation drying in the transition from seed development to germination. II. Post-germinative enzyme production and soluble protein synthetic pattern changes within the endosperm of *Ricinus communis* L. seeds. J. Exp. Bot.

Kermode, A.R., Dasgupta, J., Misra, S. and Bewley, J.D. In press. The transition from seed development to germination: A key role for desiccation? Hort. Sci. (special suppl.).

Kermode, A.R., Gifford, D.J. and Bewley, J.D. In press. The role of maturation drying in the transition from seed development to germination. III. Insoluble protein synthetic pattern changes within the endosperm of *Ricinus communis* L. seeds. J. Exp. Bot.

King, R.W. 1976. Abscisic acid in developing wheat grains and its relationship to grain growth and maturation. Planta 132: 43-51.

Klein, S. and Pollock, B.M. 1968. Cell fine structure of developing lima bean seeds related to seed desiccation. Am. J. Bot. 55: 658-672.

Kulieva, L.K., Obrucheva, N.V. and Prokofev, A.A. 1985. Physiological maturity of the ripening seeds of fine-fibered cotton. Sov. Plant Physiol. 31: 732-736.

Lee, D.R. and Atkey, P.T. 1984. Water loss from the developing caryopsis of wheat (*Triticum aestivum*). Can. J. Bot. 62: 1319-1326.

Long, S.R., Dale, R.M.K. and Sussex, I.M. 1981. Maturation and germination of *Phaseolus vulgaris* embryonic axes in culture. Planta 153: 405-415.

McDaniel, S., Smith, J.D. and Price, H.J. 1977. Response of viviparous mutants to abscisic acid. Maize Genet. Newslett. 51: 85-86.

Meredith, P. and Jenkins, L.D. 1975. Loss of moisture from developing and ripening cereal grains. N.Z.J. Sci. 18: 501-509.

Misra, S. and Bewley, J.D. 1985. Reprogramming of protein synthesis from a developmental to a germinative mode induced by desiccation of the axes of *Phaseolus vulgaris*. Plant Physiol. 78: 876-882.

Misra, S. and Bewley, J.D. In press. The messenger RNA population in the embryonic axes of *Phaseolus vulgaris* during development and following germination. J. Exp. Bot.

Misra, S., Kermode, A. and Bewley, J.D. 1985. Maturation drying as the "switch" that terminates seed development and promotes germination. *In* Molecular Form and Function of the Plant Genome (L. van Vloten-Doting, G.S.P. Groot and T.C. Hall, eds.). NATO Advanced Studies Institute Proceedings. Plenum Press, New York and London. Pp. 113-128.

Mitchell, B., Armstrong, C., Black, M. and Chapman, J.M. 1978. Physiological aspects of sprouting and spoilage in developing *Triticum aestivum* L. (wheat grains). *In* Seed Production (P.D. Hebblethwaite, ed.). Butterworths, London, Boston. Pp. 339-356.

Mori, T., Wakabayashi, Y. and Takagi, S. 1978. Occurrence of mRNA for storage protein in dry soybean seeds. J. Biochem. 84: 1103-1111.

Muto, S. and Beevers, H. 1974. Lipase activities in castor bean endosperm during germination. Plant Physiol. 54: 23-28.

Nechiporenko, G.A. and Rybalova, B.A. 1983. Water exchange of vegetable marrow fruits during ripening. Sov. Plant Physiol. 30: 223-226.

Norman, H.A., Black, M. and Chapman, J.M. 1982. Induction of sensitivity to gibberellic acid in aleurone tissue of developing wheat grains. II. Evidence for temperature-dependent membrane transitions. Planta 154: 578-586.

Obendorf, R.L. and Wettlaufer, S.H. 1984. Precocious germination during in vitro growth of soybean seeds. Plant Physiol. 76: 1023-1028.

Payne, P.I. 1976. The long-lived messenger ribonucleic acid of flowering plant seeds. Biol. Rev. 51: 329-363.

Payne, P.I., Gordon, M.E., Barlow, P.W. and Parker, M.L. 1977. The subcellular location of the long-lived messenger RNA of rape seed. *In* Translation of Natural and Synthetic Polynucleotides (A.B. Legocki, ed.). Agric. Univ. Press, Poznań. Pp. 224-227.

Peumans, W.J. and Carlier, A.R. 1977. Messenger ribonucleoprotein particles in dry wheat and rye embryos. *In vitro* translation and size distribution. Planta 136: 195-201.

Peumans, W.J., Stinissen, H.M. and Carlier, A.R. 1982. Lectin synthesis in developing and germinating wheat and rye embryos. Planta 156: 41-44.

Quebedeaux, B., Sweetser, P.B. and Rowell, J.C. 1976. Abscisic acid levels in soybean reproductive structures during development. Plant Physiol. 58: 363-366.

Robichaud, C.S., Wong, J. and Sussex, I.M. 1980. Control of *in vitro* growth of viviparous embryo mutants of maize by abscisic acid. Dev. Genet. 2: 325-330.

Rogerson, N.E. and Matthews, S. 1977. Respiratory and carbohydrate changes in developing pea (*Pisum sativum* L.) seeds in relation to their ability to withstand desiccation. J. Exp. Bot. 28: 304-313.

Senaratna, T. and McKersie, B.D. 1983. Characterization of solute efflux from dehyration injured soybean (*Glycine max* L. Merr) seeds. Plant Physiol. 72: 911-914.

Senaratna, T., McKersie, B.D. and Stinson, R.H. 1984. Association between membrane phase properties and dehydration injury in soybean axes. Plant Physiol. 76: 759-762.

Senaratna, T., McKersie, B.D. and Stinson, R.H. 1985. Antioxidant levels in germinating soybean seed axes in relation to free radical and dehydration tolerance. Plant Physiol. 78: 168-171.

Smith, J.D., McDaniel, S. and Lively, S. 1978. Regulation of embryo growth by abscisic acid in vitro. Maize Genet. Coop. Newslett. 52: 107-108.

Sprague, G.F. 1936. The relation of moisture content and time of harvest to germination of immature corn. J. Am. Soc. Agron. 28: 472-478.

Stinissen, H.M., Peumans, W.J. and De Langhe, E. 1984. Abscisic acid promotes lectin biosynthesis in developing and germinating rice embryos. Plant Cell Rep. 3: 55-59.

Sussex, I.M. 1975. Growth and metabolism of the embryo and attached seedling of the viviparous mangrove, *Rhizophora mangle*. Amer. J. Bot. 62: 948-953.

Sussex, I.M. and Dale, R.M.K. 1979. Hormonal control of storage protein synthesis in *Phaseolus vulgaris*. *In* The Plant Seed. Development, Preservation and Germination (I. Rubenstein, R.L. Phillips, C.E. Green and B.G. Gengenbach, eds.). Academic Press, New York. Pp. 129-141.

Symons, S.J., Angold, R.E., Black, M. and Chapman, J.M. 1983. Changes in the growth capacity of the developing wheat embryo. I. The influences of the enveloping tissues and premature drying. J. Exp. Bot. 34: 1541-1550.

Tilden, R.L. and West, S.H. 1985. Reversal of the effects of aging in soybean seeds. Plant Physiol. 77: 584-586.

Triplett, B.A. and Quatrano, R.S. 1982. Timing, localization and control of wheat germ agglutinin synthesis in developing wheat embryos. Dev. Biol. 91: 491-496.

Tully, R.E. and Beevers, H. 1976. Protein bodies of castor bean endosperm. Isolation, fractionation and the characterization of protein components. Plant Physiol. 58: 710-716.

Tully, R.E. and Beevers, H. 1978. Proteases and peptidases of castor bean endosperm. Enzyme characterization and changes during germination. Plant Physiol. 62: 746-750.

Wellington, P.S. 1956. Studies on the germination of cereals. I. The germination of wheat grains in the ear during development, ripening, and after-ripening. Ann. Bot. 20: 105-120.

Youle, R.J. and Huang, A.H.C. 1981. Occurrence of low molecular weight and high cysteine-containing albumin storage proteins in oil seeds of diverse species. Am. J. Bot. 68: 44-48.

5 Loss of Desiccation Tolerance during Seed Germination: A Free Radical Mechanism of Injury

Tissa Senaratna
Bryan D. McKersie

New generations of plants begin from seed, which serves to disperse the new generation in time and space, to provide a form which is tolerant of adverse environmental conditions, and to supply the new individual with food reserves until a self-sufficient, autotrophic organism can be established. Seeds possess several unique physiological properties which contribute to these ends. One is tolerance of desiccation.

Seeds of most species remain viable at extremely low water contents for prolonged periods of time. When exposed to moist conditions, with germination, the seed loses its tolerance of desiccation, exhibiting a physiological switch from tolerance to susceptibility. The biochemical factors which impart desiccation tolerance to the mature seed and the events which occur in the germinating seed as this tolerance is lost have only recently been studied. In this review, the available experimental evidence which has led to the development of a model for the mechanism of desiccation injury in seeds is summarized, and a hypothetical mechanism is suggested which might provide desiccation tolerance in seeds.

Development and Loss of Desiccation Tolerance

After fertilization of the ovule, seed development proceeds with tissue differentiation and reserve accumulation. Water is lost in the final stages of seed

The authors thank John Warkentin and Rajini Senaratna, Department of Chemistry, McMaster University, Hamilton, Canada, for very useful discussions.

maturation. Details of these complex changes are presented in Chapter 4. Embryos develop desiccation tolerance only in late stages of seed development. For example, *Phaseolus vulgaris* seeds were unable to germinate after being dried at 22 days after anthesis, but if the same drying stress was applied at 26 days, germination proceeded upon rehydration (Dasgupta et al., 1982). Therefore, at a very precise stage in their development, these seeds acquired tolerance of desiccation.

As the dry seed imbibes water and begins to germinate, it loses its tolerance of desiccation. This characteristic has been demonstrated in various seed systems, although the precise time at which the switch from tolerance to intolerance occurs varies among the species (Crevecoeur et al., 1976; Akalehiywot and Bewley, 1980; McKersie and Tomes, 1980; Becwar et al., 1982). Desiccation tolerance is not lost immediately after imbibition. During the early stages of germination and prior to radicle emergence in most species, the seed can be dehydrated to its original water content at maturity (<10% moisture) without causing injury (Hegarty, 1978). In contrast, desiccation at later stages, usually after the initiation of radicle elongation, blocks subsequent development or dramatically reduces seed vigour. Comparison of physiological and biochemical properties of the seed at these two stages—the desiccation tolerant (at early stages of germination) and the desiccation intolerant (at latter stages of germination)—has been conducted in our laboratory in an attempt to define the mechanisms of desiccation injury and tolerance.

Cellular Site and Nature of Desiccation Injury

Ultrastructural studies of *Zea mays* seeds have attributed at least some of the loss of desiccation tolerance during germination to a general disruption of all cellular membranes after a 72-hr imbibition/24-hr dehydration cycle (Crevecoeur et al., 1976). In embryos dried after 24 hr imbibition, cell fine structure was preserved, and the embryos resumed growth upon reimbibition. In embryos dried after 72 hr imbibition, cell fine structure was quite disorganized, and rehydration led to further loss of membrane integrity, particularly disruption of nuclear and mitochondrial envelopes and mitochondrial cristae. The injury was irreversible, since rehydration for 48 hr did not restore cellular integrity, and the embryos failed to resume growth.

During the rehydration of dry tissues there is a characteristic leakage of cytoplasmic solutes (Simon, 1974; Simon and Mills, 1983). In general, seeds with low viability or low vigour leak solutes more rapidly and in larger quantities than high-vigour seeds. This statistically significant negative correlation is the basis for commercial viability and vigour testing in several crop and horticultural species. Seeds and isolated embryonic axes which have been imbibed for specific periods of time and then dried show increased rates of solute leakage during subsequent rehydration in a manner that is highly and negatively correlated with the ability of the seed to germinate subsequently and with the ability of the axis to elongate or to reduce the vital stain, tetrazolium chloride (Senaratna and McKersie, 1983a). This observation provides further evidence that the integrity of cellular membranes which limit solute efflux from these seeds has been damaged. Since the changes in membrane integrity can be detected as soon as rehydration commences, injury to the membranes apparently occurs while the tissue is in the dried state; thus cellular membranes are an early site of injury, if not the primary one. Similar experimental observations have implicated cellular membranes as the site of injury for most other environmental stresses, such as freezing (Steponkus, 1984), chilling (Lyons, 1973), and damage by air pollutants (McKersie et al., 1982).

Two hypothetical but mechanistically different types of disruption may be envisaged to account for this loss of membrane integrity. The plasmalemma and other cellular membranes may have physically ruptured as a consequence of the strain induced by the rapid fluxes of water during drying or rehydration. Alternatively, the membrane may experience changes in composition or structure, resulting in its permeability and its altered appearance in electron micrographs.

Several lines of evidence support the latter alternative. Leakage of cytoplasmic solutes from seeds during imbibition is biphasic (Simon, 1974). A rapid initial leakage of solutes from sites in the apparent free space is followed by a slower, linear efflux from cytoplasmic compartments presumably across rate-limiting membranes. This biphasic pattern has been cited as indicating a reorganization of membrane structure during rehydration (Simon, 1974) but more likely results from the differential rates of leakage from different compartments of a spherical seed during rehydration (McKersie and Senaratna, 1983). When soybean axes were desiccated in both the tolerant and intolerant stages of germination, the kinetics of electrolyte leakage during rehydration followed this biphasic pattern and indicated an increased rate of efflux for both phases (Figure 5-1). If, however, cell elongation was prevented by treatment with polyethylene glycol or cycloheximide, the initial rate of leakage was lower, and only in the linear rate of efflux during phase 2 was the rate the same as in the control. Axes dehydrated after the latter treatments failed to continue growth after rehydration and thus may be considered to have become intolerant of desiccation. These results are inter-

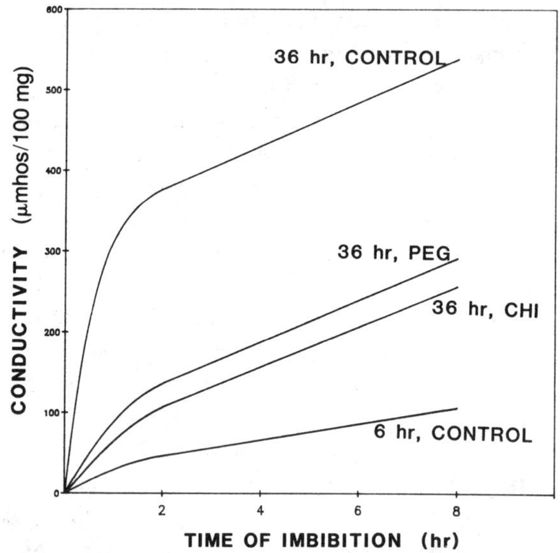

Figure 5-1. Time profile of electrolyte leakage during reimbibition of soybean embryonic axes after 6 or 36 hr germination. Seeds were prepared as described in Table 5-1. After treatment, the embryonic axes were removed and air dried at 25°C to 10% moisture. Ten axes were reimbibed in 10 ml distilled water. Conductivity readings of the imbibing solution were made at 15-min intervals for hour 1 and 60-min intervals to 8 hr (adapted from Senaratna and McKersie, 1983a).

preted as indicating that dehydration of soybean axes in the intolerant state (36 hr) leads to a increased incidence of cell rupture and/or an increase in membrane permeability coincident with the loss of viability.

The rate of dehydration of axes in the intolerant state also influences the amount of cell rupture during rehydration but does not have any significant effect on membrane permeability or viability. Membranes from desiccation-injured axes retain some elements of selective permeability. For example, if the pH of the bathing solution is varied from 5.5 to 8.0, the rate of K^+ efflux decreases, whereas sugar efflux increases (Figure 5-2). This pattern is observed both before and after desiccation of intolerant axes even though the rate of efflux after desiccation is increased 10-fold. Other experiments on the kinetics and quality of solute efflux are also inconsistent with the hypothesis that the major lesion to cellular membranes is physical rupture (Senaratna and McKersie, 1983b). Therefore, desiccation injury appears to be associated with membrane dysfunction manifested as increases in membrane permeability resulting from biochemical or biophysical changes in the membrane and not from simple physical rupture.

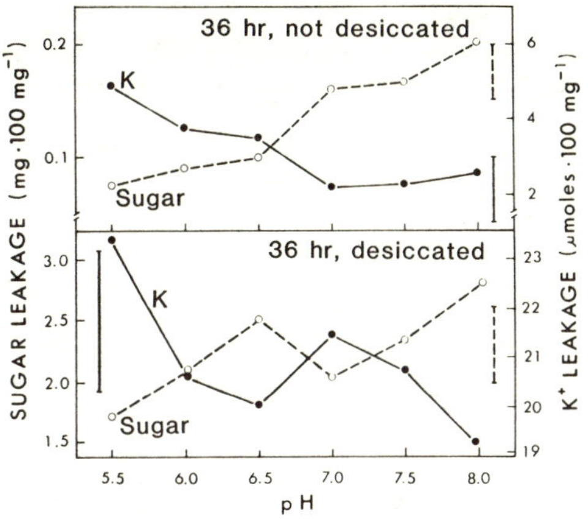

Figure 5-2. Leakage of potassium (●——●) and sugar (○——○) from desiccation-intolerant soybean embryonic axes in response to changes to external pH. Axes were prepared from soybean seeds imbibed for 36 hr as described in Table 5-2. Axes were either not desiccated (a) or were dried to 10% moisture and reimbibed for 2 hr in distilled water to full rehydration (b). All treatments were soaked for 4 hr in 25 ml of 50 mM MES-HEPES buffer. Values represent the change in solute concentration in the bathing solution between 1 and 4 hr. Vertical bars represent least significant differences at $p = 0.05$ among pH treatments (adapted from Senaratna and McKersie, 1983b, courtesy of *Plant Physiology*.).

To define the nature of this lesion to the membrane, the physical structure of the lipid bilayer of microsomal membranes from tolerant and intolerant axes was investigated using wide angle x-ray diffraction (Table 5-1). Microsomal membranes were selected because this fraction represents a cross section of cellular membranes, and microscopy has indicated that all cellular membranes exhibited some symptoms of injury. Diffraction was chosen because of its previous success in identifying changes in lipid packing during senescence of several plant tissues (McKersie et al., 1976; McKersie and Thompson, 1977, 1978). A lipid-phase transition was observed at approximately 7°C in microsomal membranes from tolerant axes (6 hr imbibition) before and after dehydration and in those from intolerant axes (36 hr imbibition) before dehydration. The transition temperature increased to 47°C in the same membrane fraction from intolerant axes after dehydration and 2 hr rehydration. These latter membranes apparently suffered considerable phospholipid degradation as evidenced by a 2-fold decrease in the phospholipid:sterol ratio and a 10-fold increase in the free fatty acid:phospholipid ratio. There were only minor changes in the fatty acid saturation of the total membrane lipid and in the composition of the sterol fraction (Senaratna et al., 1984).

Table 5-1
Lipid phase transition temperature, molar phospholipid:sterol ratio, and molar free fatty acid:phospholipid ratio of smooth microsomal membranes isolated from soybean embryonic axes

Imbibition time (hr)	Treatment	Phase transition temperature (°C)	PL:sterol	FFA:PL ($\times 10^{-3}$)
6	Control	7 b	46 a	42 b
	Desiccated	7 b	38 a	38 b
36	Control	9 b	38 a	34 b
	Desiccated	47 a	16 b	343 a

Note: Values within a column followed by the same letter are not significantly different at $p = 0.05$, according to analysis of variance. Soybean seeds were imbibed and embryonic axes dried as described for Figure 5-2. Axes were reimbibed for 2 hr to rehydrate the tissue fully. Smooth microsomal membranes were isolated by differential centrifugation. The membrane pellet was analyzed by wide-angle x-ray diffraction to determine the phase transition temperature, which is operationally defined by this technique as the highest temperature at which gel-phase lipid (4.15 Å band) could be detected to within 2°C. Lipids were extracted with hot isopropanol and chloroform-methanol. Free fatty acids and sterols were separated by thin-layer chromatography and quantified by gas chromatography. Phospholipid was quantified as Pi after perchloric acid digestion (adapted from Senaratna et al., 1984, courtesy of *Plant Physiology*).

The presence of gel-phase domains in cellular membranes could conceivably contribute to their loss of function and consequently to loss of cell viability in response to desiccation. The formation of these domains would indicate a lateral phase separation of phospholipid in the bilayer. Membrane proteins are known to be squeezed out of phospholipids as they form gel-phase lipid domains (Shechter et al., 1974; Shinitzky and Inbar, 1976). As shown in Chapter 11, the pattern of protein organization is altered, affecting enzyme activity (Sandermann, 1978) and transport (Baldassare et al., 1979). The presence of gel-phase lipid also renders the lipid bilayer leaky (Barber and Thompson, 1980), which may contribute to the increased leakage observed in the dehydration-injured tissue.

The lateral phase separation is apparently induced by the accumulation of phospholipid degradation products in the membranes. In senescent tissues, the phospholipids are induced to form discrete gel and liquid crystalline domains by the presence of several components in the neutral lipid fraction (McKersie and Thompson, 1979). At least one of these components is free fatty acids, which may remain in the membrane after the cleavage of the ester linkage between the glycerol backbone and fatty acid tail of a phospholipid molecule. The phospholipids from microsomal membranes of soybean axes in the intolerant state can be ordered into gel-phase domains by the inclusion of saturated free fatty acids prepared from these extracts. Unsaturated fatty acids do not have the same effect.

Factors Initiating Membrane Injury

Two possible causes for the observed phospholipid deesterification can be envisaged. Phospholipid deesterification can occur as a result of enhanced activity of an enzyme such as acyl hydrolase (phospholipase A_2). Alternatively, degradation may be strictly chemical. For example, nucleophilic attack by the free radical superoxide on the ester linkage of the phospholipid may promote deesterification (Niehaus, 1978). It is difficult to distinguish between these alternatives, but a time-course study of phospholipid degradation and free fatty acid accumulation in the membranes of soybean axes during dehydration stress appeared to be inconsistent with what might be expected from a strictly enzymatic process (Senaratna et al., 1985b).

The injury symptoms previously observed in microsomal membranes from dehydration-injured soybean axes could be closely simulated by exposing isolated membranes from dehydration-intolerant axes to free radicals in vitro. Free radical treatment using the xanthine-xanthine oxidase system caused a rise in lipid phase transition temperatures and promoted phospholipid deesterification (Table 5-2). Membranes from axes in the dehydra-

Table 5-2

Effect of free radical treatment on lipid phase transition temperature and phospholipid:sterol and free fatty acid:phospholipid ratios of smooth microsomal membranes isolated from desiccation-tolerant and desiccation-intolerant embryonic axes of soybean

		Microsomal membrane			
Imbibition time (hr)	Free radical treatment	Phase transition temperature (°C)	PL:sterol	FFA:PL ($\times 10^{-3}$)	Liposome phase transition temperature (°C)
6	Control	7	57	52	3
	Treated	14	51	88	8
36	Control	9	52	43	6
	Treated	40	30	212	31

Note: Microsomal membranes were isolated as described in Table 5-1 from embryonic axes not previously subjected to any desiccation treatment. The membrane pellet was resuspended in 10 ml 0.02 M xanthine, 50 mM $NaHCO_3$, 0.5 units xanthine oxidase for 30 min. The membranes were reisolated by centrifugation and analyzed as previously described.

Liposomes were prepared from the total lipid extract from smooth microsomal membranes. The dried lipid fraction was hydrated in 10 ml 100 mM NaCl and isolated by centrifugation. The liposomes were exposed to free radicals as detailed above (adapted from Senaratna et al., 1985a, 1985b, courtesy of *Plant Physiology*).

tion-tolerant stage (6 hr) are relatively more tolerant of this free radical treatment. This similarity in injury symptoms and the coincidence during germination of the loss of desiccation tolerance and the reduced tolerance to free radicals suggests that desiccation stress may induce a free radical-mediated deesterification of membrane phospholipid, leading to the accumulation of degradation products such as free fatty acids in membranes (Senaratna et al., 1985a).

The source of these free radicals remains obscure, although the occurrence of free radicals in plant systems is well documented (Elstner, 1982; Leshem, 1981). The oxy free radicals are produced by the univalent reduction of oxygen. Excited electrons are easily accepted by oxygen which has vacancies in its outer molecular orbital. When one electron is accepted, oxygen becomes a superoxide free radical ($O_2^{\overline{\cdot}}$). Superoxide in turn may produce a hydroxyl (·OH) free radical, the most potent oxidant known, and organic free radicals. Disruption of normal metabolism due to dehydration may transiently perturb enzyme function and may give rise to the production of free radical species. For example, inhibition of mitochondrial electron transport by cyanide has induced phospholipid deesterification in membranes of isolated mitochondria from alfalfa roots, increasing the free fatty acid contents of the membranes (Illman, 1985).

The presence of organic free radicals in soybean seeds has been documented in respect to natural and artificial ageing of seeds (Buchvarov and Grantcheff, 1984; Priestley et al., 1985). No attempts have been made to compare free radicals in desiccation-tolerant and intolerant stages during germination. These studies establish two relevant points. First, the levels of organic free radicals increase with the loss of moisture (Figure 5-3). The

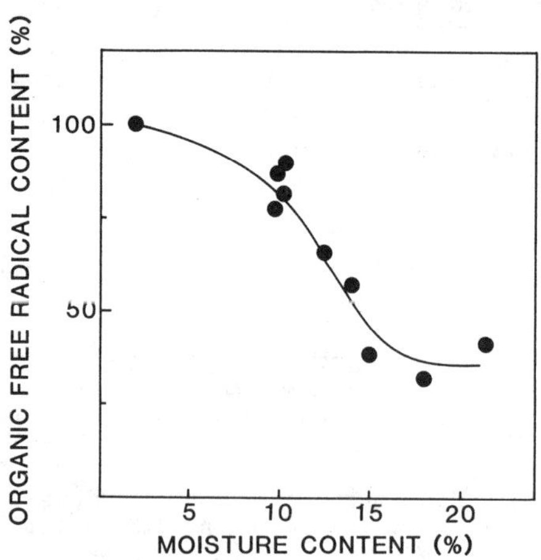

Figure 5-3. The effect of hydration on the organic free radical content of powdered soybean embryonic axes. Results are expressed on the basis of the number of free radicals detected by ESR spectroscopy per dry weight of tissue relative to the maximum observed. (Modified from Priestley et al., 1985, courtesy of *Physiologia Plantarum*.)

threshold for the accumulation of free radicals is about 15% water, rising to a maximum at 10% water (Priestley et al., 1985). Injury from desiccation in this and other species starts at 20% and increases in severity with further drying (McKersie and Tomes, 1980; Senaratna and McKersie, 1983a). Priestley et al. (1985) interpreted their data as indicating that free radicals are released from a trapped state as the tissue is hydrated. Second, there is some indication that free radicals are more readily produced in the embryonic axis than in the cotyledons (Buchvarov and Grantcheff, 1984), but other workers have reported similar accumulations in both tissues (Priestley et al., 1985). It is an interesting indication, however, because desiccation injury at this stage occurs more readily in the axes (Senaratna and McKersie, 1983a). Thus, it is reasonable to assume that production and accumulation of free radicals occur during desiccation.

Mechanisms of Free Radical Injury to Membranes

Free radical-mediated phospholipid degradation is commonly thought to involve a peroxidation of unsaturated fatty acids. The divinyl methane structure present in all polyunsaturated fatty acids is particularly susceptible to hydrogen abstraction, resulting in the formation of a fairly stable free radical; in the presence of oxygen, this process initiates a free radical chain reaction leading to autooxidation (Mead, 1976). Peroxide and hydroperoxide intermediates produced during this reaction decompose. A product of the decomposition is malondialdehyde, which can be measured to indicate free radical-mediated lipid peroxidation. Thus, double bonds of unsaturated fatty acids have been assumed to be the primary site of free radical attack (Figure 5-4a).

The allelic hydrogen (hydrogen attached to a carbon atom adjacent to a double bond) is easier to abstract than a tertiary or secondary hydrogen. If free radical-mediated injury proceeds by such a mechanism, the degree of fatty acid unsaturation would be expected to change. However, the desiccation of plant tissues does not normally induce changes in fatty acid saturation even if the stress is lethal (Stewart and Bewley, 1982; Senaratna et al., 1984).

Alternatively, free radicals generated in aqueous environments have been shown to induce deesterification in pure phospholipid (Niehaus, 1978) and in liposomes prepared from phospholipids of microsomal membranes from seed axes (Senaratna et al., 1985b). In addition, similar deesterification has been observed in phospholipid/water mixtures exposed to gamma radiation, which produces free radicals (Kendall et al., 1985b).

Deesterification may proceed by one of two mechanisms. The fatty acid

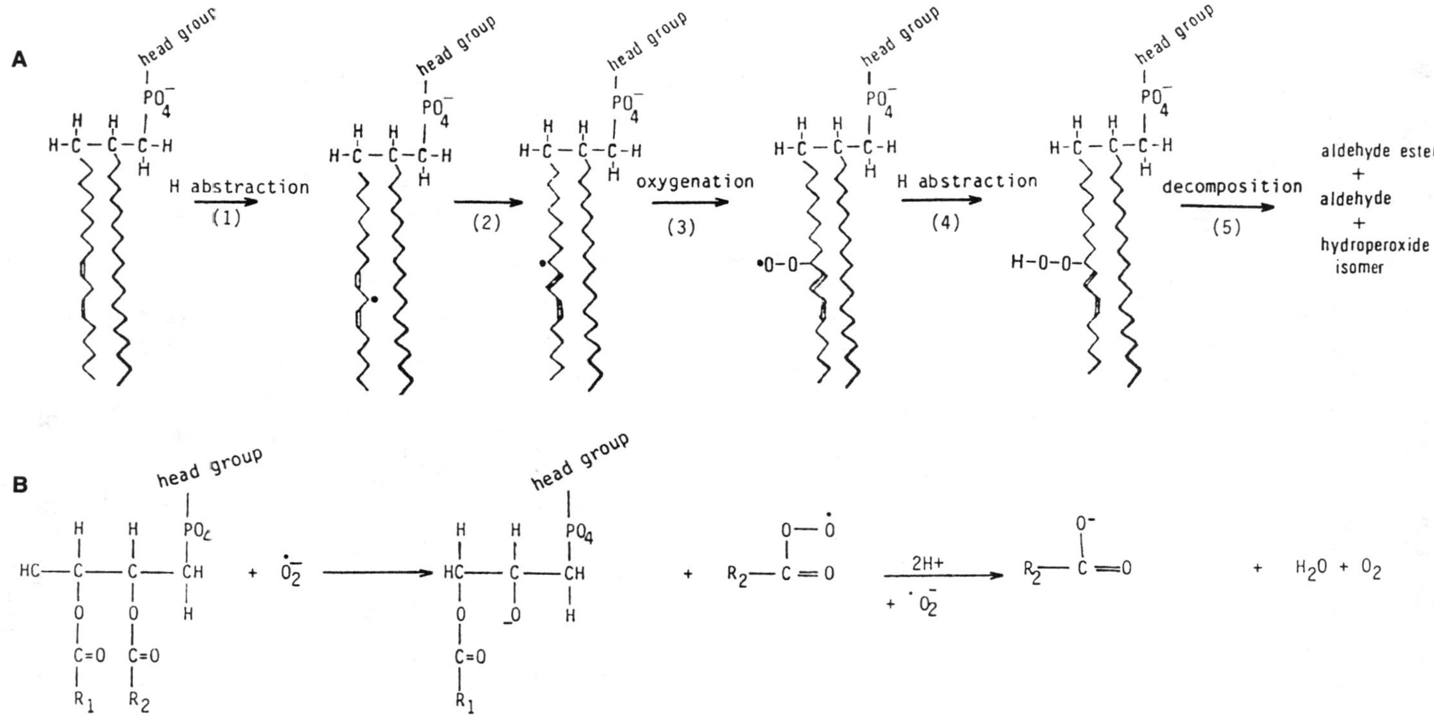

Figure 5-4. (a) Mechanism of phospholipid peroxidation. Outline of the peroxidation reaction of the unsaturated fatty acid in a lipid molecule. *Step 1*: A free radical initiates the reaction by abstracting H from the allelic carbon. *Step 2*: Electron rearrangement. At this step, two radicals can conjugate together, terminating the reaction and producing polymers. *Step 3*: Oxygen attachment to the radical produces peroxide radical. *Step 4*: H abstraction from another molecule to produce hydroperoxide, and this molecule becomes a free radical. This step perpetuates a chain reaction. *Step 5*: Hydroperoxide decomposes to produce various products. The first degradation is at the bond between the two oxygen atoms to yield OH radical. Subsequent cleavages occur between C—O bonds or C—C bonds (for details see Frankel, 1980; Mead, 1976).

(b) Nucleophilic addition of superoxide (for details, see San Filippo et al., 1976).

C

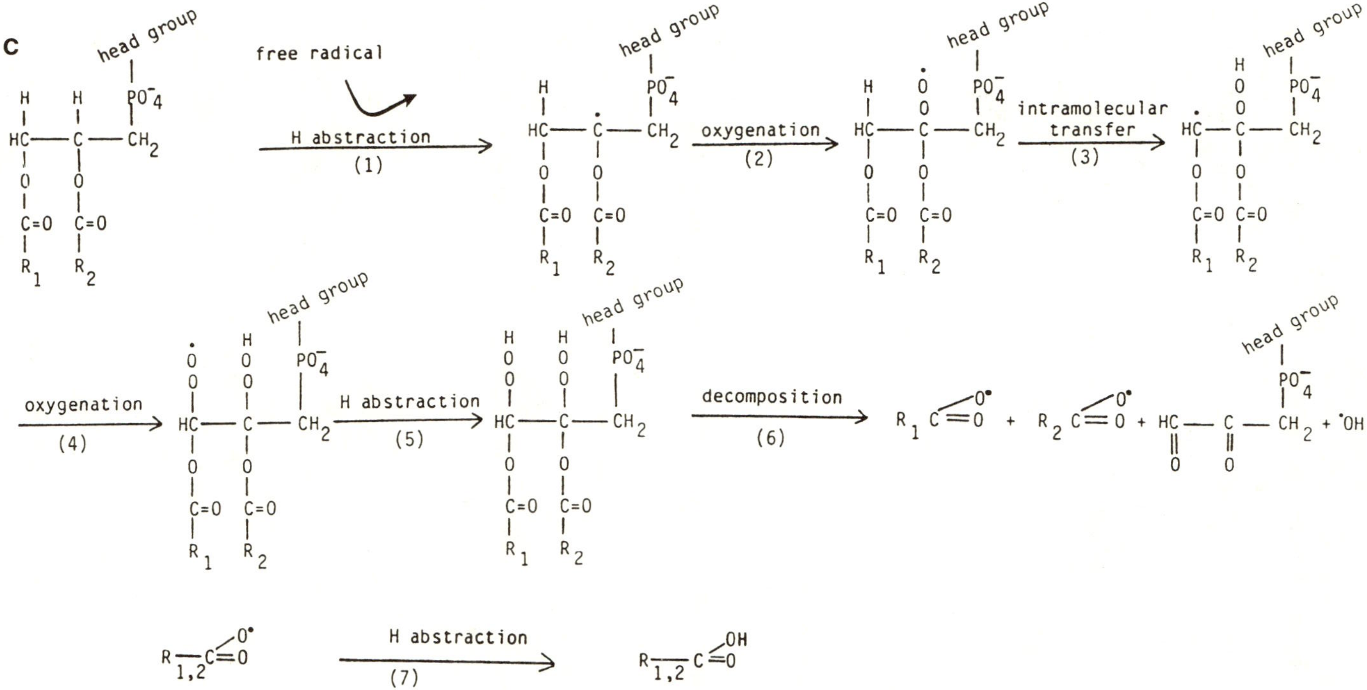

(c) Proposed mechanism of phospholipid deesterification. *Step 1*: A free radical initiates the reaction by abstracting H from the second carbon of the glycerol backbone, yielding a radical. *Step 2*: Oxygen attachment yielding a peroxide radical. *Step 3*: Peroxide radical abstracts hydrogen from the first carbon of the glycerol backbone. *Step 4*: Oxygen attachment to the first carbon. *Step 5*: H abstraction from another molecule, which may be the starting substance leading to a chain reaction. The product is a dihydroperoxide. *Step 6*: Decomposition of dihydroperoxide between O—O to produce ·OH radical. Subsequent breakdown of ester bond O and glycerol to produce acyloxy radicals. Other products, such as aldehydes, ketones, ketoaldehydes, and hydrocarbons can also be anticipated. *Step 7*: The acyloxy radical abstracts H from another lipid molecule, perpetuating the chain reaction and becoming a free fatty acid.

ester linkage can be broken as a result of nucleophilic addition of superoxide radical at the carbonyl carbon (Figure 5-4b). The reaction scheme for this cleavage of the ester bond has been proposed by San Filippo, Jr., et al. (1976). A similar mechanism can be involved in superoxide-mediated deesterification of phospholipid (Niehaus, 1978).

Alternatively, deesterification can be initiated by abstracting a hydrogen from the tertiary carbon (middle carbon of the glycerol backbone of the phospholipid) by a free radical, forming an organic radical (Figure 5-4c). When oxygen is present, there follows an autooxidation reaction. Peroxides, hydroperoxides, and dihydroperoxides are likely intermediates in this process. Dihydroperoxides may decompose, producing an acyloxy radical and keto aldehydes. An acyloxy radical can abstract H from another phospholipid molecule, reducing into a fatty acid, and as a consequence may perpetrate a chain reaction promoting phospholipid deesterification.

Mechanisms of Tolerance

If irreversible injury to cellular membrane systems is initiated during desiccation via a free radical-mediated mechanism, as suggested above, then tolerance in seed tissues in the early stages of germination may be provided by a free radical tolerance mechanism. Tolerance to free radicals in plant cells may be provided by enzymes as well as antioxidants (Chance et al., 1979; Pryor, 1976). Three enzymes, superoxide dismutase (SOD), catalase, and peroxidase, are known to function in the cell, preventing the buildup of reactive oxyfree radicals. Another contributing enzyme is glutathione reductase. This enzyme reduces oxidized glutathione which has already reduced a free radical, thus allowing a glutathione cycle to operate (Pryor, 1976). These enzymes, however are not membrane bound and would not explain the observed free radical tolerance of isolated membranes from dehydration-tolerant tissue (Senaratna et al., 1985b). Enzymes such as SOD may not have a major role in free radical scavenging when tissues are in the dry state but may function during the drying process prior to the completion of desiccation and associated inactivation of enzymes. In soybean seeds, SOD activity was minimal in the dry state and was low during the first 2 hr of imbibition. Maximum activity was recorded between 3 and 4 hr after imbibition and declined to minimum levels at 5 hr and remained low thereafter (Stewart and Bewley, 1980). This observation does not correlate in any way with the loss of desiccation tolerance.

To identify the agent(s) imparting desiccation tolerance to the soybean axis, microsomal membranes were separated, and artificial membranes (lipo-

Table 5-3
Lipid-soluble antioxidant content of soybean embryonic axes

Imbibition period (hr)	Treatment	Tocopherol equivalents ($mg \cdot g^{-1}$)
0	Control	50.0
6	Control	57.0
6	Desiccated	52.0
36	Control	0.7
36	Desiccated	0.1

Note: Soybean embryonic axes were isolated from control and desiccation-stressed seeds as previously described. Samples were extracted with chloroform-methanol. The relative quantity of antioxidants was determined by the inhibition of linoleic acid oxidation (McKersie et al., 1982). Values are expressed as the equivalent degree of inhibition by tocopherol per g total lipid extract, determined gravimetrically (adapted from Senaratna et al., 1985b, courtesy of *Plant Physiology*).

somes) were constructed from the lipid portion of the membrane. These membranes are thus devoid of membrane protein and contaminating cytoplasmic protein. Liposomes prepared from membrane lipids of tolerant soybean axes are relatively more tolerant of free radical treatment than those from axes in the intolerant stage (Table 5-3). Thus, free radical scavenging capacity may be presumed to be present in the lipid-soluble fraction of the membrane. Soybean axes tolerant of desiccation contain large quantities of lipid-soluble antioxidants. In contrast, those in the intolerant stage (36 hr) do not contain significant quantities of lipid-soluble antioxidants. The desiccation tolerance of the seed appears to be associated with free radical scavengers and particularly with lipid soluble antioxidants in cytoplasmic membranes. During germination these pools are depleted, making the cellular membranes more susceptible to free radical-mediated desiccation injury.

The proteins (cytoplasmic or membrane bound) which may mediate the synthesis of these antioxidants and thus contribute to tolerance are not defined as yet. The involvement of enzymatic free radical scavengers is also not clearly defined at this time.

A Model for Dehydration Injury to Membranes

A working hypothesis of desiccation damage which is consistent with the previous experimental data is presented schematically in Figure 5-5.

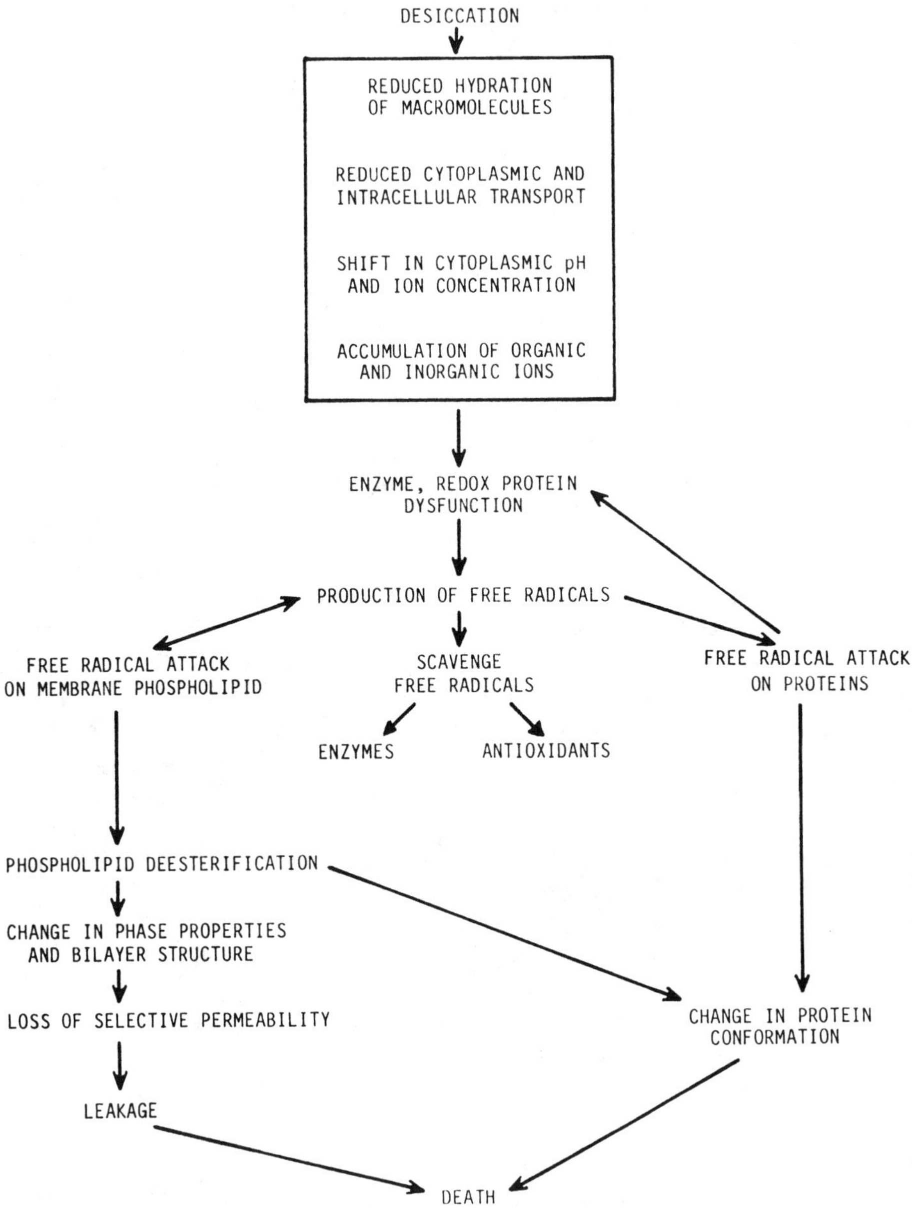

Figure 5-5. Model of the sequence of events in desiccation injury as mediated by free radicals.

Desiccation induces several changes in the cellular environment such as: (1) reduced hydration of macromolecules and consequent conformational changes; (2) reduced cytoplasmic and intracellular transport; (3) shifts in cytoplasmic pH and ion concentrations; and (4) accumulation of organic and inorganic ions. All or any of these changes might be expected to cause transient dysfunctions in specific enzymes and/or electron transport chains. Such dysfunction may lead to the production of free radicals or promote chemical reactions which would not normally occur in a fully hydrated system. Under normal metabolic conditions, free radicals would be scavenged by enzymatic mechanisms such as superoxide dismutase, but because of the reduced hydration these mechanisms may not function. The likelihood that free radicals are produced by electron transport chains and possibly other membrane-bound enzymes implies that they may be produced in the membrane, thus finding the ester linkage of phospholipids a convenient reaction site.

The free radical-mediated deesterification of phospholipid is self-perpetuating and may even be amplified with time, so that the reactions continue, once started, until two radicals conjugate together pairing the electrons, or until the radical reacts with a scavenger. Functional antioxidants may include tocopherol or coincidental scavengers such as carbohydrates or proteins which, although degraded by the reaction, may not produce free radical products. In desiccation-tolerant tissue, high levels of lipid-soluble and presumably membrane-associated antioxidants may serve to scavenge free radicals and prevent the chain reactions from continuing. In susceptible tissue, these reactions continue, leading to the degradation of membrane phospholipid and to the accumulation of substantial amounts of free fatty acids and other degradation products in the membrane. The phospholipid and protein distribution within the bilayer would be reorganized to accommodate this accumulation, causing the formation of gel-phase regions and further perturbing membrane bound enzymes. The permeability of the bilayer would increase, and loss of compartmentalization upon rehydration would be expected, and thus these events would culminate in cell death.

REFERENCES

Akalehiywot, T., and Bewley, J.D. 1980. Desiccation of oat grains during and following germination and its effect on protein synthesis. Can. J. Bot. 58: 2349-2355.

Baldassare, J., Saito, Y., and Silbert, D. 1979. Effect of sterol depletion on LM cell sterol mutants. Changes in lipid composition of the plasma membrane and their effects on 3-0-methyl-glucose transport. J. Biol. Chem. 254: 1108-1113.

Barber, R.F., and Thompson, J.E. 1980. Senescence-dependent increase in the permeability of liposomes prepared from bean cotyledon membranes. J. Exp. Bot. 31: 1305-1313.

Becwar, M.R., Stanwood, P.C. and Ross, E.E. 1982. Dehydration effects on imbibitional leakage from desiccation sensitive seeds. Plant Physiol. 69: 1132-1135.

Buchvarov, P. and Grantcheff, Ts. 1984. Influence of accelerated and natural ageing on free radical levels in soybean seeds. Physiol. Plant. 60: 53-56.

Chance, B., Sies, H. and Boveris, A. 1979. Hydroperoxide metabolism in mammalian organs. Physiol. Rev. 59: 527-605.

Crevecoeur, M., Deltour, R. and Bronchart, R. 1976. Cytological study of water stress during germination of *Zea mays*. Planta 132: 31-41.

Dasgupta, J., Bewley, J.D. and Yeung, E.C. 1982. Desiccation- tolerant and desiccation-intolerant stages during the development and germination of *Phaseolus vulgaris* seeds. J. Exp. Bot. 33: 1045-1057.

Elstner, E.F. 1982. Oxygen activation and oxygen toxicity. Ann. Rev. Plant Physiol. 33: 73-96.

Frankel, E.N. 1980. Lipid oxydation. Prog. Lipid Res. 19: 1-22.

Hegarty, T.W. 1978. The physiology of seed hydration and dehydration and the relation between water stress and control of germination. A review. Plant Cell Environ. 1: 109-119.

Illman, D.D. 1985. Isolation Injury to Alfalfa (*Medicago sativa* L.) Mitochondrial Membranes. M.S. thesis. University of Guelph.

Kendall, E.J., Katsaras, J., Stinson, R.H. and McKersie, B.D. 1985. Phase transitions in freeze-damaged membranes. Plant Physiol. (suppl.) 77: 736.

Leshem, Y.Y. 1981. Oxy free radicals and plant senescence. What's New in Plant Physiol. 12: 1-3.

Lyons, J.M. 1973. Chilling injury in plants. Ann. Rev. Plant Physiol. 24: 445-456.

McKersie, B.D., Hucl, P. and Beversdorf, W.D. 1982. Solute leakage from susceptible and tolerant cultivars of *Phaseolus vulgaris* following ozone exposure. Can. J. Bot. 60: 73-78.

McKersie, B.D. and Senaratna, T. 1983. Membrane structure in germinating seeds. *In* Recent Advances in Phytochemistry. Mobilization of Reserves in Germination. C. Nozzolillo, P.J. Lea, F.A. Loewus, eds. Pp. 29-52. Plenum, New York.

McKersie, B.D. and Thompson, J.E. 1977. Lipid crystallization in senescent plant membranes from cotyledons. Plant Physiol. 59: 803-807.

McKersie, B.D. and Thompson, J.E. 1978. Phase behaviour of chloroplast and microsomal membranes during leaf senescence. Plant Physiol. 61: 639-643.

McKersie, B.D. and Thompson, J.E. 1979. Phase properties of senescing plant membranes. Role of the neutral lipids. Biochim. Biophys. Acta 550: 48-58.

McKersie, B.D., Thompson, J.E. and Brandon, J.K. 1976. X-ray diffraction evidence for decreased lipid fluidity in senescent membranes from cotyledons. Can. J. Bot. 54: 1074-1078.

McKersie, B.D. and Tomes, D.T. 1980. Effect of dehydration treatment on germination, vigor, and cytoplasmic leakage in wild oats and birdsfoot trefoil. Can. J. Bot. 58: 471-475.

Mead, J.F. 1976. Free radical mechanisms of lipid damage and consequences for cellular membranes. *In* Free Radicals in Biology, Vol. I. W.A. Pryor, ed. Pp. 51-68. Academic Press, New York.

Niehaus, Jr., W.J. 1978. A proposed role of superoxide anion as a biological nucleophile in the deesterification of phospholipid. Bioorg. Chem. 7: 77-84.

Priestley, D.A., Werner, B.G., Leopold, A.C. and McBride, M.B. 1985. Organic free radical levels in seeds and pollen. The effects of hydration and ageing. Physiol. Plant. 64: 88-94.

Pryor, W.A. 1976. The role of free radical reactions in biological systems. *In* Free Radicals in Biology. Vol. I. W.A. Pryor, ed. Pp. 1-49. Academic Press, New York.

Sandermann, Jr., H. 1978. Regulation of membrane enzymes by lipids. Biochim. Biophys. Acta 515: 209-237.

San Filippo, Jr., J., Romano, L.J., Chern, C. and Valentine, J.S. 1976. Cleavage of esters by superoxide. J. Org. Chem. 41: 586-588.

Senaratna, T. and McKersie, B.D. 1983a. Dehydration injury in germinating soybean (*Glycine max* L. Merr.) seeds. Plant Physiol. 72: 620-624.

Senaratna, T. and McKersie, B.D. 1983b. Characterization of solute efflux from dehydration injured soybean (*Glycine max* L. Merr.) seeds. Plant Physiol. 72: 911-914.

Senaratna, T., McKersie, B.D. and Stinson, R.H. 1984. The association between membrane phase properties and dehydration injury in soybean axes. Plant Physiol. 76: 759-762.

Senaratna, T., McKersie, B.D. and Stinson, R.H. 1985a. Simulation of dehydration injury to membranes from soybean axes by free radicals. Plant Physiol. 77: 472-474.

Senaratna, T., McKersie, B.D. and Stinson, R.H. 1985b. Antioxidant levels in germinating soybean seed axes in relation to free radical and dehydration tolerance. Plant Physiol. 78: 168-171.

Shechter, E., Letellier, L. and Gulik-Krzywicki, T. 1974. Relations between structure and functions in cytoplasmic membrane vesicles isolated from E. coli fatty acid auxotroph. Eur. J. Biochem. 49: 61-76.

Shinitzky, M. and Inbar, M. 1976. Microviscosity parameters and protein mobility in biological membranes. Biochim. Biophys. Acta 433: 133-149.

Simon, E.W. 1974. Phospholipids and plant membrane permeability. New Phytol. 73: 377-420.

Simon, E.W. and Mills, L.K. 1983. Imbibition, leakage and membranes. *In* Advances in Phytochemistry. Mobilization of Reserves in Germination. C. Nozzolillo, P.J. Lea, F.A. Loewus, eds. Pp. 9-27. Plenum, New York.

Steponkus, P.L. 1984. Role of the plasma membrane in freezing injury and cold acclimation. Ann. Rev. Plant Physiol. 35: 543-548.

Stewart, R.R.C. and Bewley, J.D. 1980. Lipid peroxidation associated with accelerated ageing in soybean axes. Plant Physiol. 65: 245-248.

Stewart, R.R.C. and Bewley, J.D. 1982. Stability and synthesis of phospholipids during desiccation and rehydration of a desiccation-tolerant and a desiccation-intolerant moss. Plant Physiol. 69: 724-727.

6 Water Content in Relation to Stress in Pollen

Folkert A. Hoekstra

Pollen, the male gametophytes of higher plants, delivers the sperm cells to the ovules in order to bring about fertilization. The size of pollen grains is 5-200 μm, with the majority having a diameter of 30 μm (Stanley and Linskens, 1974). In the course of the development of the male angiosperm gametophyte, two nuclear divisions take place. The first one leads to a vegetative and a generative cell; in the second one the generative cell gives rise to two sperm cells. Depending on whether dehiscence occurs after the first or the second mitosis, pollen is bicellular or tricellular, erroneously designated binucleate and trinucleate, respectively (cf. Brewbaker, 1959). Trinucleate plant families, comprising about 30% of the angiosperms, are considered to have evolved polyphyletically from binucleate families (Brewbaker, 1959, 1967).

At dehiscence pollen undergoes rapid drying; from a maximum of 80% moisture content inside the closed anther (Keijzer, 1983), they drop to a value of 60% or much less in the free state. Pollen is then dispersed by wind, insects, or vertebrates to meet a receptive stigma. Hydration is a necessary prelude to germination, producing a filiform structure or pollen tube which grows through the style towards the ovules by a process much like a parasitic invasive action.

Two types of stigmas are encountered in angiosperms, those which are "dry" at maturity, having a proteinaceous pellicle but no liquid secretion, and those which are "wet" (i.e. bearing a fluid secretion; Heslop-Harrison and Shivanna, 1977). The dry type is often associated with trinucleate families, whereas in binucleate families both dry and wet stigmas occur. The distinction is of considerable importance because it means that pollen belonging to plants having wet stigmas must be resistant to instantaneous rehydration, whereas those with dry stigmas will experience a more subtle rehydration. Such differences are reflected in the extent to which pollen can successfully be germinated in vitro: trinucleate pollen species are considered difficult to germinate in vitro, requiring high concentrations of sugar in the germination media (Brewbaker, 1959, 1967). Controlled hydration, using media that are solidified with 2% agar and then partially dried, seems crucial for the successful germination of a great number of trinucleate pollen species (Bar-Shalom and Mattsson, 1977). In contrast, binucleate species generally germinate well in sucrose concentrations of low molarity, with small additions of boric acid and some Ca-, Mg-, and K-salts (Brewbaker, 1959; Brewbaker and Kwack, 1963). That the condition of the stigma rather than the number of nuclei is important for the ease with which germination in vitro can be accomplished is further supported by data on the behaviour of the trinucleate pollen of *Cactaceae* and *Aizoaceae*. These families have wet stigmas, and being exceptions to the trend, their trinucleate pollen germinates easily in liquid media (Hoekstra, unpublished results).

Whereas wet stigmas easily trap pollen in their sticky fluid, adhesion onto the dry-type stigma is dependent either on postrecognition reactions or on electrostatic force and the presence of filament branches (Woittiez and Willemse, 1979) on which pollen readily germinates. As mentioned earlier, trinucleate families are conspicuous in having dry stigmas. In this context it is striking that trinucleate pollen species tend to germinate rapidly, within a few minutes, thus becoming attached to the stigma by their own pollen tube (Hoekstra and Bruinsma, 1978). In contrast, among their binucleate counterparts, both rapid and slow (1 hr) types occur.

In angiosperm evolution, recurrent competition of pollen tubes in the style for the available ovules is suggested to have led to faster-growing pollen tubes and a more rapid start of tube emergence upon the first contact with the stigma (cf. Hoekstra, 1983). A pollen grain which advances further in maturation in the anther should be able to reach the ovule faster and should therefore have a competitive advantage. It has been realized that such is the case with rapidly germinating trinucleate pollen species, in which the second nuclear division occurs during maturation in the anther. Furthermore, the pollen mitochondria are in an advanced stage of development at maturity (Hoekstra, 1979), and tube growth may proceed independently from protein synthesis (Hoekstra and Bruinsma, 1979). Because such pollen condition is closely associated with the occurrence of dry stigmas (Heslop-Harrison and Shivanna, 1977), there has probably been an evolutionary change in the water relations during the early pollen-stigma interaction as well. Pollen longevity and speed of tube emergence were found to be associated with specific compositions of fatty acids in the total lipids (Hoekstra and van Roekel, 1985), and in the following sections special emphasis is therefore placed on membrane-related aspects of dehydration, rehydration, and dry storage.

Effects of Rain and High Humidity

In contrast to seed, pollen is not designed for long-term survival, but for a short period it, like seed, must cope with such rather hostile conditions as extreme drought or humidity and extreme temperatures. Upon dehiscence, liquid water is rather detrimental, as it leads to loss of vitality, precocious germination, or bursting (Lidforss, 1896). In some cases pollen may even germinate inside the closed anther when dehydration is insufficient, particularly in humid weather (Pacini and Franchi, 1982). Rain interferes with the daily anthesis of *Compositae* flowers and affects pollen viability in the closed anther such that pollen of very poor quality is set free after the weather conditions have improved (Hoekstra, 1983). Such pollen has not been in direct contact with rain water.

Pollen germinating in rain water will not substantially contribute to pollination and fertilization, unless it is floating in the vicinity of a receptive stigma. Alternatively, it may dry out after being soaked in rain water and may be carried to a receptive stigma. Drying experiments done on six pollen species that were soaked in distilled water show that, except in the gymnosperm *Pinus*, germinability is considerably impaired (Hoekstra, 1983). Survival is much better when the soaking is done in a liquid germination medium rather than in distilled water. Consequently, there may result lesser solute losses because sucrose, boric acid, and salts are present in the medium (Dickinson, 1967; cf. Čapková-Balatková et al., 1980). Retention of vitality is possible only when redrying occurs prior to outgrowth of tubes. Rapidly germinating pollen types, therefore, are particularly sensitive to the soaking and drying treatments (Hoekstra, 1983).

Exposure to a water-saturated atmosphere leads to water uptake, up to as much as 60% of the air-dry weight of the pollen (Hoekstra and Bruinsma, 1980). Such uptake of water vapor is generally insufficient to sustain tube growth, although the formation of probe tubes of 1-3 pollen diameters has been noticed (Ferrari et al., 1981; Hoekstra, unpublished results). There is general agreement among the early workers of pollen physiology that high relative humidity is unfavorable for storage. *Compositae* pollen loses vitality very rapidly, in 3 hr on the average (Hoekstra and Bruinsma, 1975a), exhibiting high rates of respiration (Hoekstra and Bruinsma, 1975b). With inhibitors of oxidative phosphorylation, it was calculated that about 75% of the energy

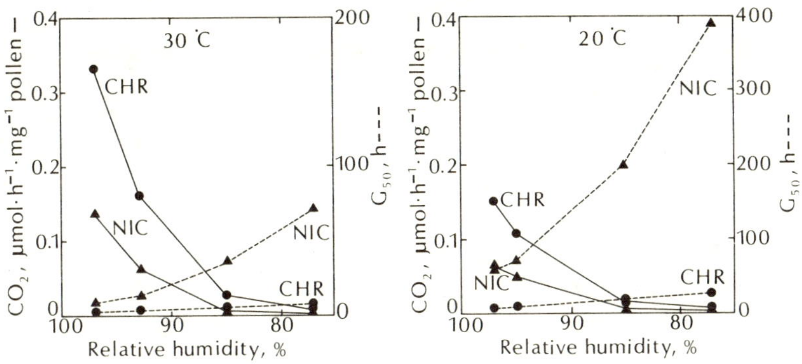

Figure 6-1. Effect of relative humidity (over saturated salt solutions) and temperature on respiration and longevity of pollen of *Chrysanthemum* (CHR) and *Nicotiania* (NIC). Longevity is expressed in hours until 50% of the initial germinative capacity (G_{50}). CO_2 was analysed by GC. (From Hoekstra and Bruinsma, 1975b, courtesy of *Physiologia Plantarum*.)

generated by this respiration is trapped as ATP for inappropriate syntheses (Hoekstra and Bruinsma, 1980). The inappropriate syntheses rather than loss of respiratory substrate are assumed to be detrimental, as respiration continues long after vitality is lost, suggesting that membrane integrity is still maintained under these humid conditions.

Pollen species with longer lag periods generally display lower rates of respiration under humid conditions and have considerably better longevity (Hoekstra and Bruinsma, 1975b). Figure 6-1 gives an example of the influence of relative humidity and temperature on respiration and longevity. The two species tested differ in that the trinucleate pollen of *Chrysanthemum cinerariaefolium* rapidly germinates on a dry stigma in few minutes, whereas it takes about 1 hr for the binucleate *Nicotiana alata* pollen to start tube emergence on the wet stigmatic surface. In most cases respiration in humid air falls after 15-20% of the pollen dry weight has been consumed (calculated from data of Hoekstra and Bruinsma, 1975b). More research is required to determine whether cessation of respiratory activity in moist air is linked to loss of membrane integrity or whether exhaustion of respiratory substrate is involved.

Dehydration of Fresh Pollen

Dehydration-Tolerant Pollen

The early workers on pollen physiology in the late 19th and early 20th century found that, in general, drying of pollen is essential for successful storage (see extensive review by Binder et al., 1974). Pfundt (1910) stored some 85 species for extended periods of time above concentrated sulfuric acid, so that relative humidity was approximately zero. During dehydration the shape of the originally hydrated grains gradually changed from round or oval to half moon or banana shape because of shrivelling. This decrease in volume increases the concentration of endogenous substances and ions such as potassium to such an extent that it can inhibit protein synthesis (Bashe and Mascarenhas, 1984). Respiration falls to undetectable rates below 20% moisture content, and adenylate energy charge values drop considerably (Hoekstra, unpublished results), suggesting an inability to generate ATP.

The general resistance to dehydration suggests a special condition or protection of pollen membranes. In a number of other anhydrobiotic organisms, trehalose is formed upon dehydration in amounts as high as 20% of the dry weight (Crowe et al., 1984; Chapters 10-12). In the pollen literature the occurrence of trehalose is not mentioned (Stanley and Linskens, 1974). In contrast, sucrose, fructose, and glucose are generally present in amounts that may reach 30% of the pollen dry weight (McLellan, 1977). Although not as effective as trehalose, these compounds are able to preserve the structure and function of lobster muscle microsomal membranes during dehydration (Chapter 11). Another compound that is rather common in pollen is proline, present in the amount of 1-2% of the dry weight (Linskens and Schrauwen, 1969). Proline seems to increase resistance to high temperatures during germination in vitro (Zhang and Croes, 1983). On the other hand, dehydration-sensitive corn pollen does contain considerable amounts of reducing and nonreducing sugars (Stanley and Linskens, 1974) and of proline (Linskens and Pfahler, 1973; Pálfi and Köves, 1984). It may be that a combination of sugars serves to provide dehydration resistance, as suggested in Chapter 1.

Simon (1974) postulated that phospholipids in dehydrating seed membranes pass from the bilayer into the hexagonal phase much as brain lipid does. Upon rehydration, the reverse process might be responsible for the initial leakage during the early stages of seed imbibition. Using Simon's suggestion, Heslop-Harrison (1979a, 1979b), and Shivanna and Heslop-Harrison (1981) suggest that such events are also common during dehydration and rehydration processes in pollen. Electron micrographs of ungerminated grains generally show poorly resolved membranes. It is more likely, I suggest, that a slow developmental change of the membranes takes place during germination (Hoekstra and van Roekel, 1983). Evidence that, in fresh *Typha latifolia* pollen, membrane phospholipids are arranged not in the hexagonal state but instead in the bilayer structure comes from ^{31}P-NMR spectra on intact pollen (Priestley and de Kruijff, 1982). At least down to 11% moisture content, the lamellar is the sole configuration present in the pollen phospholipids.

Dehydration-Sensitive Pollen

In their search for methods to extend pollen viability, pioneers of pollen physiology came across the strange behaviour of Gramineae pollen (Pfundt, 1910; Knowlton, 1922; Holman and Brubaker, 1926). Apart from finding that high temperatures are a considerable stress to grass pollen, Andronescu in 1915 was the first to show that desiccation is the cause of death of corn pollen (see references cited by Knowlton, 1922; Goss, 1968). Although sensitivity to desiccation is quite rare, its occurrence in the agriculturally important family of the Gramineae obstructs the progress of breeding programmes considerably, as flowering periods of hybrid parents must coincide for successful breeding work. Best longevity, although it is still limited to several days or at most 2 weeks, can be obtained by storage at 0-2°C in high relative humidity, as reported for corn (Sartoris, 1942; Walden, 1960), and sugar cane (Sartoris, 1942; Moore, 1976). By reducing pollen moisture content and, at the same time, avoiding irreversible loss of vitality, Barnabás and Rajki (1976, 1981) succeeded in the successful storage of corn pollen at −196°C, tested in vitro as well as in vivo. An exception to the generally poor desiccation tolerance is the pollen of the grass *Pennisetum typhoides*, which remains viable for extended periods of time when stored over concentrated sulfuric acid (Vasil, 1960).

Investigations by Heslop-Harrison (1979a) and Shivanna and Heslop-Harrison (1981) suggest that the injury from desiccation originates as impairment of the plasma membrane integrity. This comment was based on their results using the vital stain, fluorescein diacetate (FDA), an ester that easily penetrates into the cytosol where esterases give rise to the fluorescent dye, fluorescein. Leaky plasma membranes cannot retain the dye inside the grain, whereas the intact ones allow for an accumulation of fluorescein (Heslop-Harrison and Heslop-Harrison, 1970; Heslop-Harrison et al., 1984). It must be emphasized that fluorescein diacetate is applied via a liquid carrier such as a germination medium. This form of application may induce imbibitional damage in the dried grains, as discussed in the section on rehydration. However, Shivanna and Heslop-Harrison (1981) carefully avoided this possible damage by rehydrating the pollen in humid air prior to the test. If we assume that the esterase activity is not altered, their data indicate that dehydration of rye pollen irreversibly damages the capacity of the plasmalemma to retain fluorescein. As a cell with leaking membranes cannot become turgid, germination will be impossible. Conflicting with such an explanation of the loss of vitality in corn pollen is the observation that drying is followed by a strongly increased incidence of bursting in germination tests in vitro (Pfahler and Linskens, 1973). Incidence would never have increased if osmotically active substances had been leaking out freely.

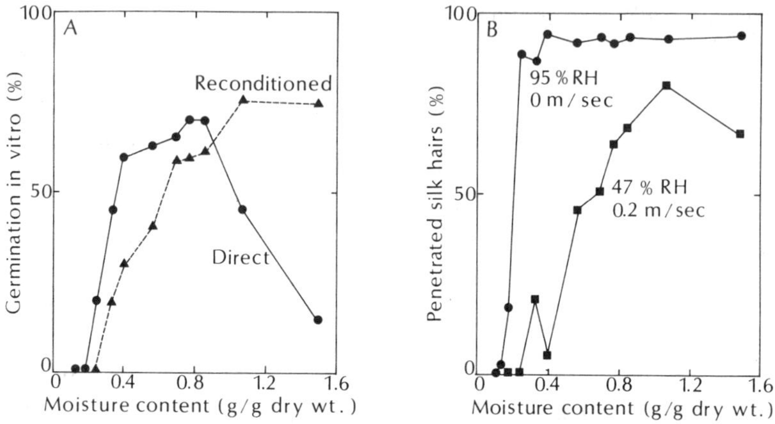

Figure 6-2. Effect of desiccation of corn pollen at room temperature on germinability in vitro (a) and in vivo (b). During drying (for 6 hr), pollen was mixed every 5 min to maintain homogeneity of the moisture content in the sample. Water contents were determined by measuring the weight losses after 4 hr at 110°C. (a) Pollen was either spread directly on the medium (according to Pfahler, 1966) or reconditioned for 18 hr at 2°C in humid air. (b) Pollination was carried out in climate-controlled rooms at 22°C. Tube penetration was analysed on 10 silks per treatment using aniline blue and counting the penetrated silk hairs under a UV microscope.

Hoekstra and Bal (unpublished) analysed the sensitivity of corn pollen to drying. They found that the water content of the pollen was 1.6-1.8 g/g dry weight at the time of shedding from tassels collected in the field. The effect of drying on the germinative capacity is outlined in Figure 6-2. During the early period of drying, the percentage of germination in vitro rose. This phenomenon of improved germinability with partial drying has been noticed by several investigators (Pfahler and Linskens, 1973; Heslop-Harrison, 1979a). The promotive effect of 1 day of storage in the refrigerator, as observed by Frova and Feder (1979), is also due to some water loss. When the moisture content falls below 0.4 g/g dry weight, subsequent germinative ability in vitro starts to decline until, below 0.2 g/g, vitality is completely lost. The drying procedure took about 6 hr, during which a nondried control fully retained its viability. Reconditioning in humid air for 18 hr at 2°C after the partial drying, prior to the germination test, did not improve the capacity of germination in vitro. Tests on germinability in vivo under conditions of high relative humidity were carried out with the same pollen and indicate that viability of pollen in situ is also lost below a moisture content of 0.2 g/g dry weight. A dry wind (47% RH; 2 m/sec) during pollination, which leads to

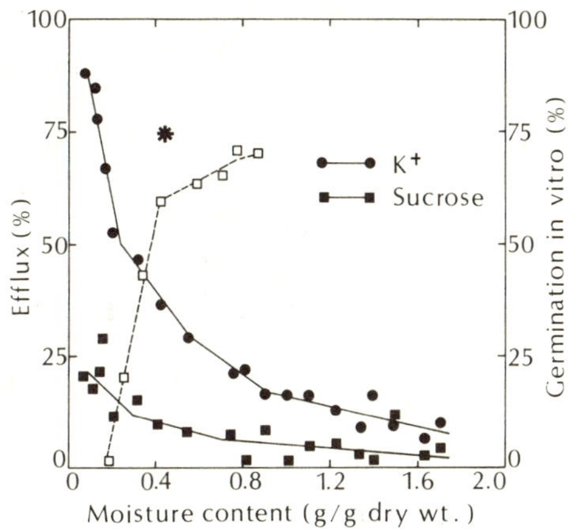

Figure 6-3. Effect of desiccation of corn pollen on germination in vitro and on the efflux of K^+ and sucrose from the grains during a 2-min wash in a liquid medium containing 0.5 M mannitol, 0.8 mM H_3BO_3, and 2.1 mM $Ca(NO_3)_2 \cdot 4H_2O$. Drying procedures were as in Figure 6-2. The asterisk represents the K^+-efflux after a previous rehydration of dry pollen (0.07 g/g dry weight) in humid air. The open symbols represent % germination.

further desiccation of the pollen, considerably impaired germinability. This finding is in accordance with the observation that, under field conditions, dry, hot winds strongly reduce seed set because of both pollen and silk desiccation (Lonnquist and Jugenheimer, 1943).

As conductivity measurements failed to indicate increased leakage of electrolytes during drying, we measured the efflux of a specific ion such as K^+ and of the osmotically important component, sucrose. Figure 6-3 shows that the decrease in viability coincided with a rapid increase of K^+ leakage out of the grains. Sucrose, however, did not leak to a large extent, which may explain the maintenance of turgor pressure in the nonviable grains. It is evident that a deteriorative change in membrane integrity has taken place with drying. However, the extent to which membrane damage and loss of vitality are causally related is far from clear. It seems to me that initially the membrane damage is rather subtle.

Mitochondrial membranes also undergo a change during drying, as it is increasingly difficult to isolate intact, functioning mitochondria from corn pollen with drying (Hoekstra, unpublished results). In an attempt to gain insight into the nature of these membrane changes, the microviscosity of the purified mitochondrial membrane fraction was determined by measuring the mobility of the dye, 1,6-diphenylhexatriene, during its fluorescence lifetime after incorporation into the lipophylic domain of the mitochondrial membrane. From the fluorescence polarization value p, the microviscosity was calculated (Shinitsky and Barenholz, 1978). Viscosity values were essentially unchanged (range 2.05-2.18 poise) over the range of 9-60% water contents, indicating no important change in microviscosity.

Rehydration of Dry Pollen

In their studies on pollen longevity, Holman and Brubaker (1926) and Nebel and Ruttle (1936) noticed that, during dry storage, germination was obtained from samples which had previously failed to germinate in vitro. Nebel and Ruttle (1936) and later Pfeiffer (1938, 1939) found that stored pollen germinates markedly better when it is exposed to 80% RH during 1 day prior to the test. These authors did not realize the importance of their finding. But Duffield and Snow (1941), having success with 12 hr pretreatment in 75% RH, advised prehumidification as a general method in studies on pollen viability. The erratic germination mentioned earlier was most likely due to variation in relative humidity in the laboratory.

Prehumidification in water-saturated atmosphere (Visser, 1955; Lichte, 1957) is superior to that at 75-80% RH and is now widely accepted as pretreatment to germination in vitro. Lichte (1957) classified species as to the prehumidification time required for optimal germination in vitro. Some species only require 10-15 min, whereas others do not germinate well until after 3-hr pretreatment in humid air. An excellent example of the sensitivity of dry pollen to immediate hydration in liquid germination medium is shown in Figure 6-4. The restorative effect of pretreatment in humid air is striking.

Lichte (1957) published micrographs showing a difference in size between germinable and nongerminable pollen. Apparently the nongerminative

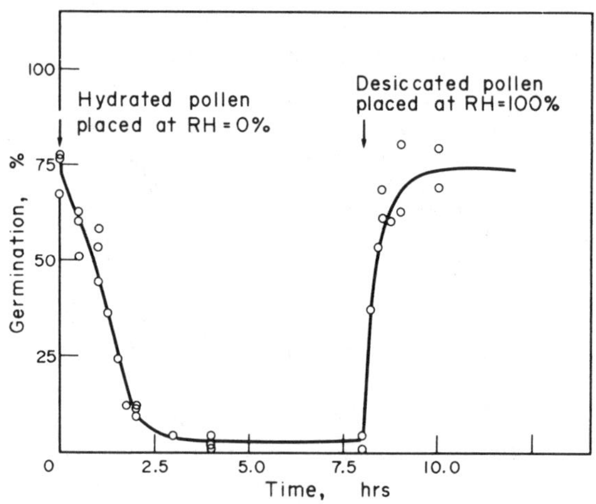

Figure 6-4. Influence of drying fresh *Petunia* pollen over P_2O_5, and of placing dry pollen in a humid atmosphere on the germinative capacity in a liquid medium. (From Gilissen, 1978, courtesy of *Environ. Expt. Bot.*, copyright 1978, Pergamon Press, Ltd.)

condition coincides with an inability to swell. Some similar observations have been made by Gilissen (1977), who attributes the difference in swelling ability to different rigidities of the pollen grain wall, caused by different hydration states. However, Shivanna and Heslop-Harrison (1981) regard the inability to swell as resulting from the loss of osmotic substances due to impaired plasma membrane integrity. They assert that turgescence is an absolute prerequisite for germination, and a cell with leaking membranes cannot become turgid. Testing the desiccation resistance of eight pollen species, these authors showed that, with the exception of Gramineae, humid air pretreatment restores both the ability of germination in vitro and cell fluorescence in the FDA test. Some pollen species are more resistant to drying and subsequent immersion in germination medium than others.

Apart from the initial water content of pollen, another factor determines the extent to which the plasmalemma can be injured, namely the imbibition temperature (Hoekstra, 1984). Figure 6-5 shows that air-dry *Typha latifolia* pollen (7% moisture content) germinates well when imbibition occurs at room temperature or higher. However, imbibition for 3 min at temperatures below 20°C followed by incubation at 23°C reduced germinability; there was a concomitant decrease in respiration and a rise in leakage of NAD^+ out of the grains.

The cause of this low respiratory activity upon imbibitional chilling has been studied in detail (Hoekstra, 1984). Mitochondrial activity in situ is strongly impaired by the loss of cytochrome *c* from its site of action. Furthermore, lesions at the substrate side of ubiquinone determine the rate of O_2 consumption. This may result from the leakage of NAD^+, adenylate phosphates, and Krebs-cycle intermediates from the grains. These results are interpreted as evidence that loss of membrane integrity is the primary cause of both the respiratory failure and the loss of viability. Like imbibitional damage at room temperature, the effects of cold imbibition can be prevented by increasing the initial moisture content of the pollen. Evidence for membrane damage at the ultrastructural level has been obtained for cold imbibition treatment of *Typha* pollen (Figure 6-6). The damaged cells show abundant osmiophylic globules associated with most membranes and a higher degree of vesiculation. Furthermore, electron-dense material accumulates between the intine and plasmalemma, most likely due to the loss of turgor pressure (see also Sack et al., 1983).

Injury is a general phenomenon in many dry pollen species upon cold imbibition, although some resistant types have been encountered (Hoekstra and van der Wal, in preparation). When the initial moisture content is reduced further, from 8% to 5% or lower, the excellent germinability at imbibition temperatures of 15-30°C becomes a poor performance. Potassium leak-

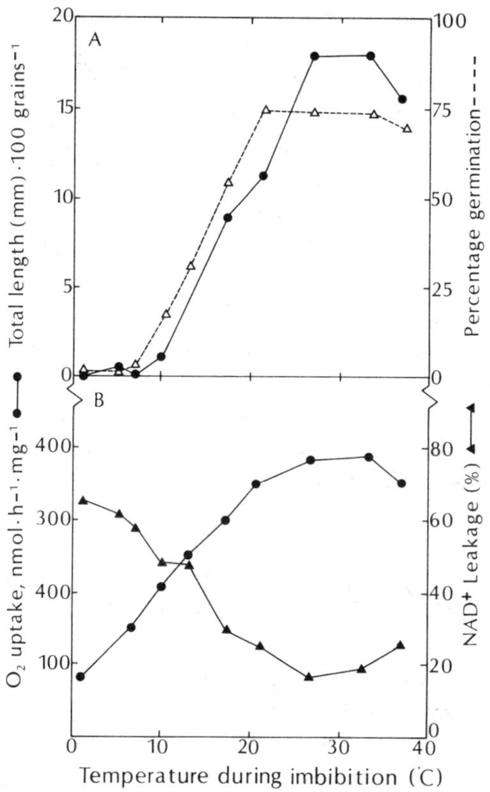

Figure 6-5. Effect of the temperature during imbibition of dry *Typha* pollen (moisture content 7%) on germination in vitro and pollen tube length (a), and respiration and NAD^+-leakage (b). After imbibition for 3 min, pollen was further incubated at 23°C. (From Hoekstra, 1984, courtesy of *Plant Physiology*.)

age is highly negatively correlated with the germinability at all temperatures tested, indicating that imbibitional stress is not essentially different from imbibitional chilling stress as far as the membrane damage is concerned.

The beneficial effect of prehumidification can be completely replaced by letting the imbibition occur at a high temperature (Hoekstra, 1984; Hoekstra and van der Wal, in preparation). When warm imbibition medium is exchanged after 1 min for an ice cold medium, pollen do not suffer any damage, which means that successful rehydration occurs in seconds. The foregoing discussion suggests that an elevated initial moisture content may be increasing the mobility of the phospholipids and lowering the transition temperature. The massive entry of water can only be withstood when the membrane is sufficiently flexible (i.e. has a lower viscosity through higher tem-

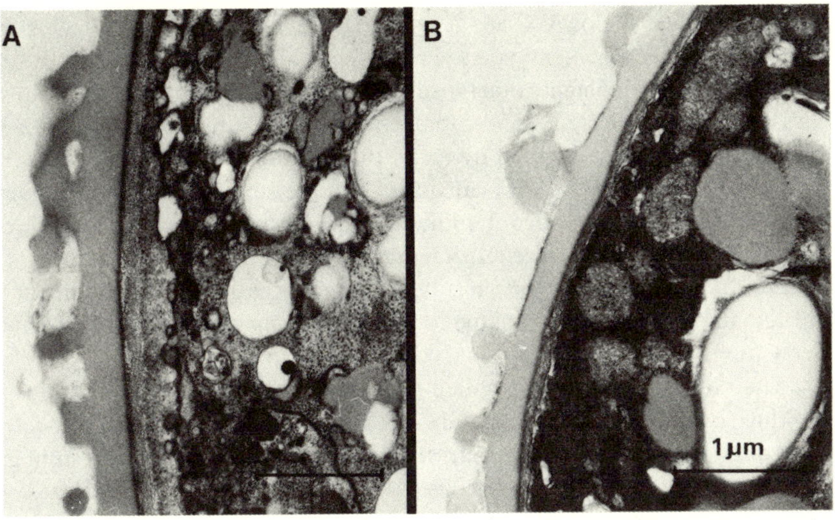

Figure 6-6. Electron micrographs of *Typha latifolia* pollen imbibed in liquid germination medium for 5 min at 0-1°C and further incubated at 22°C for 20 min until fixation in cacodylate buffer containing 2% glutaraldehyde and 1% formaldehyde and further processing for electron microscopy. (a) Initial moisture content 7% (injured); (b) initial moisture content 31% (viable). (Data from Sack et al., unpublished; micrographs courtesy of A. C. van Aelst, Dept. Plant Cytol. Morphol., Agric. Univ. Wageningen.)

peratures at imbibition, and/or a higher mobility of phospholipids through a higher initial hydration level; Hoekstra, 1984).

Dry Storage and Longevity

The extensive literature on pollen storage indicates that there is considerable variation in longevity among the various pollen species (Pfundt, 1910; Knowlton, 1922; Holman and Brubaker, 1926; Binder et al., 1974). As has been discussed earlier, this variation may be partly due to rehydration difficulties involving initial moisture content of the pollen and temperature of imbibition.

In storage experiments over very strong desiccants such as concentrated H_2SO_4 and anhydrous $CaCl_2$, the omission of prehydration in humid air may have led to the idea that best storage occurs at a relative humidity a little above zero percent (cf. Binder et al., 1974). Brewbaker (1959) has discovered that the short-lived characteristic tends to be associated with the trinucleate constitution, characterized by rapid start of tube emergence (Hoekstra and Bruinsma, 1978).

Temperatures below 0°C are generally optimal for storage, preserving the life of dry pollen for many years. Freezing should be done when the pollen has low moisture content, otherwise physical freezing damage will lead to loss of viability (Visser, 1955). Storage at the temperature of liquid nitrogen has been successfully applied by many investigators. From decay curves at 0°C, Bredemann et al. (1947) calculated that the survival period of *Lupinus* pollen at −196°C exceeded 1 million years. There are several papers on successful storage in a vacuum (Jensen, 1964; cf. Binder et al., 1974). Freeze drying may give good results as long as the initial water content is sufficiently reduced to prevent the formation of damaging ice crystals (Ching and Ching, 1964).

The observation that dry pollen store well in apolar organic solvents is striking (Iwanami and Nakamura, 1972). Heslop-Harrison (1979a) wondered why this treatment would not extract all the lipids from pollen. From many species we obtained similar amounts of neutral and polar lipids by disruption with French pressure cell, irrespective of a previous hexane wash of the pollen (unpublished data). Also a hexane-soluble inhibitor from *Impatiens* pollen could be obtained only by methanol extraction and not by direct hexane washing, again suggesting that the apolar solvents do not enter the cytoplasm.

It has been suggested that metabolic processes and respiration underlie the loss of pollen viability (Knowlton, 1922; Hoekstra and Bruinsma, 1975b). On the other hand, dry pollen containing approximately 7% water does not display detectable respiratory activity. However, in the presence of tritiated H_2O in a saturated magnesium nitrate solution, dry *Pinus* pollen of 8.6% moisture content slowly incorporated tritiated vapor into several metabolites (Wilson et al., 1979).

In an attempt to analyse the causes of viability loss during dry storage, Hoekstra and van Roekel (1985) examined nine pollen species (5-6% moisture content) with widely divergent longevity characteristics. At 22°C half of the initial germinative capacity was lost in periods of 6-170 days. Regardless of storage longevity, decrease of vitality was always accompanied by elevated K^+ leakage into the germination medium, suggesting a relation between loss of vigor and injury of the plasma membrane. Imbibitional damage was carefully avoided by prehydration in humid air. Analysis of the fatty acid compositions of total lipids of the nine species revealed palmitic (C16:0), linoleic (C18:2), and linolenic acid (C18:3) to be the major components. Those pollen with short longevity tend to have a high linolenic acid content, and vice versa (Figure 6-7a).

These nine pollen species were selected on account of their widely divergent lag periods (i.e. the time required from immersion in the medium until

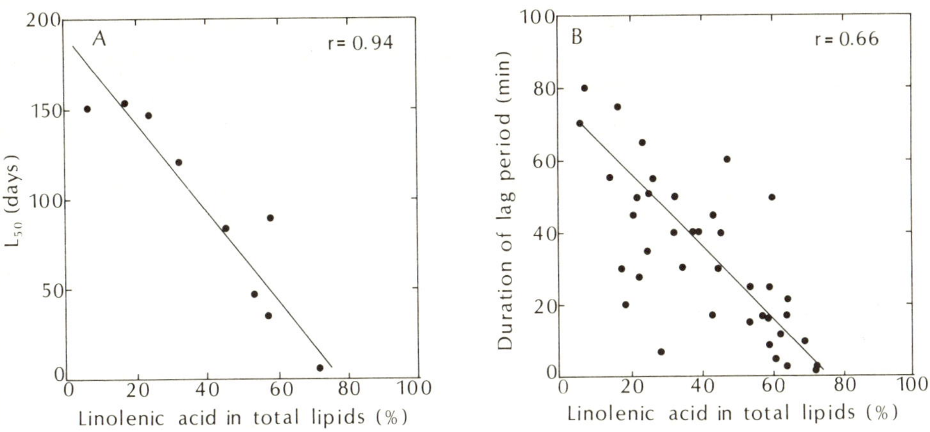

Figure 6-7. Correlation of linolenic acid content in total lipids and (a) pollen longevities (in days until half of original viability, L_{50}), and (b) duration of lag periods (period from immersion in the medium until appearance of the first pollen tubes). (a) Storage of nine pollen species belonging to nine different plant families was performed in an incubator at 22°C; moisture content of the pollen was 5-6%. Prior to germination in vitro, pollen was prehydrated in humid air. (Data from Hoekstra and van Roekel, 1985.) (b) Lengths of lag period in vitro were determined on 38 pollen species belonging to 35 genera of 28 plant families.

the appearance of the first tubes). Our attention is thus drawn to the significance of a high linolenic acid content in certain characteristics of pollen tube growth and development. It was shown earlier that faster growth of pollen tubes is the evolutionary result of gametophytic competition (Mulcahy, 1974; Ottaviano et al., 1980). As suggested by Hoekstra (1983), this fast tube growth would lose its competitive value without an equally rapid start of tube emergence. I assume that high fluidity of phospholipids must allow for the rapid fusion and rearrangements required for the extending plasma membrane of the growing tube. Highly unsaturated fatty acids in the lipids, such as linolenic acid, bring about a high fluidity. That such a trend exists is shown in Figure 6-7b, which displays the negative correlation between linolenic acid content and lag period for pollen germination. The correlation was made with 38 species belonging to 35 genera of 28 families. A less significant, positive correlation was found between linoleic acid content and both length of the lag period and longevity. The correlations with linolenic acid in Figure 6-7 were chosen because of the extreme sensitivity of this fatty acid to peroxidation and its effectuation of high fluidity in the phospholipids. A causal relationship between linolenic acid content in the pollen lipids and longevity or rapid tube emergence/growth has yet to be demonstrated.

In an attempt to curtail possible peroxidation reactions, Hoekstra and

Barten (in press) infiltrated four pollen species with vitamin E and the synthetic antioxidant butylated hydroxytoluene (BHT). This was done either via hexane suspensions or via aqueous suspensions. After evaporation of the hexane or drying, pollen was stored dry. However, the antioxidant treatments did not extend longevity over that of the controls.

In general, peroxidation of the polyunsaturated fatty acids of the neutral and polar lipids causes compositional changes. Table 6-1 compares the viability and the composition of the five main fatty acids of the phospholipids

Table 6-1

Effect of dry storage at 24°C on viability, phospholipid (PL) content, and composition of fatty acids in the PL of four pollen species having widely differing longevities

Storage time (days)	Percentage germination	PL content (μg·mg⁻¹)	Mol % of PL fatty acids					
			16:0	18:0	18:1	18:2	18:3	USR
			Impatiens					
0	91	97	31.8	0.9	1.6	3.2	62.5	2.1
3	76	102	31.6	0.9	1.4	3.1	63.0	2.1
6	51	107	32.0	0.7	1.5	3.1	62.7	2.1
9	40	107	32.2	0.8	1.4	3.2	62.4	2.0
19	0	95	31.7	0.8	1.9	3.3	62.4	2.1
			Papaver					
0	72	87	26.2	0.4	1.5	22.4	49.6	2.8
14	71	82	24.7	0.6	2.0	21.1	51.5	2.9
28	47	94	23.6	0.7	1.8	21.2	52.6	3.1
55	29	92	24.3	0.6	1.8	20.7	52.6	3.0
98	9	71	28.4	0.9	1.9	17.2	51.7	2.4
			Narcissus					
0	95	98	22.4	0.5	1.5	26.4	49.2	3.4
35	81	90	20.8	0.6	1.6	25.1	52.0	3.7
84	34	97	22.1	0.4	1.7	26.8	49.0	3.4
118	2	95	22.6	0.4	1.5	25.6	49.9	3.3
			Typha					
0	91	63	27.5	1.5	0.7	64.3	6.0	2.4
33	91	56	29.1	1.9	0.8	63.1	5.1	2.2
96	79	62	27.6	0.6	0.7	63.3	7.8	2.5
130	31	57	29.6	0.1	0.3	63.9	6.0	2.4

Note: USR stands for Unsaturated-Saturated Ratio, i.e. the sum of the mol percentages of 18:1, 18:2, and 18:3 FAs divided by the sum of those of 16:0 and 18:0 FAs. PLs were separated from total lipids by TLC, scraped off, eluted from the silicagel, methylated in methanolic KOH, and the FA-methylesters analysed by GC. Internal standardization was with diheptadecanoyl phosphatidylcholine added to pollen in $CHCl_3$/MeOH (2:1) prior to French pressure cell disruption.

and the phospholipid content during dry storage of four pollen species. In addition, this table shows the unsaturated/saturated ratio (USR), a good indicator of a possible increase in fatty acid saturation during storage. The fatty acids of the phospholipids underwent only slight, nonsignificant changes in composition and saturation, although composition and longevity differed considerably in the four species. This observation renders it highly improbable that peroxidation is involved on a large scale in the deterioration of pollen membranes during dry storage. Such lack of peroxidation throws doubt upon a possible causality of linolenic acid content and longevity.

Membrane responses of aging pollen strikingly resemble those of dehydration-sensitive germinating soybean seeds. Apart from an increase in the temperature of phase transition of the lipids, these seeds also exhibit increased rates of solute leakage, a decrease in phospholipid content, and an increase in free fatty acids (FFAs), but no change in degree of fatty acid saturation of total lipids (Senaratna and McKersie, 1983; Senaratna et al., 1984; Chapter 5). FFAs in the membrane are thought to increase the temperature of phase transition, which may contribute to the formation of gel-phase domains at physiological temperatures and may cause an increase in the permeability of the membranes (cf. Senaratna et al., 1984). During dry storage of pollen there is an increase in FFAs already, prior to the decline in vitality, and as shown in Table 6-1, the degree of fatty acid saturation of the

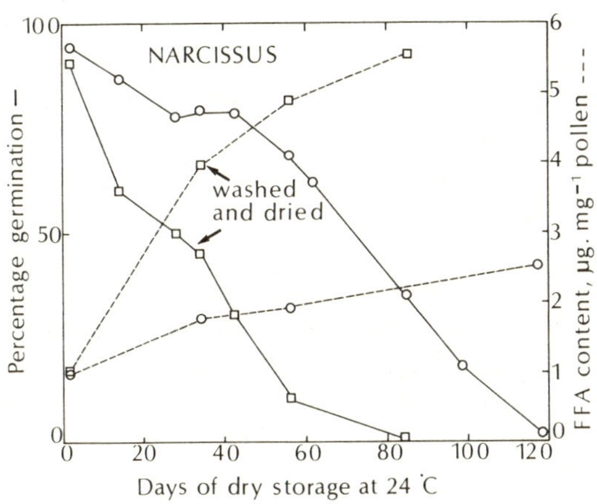

Figure 6-8. Relation of free fatty acids (FFA) and longevity. Effect of storage of *Narcissus* pollen (5.6% moisture content) at 24°C on vitality and accumulation of free fatty acids. One pollen sample was washed in germination medium for 5 min and immediately filtered and dried to a moisture content of 5-6%. Pollen was always prehydrated in humid air (2 hr) prior to germination in vitro. (Data from Hoekstra and Barten, in press.)

phospholipids remains unchanged. No data are available on the transition temperature of the lipids. During dry storage the FFAs are apparently generated by a deesterification process that cannot be blocked by exogenously applied tocopherol and BHT (Hoekstra and Barten, in press).

The washing of pollen in germination medium and subsequent drying does not significantly reduce the germinative capacity, but longevity is much reduced (Figure 6-8); there is a concomitant increased leakage of K^+. At the same time the accumulation of FFAs in the prewashed pollen is more rapid. As yet a causal relationship of rise in FFA and decline in vitality remains obscure, although the accelerated and simultaneous manifestation of both processes after medium washing suggests that one may exist. In a more recent paper, Senaratna et al. (1985) correlate the membrane injury syndrome to the strongly reduced levels of endogenous lipid-soluble antioxidants. That exogenously applied free radical scavengers are not able to prevent FFA accumulation and viability loss in pollen might be due to insufficient penetration of these compounds in the grains, and it may be that more penetrable antioxidants will prove to be effective.

Conclusions

Upon maturation, pollen may be stressed by dehydration, rehydration, and exposure to excessive humidity for extended periods of time. Exposure to high humidities at sites other than the receptive stigma invites inappropriate syntheses, resulting in loss of vitality. The picture that emerges from the present survey of the effects of dehydration and rehydration seems to indicate that in these cases membrane damage is involved. Such damage is also the most likely cause of the loss of pollen vitality during storage. We have only begun to understand that the nature of the variation in developmental processes and longevity among the different pollen species may initially be determined by differences in membrane composition.

Relatively little has been done so far in the field of membrane research in relation to pollen viability. Because of its small size, pollen offers substantial advantages over the much larger and more complex seeds. It is hoped that it will be possible to study the in situ state and functioning of membranes in pollen during desiccation and aging by means of such modern techniques as x-ray diffraction and nuclear magnetic resonance.

REFERENCES

Barnabás, B., and Rajki, E. 1976. Storage of maize (*Zea mays* L.) pollen at −196°C in liquid nitrogen. Euphytica 25: 747-752.

Barnabás, B., and Rajki, E. 1981. Fertility of deep-frozen maize (*Zea mays* L.) pollen. Ann. Bot. 48: 861-864.

Bar-Shalom, D., and Mattsson, O. 1977. Mode of hydration, an important factor in the germination of trinucleate pollen grains. Bot. Tidsskrift 71: 245-252.

Bashe, D., and Mascarenhas, J.P. 1984. Changes in potassium ion concentrations during pollen hydration and germination in relation to protein synthesis. Plant Sci. Lett. 35: 55-60.

Binder, W.D., Mitchell, G.M., and Ballantyne, D.J. 1974. Pollen viability testing, storage and related physiology. Can. Forestry Service, Pacific Forest Res. Centre, Victoria, BC, Report BC-X-105, pp. 1-37.

Bredemann, G., Garber, K., Harteck, P., and Suhr, K.A. 1947. Die Temperaturabhängigkeit der Lebensdauer von Blutenpollen. Naturwissenschaften 9: 279-280.

Brewbaker, J.L. 1959. Biology of the angiosperm pollen grain. Ind. J. Gen. Plant Breeding 19: 121-133.

Brewbaker, J.L. 1967. The distribution and phylogenetic significance of binucleate and trinucleate pollen grains in the angiosperms. Am. J. Bot. 54: 1069-1083.

Brewbaker, J.L., and Kwack, B.H. 1963. The essential role of calcium ion in pollen germination and pollen tube growth. Am. J. Bot. 50: 859-865.

Čapková-Balatková, V., Hrabětová, E., and Tupý, J. 1980. Effect of some mineral ions on pollen tube growth and release of proteins in culture. Biol. Plant. 22: 294-302.

Ching, T.M., and Ching, K.K. 1964. Freeze-drying pine pollen. Plant Physiol. 39: 705-709.

Crowe, J.H., Crowe, L.M., and Chapman, D. 1984. Preservation of membranes in anhydrobiotic organisms: the role of trehalose. Science 223: 701-703.

Crowe, L.M., Mouradian, R., Crowe, J.H., Jackson, S.A., and Womersley, C. 1984. Effect of carbohydrates on membrane stability at low water activities. Biochim. Biophys. Acta 769: 141-150.

Dickinson, D.B. 1967. Permeability and respiratory properties of germinating pollen. Physiol. Plant. 20: 118-127.

Duffield, J.W., and Snow, A.G. 1941. Pollen longevity of *Pinus strobus* and *P. retinosa* as controlled by humidity and temperature. Am. J. Bot. 28: 175-177.

Ferrari, T.E., Lee, S.S., and Wallace, D.H. 1981. Biochemistry and physiology of recognition in pollen-stigma interactions. Phytopathology 71: 752-755.

Frova, C.B., and Feder, W.A. 1979. Effect of cold treatment on the in vitro germination and day to day variability of maize (*Zea mays* L.) pollen. Ann. Bot. 43: 75-79.

Gilissen, L.J.W. 1977. The influence of relative humidity on the swelling of pollen grains in vitro. Planta 137: 299-301.

Gilissen, L.J.W. 1978. Post-X-irradiation effects of *Petunia* pollen germinating in vitro and in vivo. Environ. Exp. Bot. 18: 81-86.

Goss, J.A. 1968. Development, physiology, and biochemistry of corn and wheat pollen. Bot. Rev. 34: 333-358.

Heslop-Harrison, J. 1979a. Aspects of the structure cytochemistry and germination of the pollen of rye (*Secale cereale* L.). Ann. Bot. 44: 1-47 (supp.).

Heslop-Harrison, J. 1979b. An interpretation of the hydrodynamics of pollen. Am. J. Bot. 16: 737-743.

Heslop-Harrison, J., and Heslop-Harrison, Y. 1970. Evaluation of pollen viability by enzymatically induced fluorescence: intracellular hydrolysis of fluorescein diacetate. Stain Technol. 45: 115-120.

Heslop-Harrison, J., Heslop-Harrison, Y., and Shivanna, K.R. 1984. The evaluation of pollen quality, and a further appraisal of the fluorochromatic (FCR) test procedure. Theor. Appl. Genet. 67: 367-375.

Heslop-Harrison, Y., and Shivanna, K.R. 1977. The receptive surface of the angiosperm stigma. Ann. Bot. 41: 1233-1258.

Hoekstra, F.A. 1979. Mitochondrial development and activity of binucleate and trinucleate pollen during germination in vitro. Planta 145: 25-36.

Hoekstra, F.A. 1983. Physiological evolution in angiosperm pollen: possible role of pollen vigour. *In* Pollen: Biology and Implications for Plant Breeding. Mulcahy, D.L., and Ottaviano, E., eds. Elsevier, New York, pp. 35-41.

Hoekstra, F.A. 1984. Imbibitional chilling injury in pollen: involvement of the respiratory chain. Plant Physiol. 74: 815-821.

Hoekstra, F.A., and Barten, J.H.M. In press. Do anti-oxidants and local anaesthetics extend pollen longevity during dry storage? *In* Pollen Biotechnology and Ecology. Mulcahy, D.L., and Ottaviano, E., eds. Springer-Verlag, New York.

Hoekstra, F.A., and Bruinsma, J. 1975a. Viability of *Compositae* pollen: germination in vitro and influences of climatic conditions during dehiscence. Z. Pflanzenphysiol. 76: 36-43.

Hoekstra, F.A., and Bruinsma, J. 1975b. Respiration and vitality of binucleate and trinucleate pollen. Physiol. Plant. 34: 221-225.

Hoekstra, F.A., and Bruinsma, J. 1978. Reduced independence of the male gametophyte in angiosperm evolution. Ann. Bot. 42: 759-762.

Hoekstra, F.A., and Bruinsma, J. 1979. Protein synthesis of binucleate and trinucleate pollen and its relationship to tube emergence and growth. Planta 146: 559-566.

Hoekstra, F.A., and Bruinsma, J. 1980. Control of respiration of binucleate and trinucleate pollen under humid conditions. Physiol. Plant. 48: 71-77.

Hoekstra, F.A., and van Roekel, T. 1983. Isolation-inflicted injury to mitochondria from fresh pollen gradually overcome by an active strengthening during germination. Plant Physiol. 73: 995-1001.

Hoekstra, F.A., and van Roekel, T. 1985. Correlation of membrane properties to loss of pollen vigor during dry storage. Plant Physiol. 77: 669 (suppl.).

Holman, R.M., and Brubaker, F. 1926. On the longevity of pollen. Univ. Calif. Publ. in Bot. 13: 179-204.

Iwanami, Y., and Nakamura, N. 1972. Storage in an organic solvent as a means for preserving viability of pollen grains. Stain Technol. 47: 137-139.

Jensen, C.J. 1964. Pollen storage under vacuum. Årsskrift den Kongeliges Veterinaer og Landbo Högskole, pp. 133-146.

Keijzer, C.J. 1983. Hydration changes during anther development. *In* Pollen: Biology and Implications for Plant Breeding. Mulcahy, D.L., and Ottaviano, E., eds. Elsevier, New York, pp. 197-202.

Knowlton, H.E. 1922. Studies in pollen, with special reference to longevity. Cornell Univ. Agric. Exp. Sta. Mem. 52: 751-793.

Lichte, H.-F. 1957. Über die Physiologie von Angiospermen Pollen und ihre Bedeutung für die Pflanzenzüchtung. Angew. Bot. 31: 1-28.

Lidforss, B. 1896. Zur Biologie des Pollens. Jahrb. Wiss. Bot. 29: 1-38.

Linskens, H.F., and Pfahler, P.L. 1973. Biochemical composition of maize (*Zea mays* L.) pollen III. Theor. Appl. Genet. 43: 49-53.

Linskens, H.F., and Schrauwen, J. 1969. The release of amino acids from germinating pollen. Acta Bot. Neerl. 18: 605-614.

Lonnquist, J.H., and Jugenheimer, R.W. 1943. Factors affecting the success of pollination in corn. J. Am. Soc. Agron. 35: 923-933.
McLellan, A.R. 1977. Minerals, carbohydrates and amino acids of pollens from some woody and herbaceous plants. Ann. Bot. 41: 1225-1232.
Moore, P.H. 1976. Studies on sugarcane pollen. II. Pollen storage. Phyton 34: 71-80.
Mulcahy, D.L. 1974. Correlation between speed of pollen tube growth and seedling height in *Zea mays* L. Nature 249: 491-493.
Nebel, B.R., and Ruttle, M.L. 1936. Storage experiments with pollen of cultivated fruit trees. J. Pom. 14: 347-359.
Ottaviano, E., Sari-Gorla, M., and Mulcahy, D.L. 1980. Pollen tube growth rates in *Zea mays*: implications for genetic improvement of crops. Science 210: 437-438.
Pacini, E., and Franchi, G.G. 1982. Germination of pollen inside anthers in some non-cleistogamous species. Caryologia 35: 205-215.
Pálfi, G., and Köves, E. 1984. Determination of vitality of pollen on the basis of its amino acid content. Biochem. Physiol. Pflanzen 179: 237-240.
Pfahler, P.L. 1966. In vitro germination and pollen tube growth of maize (*Zea mays* L.) pollen. I. Calcium and boron. Can. J. Bot. 45: 839-845.
Pfahler, P.L., and Linskens, H.F. 1973. In vitro germination and pollen tube growth of maize (*Zea mays* L.) pollen. Planta 111: 253-259.
Pfeiffer, N.E. 1938. Viability of stored *Lilium* pollen. Contr. Boyce Thompson Inst. 9: 199-213.
Pfeiffer, N.E. 1939. Life of *Gladiolus* pollen prolonged by controlled conditions of storage. Contr. Boyce Thompson Inst. 10: 429-440.
Pfundt, M. 1910. Der Einfluss der Luftfeuchtigkeit auf die Lebensdauer des Blütenstaubes. Jahrb. Wiss. Bot. 47: 1-40.
Priestley, D.A., and de Kruijff, B. 1982. Phospholipid motional characteristics in a dry biological system: a ^{31}P NMR study of hydrating *Typha latifolia* pollen. Plant Physiol. 70: 1075-1078.
Sack, F.D., Hoekstra, F.A., and Leopold, A.C. 1983. Ultrastructural evidence of overall membrane damage in imbibitionally chilled *Typha* pollen. Plant Physiol. 72: 538 (suppl.).
Sartoris, G.B. 1942. Longevity of sugar cane and corn pollen: a method for long distance shipment of sugar cane pollen by airplane. Am. J. Bot. 29: 395-403.
Senaratna, T., and McKersie, B.D. 1983. Characterization of solute efflux from dehydration injured soybean (*Glycine max* L. Merr.) seeds. Plant Physiol. 72: 911-914.
Senaratna, T., McKersie, B.D., and Stinson, R.H. 1984. Association between membrane phase properties and dehydration injury in soybean axes. Plant Physiol. 76: 759-762.
Senaratna, T., McKersie, B.D., and Stinson, R.H. 1985. Antioxidant levels in germinating soybean seed axes in relation to free radical and dehydration tolerance. Plant Physiol. 78: 168-171.
Shinitzky, M., and Barenholz, Y. 1978. Fluidity parameters of lipid regions determined by fluorescence polarization. Biochim. Biophys. Acta 515: 367-394.
Shivanna, K.R., and Heslop-Harrison, J. 1981. Membrane state and pollen viability. Ann. Bot. 47: 759-770.
Simon, E.W. 1974. Phospholipids and plant membrane permeability. New Phytol. 73: 377-420.

Stanley, R.G., and Linskens, H.F. 1974. Pollen: Biology, Biochemistry, and Management. Springer-Verlag, New York, pp. 1-307.

Vasil, I.K. 1960. Pollen germination in some Gramineae: *Pennisetum typhoides*. Nature 187: 1134-1135.

Visser, T. 1955. Germination and storage of pollen. Mededelingen Landbouwhogeschool Wageningen, The Netherlands, 55: 1-68.

Walden, D.B. 1960. Preliminary studies on longevity of corn pollen and related physiological factors. Diss. Abstr. 20: 3488-3489.

Wilson, A.T., Vickers, M., and Mann, L.R.B. 1979. Metabolism in dry pollen: a novel technique for studying anhydrobiosis. Naturwissenschaften 66: 53.

Woittiez, R.D., and Willemse, M.T.M. 1979. Sticking of pollen on stigmas: the factors and a model. Phytomorphology 29: 57-63.

Zhang, H.Q., and Croes, A.F. 1983. Protection of pollen germination from adverse temperatures: a possible role for proline. Plant Cell Environ. 6: 471-476.

7 Does Trehalose Play a Role in Yeast Cells under Stress?

Ana C. Panek
E. Bernardes
Anita D. Panek

Yeast cells have presented scientists with fascinating and intriguing questions and have allowed for the development of new techniques, contributing greatly to the progress of biochemistry. Yeasts have also contributed to the well-being of mankind over the years.

This past decade has been rich in dramatic developments in yeast research. The less prominent new facts that have emerged include the striking accumulation of trehalose under conditions of stress.

An overview of the function of trehalose in microorganisms has led to the conclusion that the disaccharide acts as a storage carbohydrate. As yet, however, no definite, unequivocal functions have been attributed to trehalose, because the reported results are not conclusive enough to establish cause-and-effect relations. In any case it does not seem unreasonable to imagine a unique role for a compound such as trehalose, which displays a rather rare α-1,1-glucosidic bond with an unusually high energy of activation.

Nutrient deprivation and different stages of the life cycle can lead to variations in trehalose content, from less than 1% to over 20% of the dry weight (Küenzi and Fiechter, 1972; Panek, 1975; Lillie and Pringle, 1980). Such dramatic variations suggest that the mechanisms governing the regulation of trehalose metabolism are rather complex.

Yeast cells also accumulate glycogen in response to certain stress condi-

We are grateful to our colleagues who helped us with some of the determinations included in this paper. We also acknowledge the financial support given to our research by CNPq, FINEP, and PADCT.

tions. But trehalose, because of its special configuration, could play roles in the yeast cell apart from functioning as carbon and energy reserve. This view has been put forward by some investigators (Clegg, 1965; Lillie and Pringle, 1980; Keller et al., 1982) but has not really been pursued in yeast research.

During fermentation processes which lead to ethanol production, yeast cells are recycled several times in the fermentors and remain viable under the stress of temperatures above 30°C and high ethanol concentrations. Ethanol tolerance has been correlated with the rate of production and the rate of excretion; ethanol also depresses the maximum temperature for growth and enhances thermal death (Sá Correia and van Uden, 1982). Moreover, the tolerance depends upon the lipid composition of the membrane (Thomas et al., 1978). On the other hand, alterations in the membrane fluidity are elicited by the presence of alcohol in the medium (Ingram, 1976). Might trehalose play any role in protecting the membranes against lysis under such conditions? No investigations with that special purpose have so far been reported.

Years ago Merdinger and Kohn (1967) showed that trehalose was loosely bound to the lipid fraction of the cytoplasmic membrane in the yeastlike fungus *Pullularia pullulans*. In the fermenting yeast *Saccharomyces cerevisiae*, the bulk of trehalose accumulated during an incubation with glucose^{14}C was found in the particulate fraction (Souza and Panek, 1968). Although the results were disputed by Keller et al. (1982), who claimed that trehalose was found in the cytosol, these authors did suggest that the carbohydrate may act as a protective agent of cells under stress.

Producers of baker's yeast strive for a product with a high cellular carboyhydrate content (Suomalainen and Pfäffli, 1961). To obtain a good, active dry yeast, an elevated content of trehalose is important. Such a yeast has another advantage in that its osmoresistance is improved, and it can be used in doughs with a high sugar content (Edelman et al., 1980). Here again the facts suggest a relation between trehalose and membrane stability.

Accumulation of trehalose is a result of the activities of trehalose-6-phosphate synthase and of trehalase. Therefore, for studies of trehalose accumulation it is important to assess the activities of both enzymes. In wild type yeast the synthesizing activity predominates over the hydrolytic enzyme, which is normally found in a cryptic, inactive form, capable of being activated under certain physiological conditions.

In an attempt to get a deeper insight into the role that trehalose plays when yeast cells are faced with different kinds of environmental or metabolic stresses, we have constructed diploid and haploid strains which harbour specific lesions in the regulation of the enzymes involved in trehalose metabolism. Sporulation, germination, and nutrient deprivation were the major parameters analysed.

Biosynthesis of Trehalose

α, α-Trehalose may be formed from uridine-5′-diphosphoglucose and D-glucose-6-phosphate by the action of trehalose-6-phosphate synthase (E.C.2.4.1.15) followed by that of a phosphatase (E.C.3.1.12) (Leloir and Cabib, 1953).

The synthase shows maximal activity at pH 6.6 in the presence of 25mM Mg^{2+} (Cabib and Leloir, 1958). The partially purified trehalose-6-phosphate synthase prepared from brewer's yeast contains a specific phosphatase that cleaves trehalose-6-phosphate without hydrolysing the substrates of the reaction. Such a specific cleavage of a phosphate group ensures the equilibrium of the reaction in favor of the formation of the trehalose ester.

The effect of Mg^{2+} on yeast trehalose synthase was studied in more detail, and it was found that the ion participates in the reaction catalysed by trehalose-6-phosphate synthase in two ways with opposite effects. At lower concentrations (below 12mM) Mg^{+2} binds to glucose-6-phosphate, producing a G-6-P·Mg complex which is probably a better substrate for the enzyme than the free ester. Therefore, activation is observed. As the concentration of Mg^{+2} increases and the concentration of glucose-6-phosphate is maintained, the ions interact with the enzyme itself, producing an inhibitory effect (Oestreicher and Panek, 1982).

Elander (1969) studied trehalose-6-phosphate synthase from yeast after a 27-fold purification. He showed that uridine 5′ pyrophosphate and UTP were competitive inhibitors of UDPG. In further studies it was shown that the enzyme was inhibited by 20mM inorganic phosphate and by trehalose-6-phosphate. Free trehalose had no effect on the yeast enzyme in vitro (Elbein, 1974).

In assessing trehalose-6-phosphate synthase activity, possible allosteric effectors and catabolite repression effects should not be overlooked. During growth of cells, synthesis and breakdown could occur at similar rates, and in consequence no significant accumulation of trehalose would be seen in wild-type strains (Panek et al., 1981). Only cells harvested after the first exponential phase are able to accumulate trehalose when they are placed in non-proliferating conditions, suggesting either that trehalose synthase is under the control of catabolite repression (Panek and Mattoon, 1977) or that operative mitochondria are a requisite for synthesis or both.

Preliminary purification steps indicate that trehalose-6-phosphate synthase is a complex protein of high molecular weight (unpublished data), and as yet no studies on the regulation of this enzyme in yeast cells have been published.

We have described a specific deficiency in trehalose-6-phosphate synthase in *Saccharomyces cerevisiae* and have demonstrated that it is associated with a single nuclear gene. Strains bearing this abnormal allele (*fdp*) lack the capacity to accumulate trehalose during growth on glucose or galactose or when incubated in nonproliferating conditions. The absence of trehalose-6-phosphate synthase activity is one of the phenotypes of the pleiotropic *fdp* mutation. Another characteristic that has been described for this strain is the high ratio of dependence of glycogen synthase upon glucose-6-phosphate (van de Poll and Schamhart, 1977). This result indicates that glycogen synthase predominates in the mutant in the phosphorylated, less active form.

Another mutant (*cif*) described as similar to *fdp* (Navon et al., 1979) also lacks trehalose-6-phosphate synthase activity. Complementation tests between *cif* and *fdp* mutants indicate that the genes are allelic (Charlab et al., in press). Furthermore, a yeast mutant (*bcyl*) has been described which produces very low levels of the regulatory subunit of a protein kinase which is cAMP dependent; it exhibits high levels of kinase activity (Uno et al., 1983) and shows no trehalose-6-phosphate synthase activity as determined in vitro.

In all three mutants, albeit the genetic lesions might not be identical, the phenotypes indicate a physiological condition in which enzymes are predominantly in their phosphorylated forms. The preliminary results obtained when assaying trehalose-6-phosphate synthase in these mutant strains suggest a regulation of its activity by a phosphorylating-dephosphorylating cascade. As is the case with glycogen synthase, trehalose-6-phosphate synthase would be less active in its phosphorylated form.

Trehalose Catabolism

In *Saccharomyces* the only enzyme known to be responsible for trehalose cleavage is the hydrolytic enzyme trehalase (α, α-trehalose 1-D-glucohydrolase, E.C.3.3.1.28). For many years trehalose breakdown was considered to be a simple unregulated hydrolytic process.

An involvement of cyclic 3′5′-AMP (cAMP) in the control of trehalose metabolism in yeast was shown by van der Plaat and van Solingen (1974). The authors studied the lag phase which precedes growth and which is characterized by a rapid degradation of trehalose (Panek, 1963) and showed that the peak of trehalase activity coincided with an increase in the concentration of cAMP. They obtained a partial purification of the inactive, cryptic trehalase as well as its activating protein (van Solingen and van der Plaat, 1975). That trehalase activation might be associated with phosphorylation of the enzyme protein was further investigated by Ortiz et al. (1983), who demonstrated that the cryptic form of trehalase can be activated by incubation with ATP·Mg^{+2} in the presence of cyclic AMP. The conversion of active trehalase to its cryptic form by a protein phosphatase was demonstrated at the same time. The kinetic data presented suggest that interconversion between cryptic and active forms of trehalase is regulated by a cAMP-dependent monocyclic cascade system.

Two lines of investigation have recently converged, resulting in the independent discovery of mutations which alter trehalase activity by altering cyclic-AMP dependent protein kinase activity in *Saccharomyces cerevisiae*. Studying strains which bear the *glcl* mutation described for glycogen deficiency by Rothman-Denes and Cabib (1971), we have established also that trehalose accumulation in these strains is restricted as a result of an abnormal activation of trehalase by a protein kinase with altered binding kinetics between regulatory and catalytic subunits (Ortiz et al., 1983). Furthermore, Matsumoto et al. (1985) have shown that in these mutants trehalase is maintained in its active form because one specific phosphatase is absent. Another mutation, described by Uno et al. (1983), renders cells incapable of storing trehalose. Cells bearing the *bcyl* mutation produce an altered protein kinase because of extremely low levels of the cAMP binding protein, as mentioned before. This altered kinase activates trehalase, and as a result no trehalose is accumulated. Genetic complementation experiments have clearly established that the *bcyl* mutation involves a gene different from that altered by the *glcl* mutation (Tenan et al., 1985).

It has been suggested that cryptic trehalase is located in the cytosol and that cAMP-dependent phosphorylation of the enzyme is involved in its transfer into the vacuoles (Wiemken and Schellenberg, 1982).

On the other hand, Londesborough and Varimo (1984) have reported the existence of a nonregulated vacuolar trehalase and claim that only the cytosolic one suffers activation by phosphorylation.

Addition of glucose to yeast ascospores induces a 10-fold increase in trehalase activity, which closely correlates with an increase in cAMP levels. However, in vegetative cells nitrogen-source induced trehalase activation is not associated with an increase of the cAMP level (Thevelein, 1984).

It becomes obvious from the above discussion that many discrepancies will have to be explained and intriguing questions will have to be answered before a clear picture of the regulation of trehalose metabolism emerges. A general statement of trehalose regulation should involve, first, a control of activity of the two enzymes responsible for trehalose synthesis by phosphorylation and, second, an activation of trehalase by the same mechanism. Thus a predominance of phosphorylation would lead to trehalose breakdown whereas the action of specific protein phosphatases would lead to accumulation.

Trehalose Mobilization as a Response to Stress

Yeast cells are able to store two carbohydrates: glycogen and trehalose. Unequivocally they play a role in maintenance, differentiation, and survival. However, nature must have had some reason for providing these cells with two reserve systems. The goal of this chapter is to identify some alternative and specific roles for trehalose. Our analysis will utilize an array of yeast strains outlined in Table 7-1.

Formation of spores is a response to stress. The accumulation of trehalose under conditions leading to sporulation was reported by Roth (1970). Kane and Roth (1974) subsequently demonstrated that trehalose and glycogen are synthesized to the same extent in both sporulating and nonsporulating strains.

For the fission yeast *Schizosaccharomyces pombe* there is good evidence that both accumulation of trehalose and induction of trehalase activity during sporulation are preparatory steps for germination and are not necessarily prerequisite for sporulation itself (Inoue and Shimoda, 1981). In ascospores of *Picchia pastoris*, energy is supplied during early germination by a rapid breakdown of trehalose caused by an abrupt activation of trehalase by cyclic AMP (Thevelein et al., 1982).

By constructing diploids homozygous for a specific defect in trehalose metabolism, we have tried to assess the role this disaccharide plays during sporulation, germination, and starvation. The *fdp* mutation is characterized by the absence of trehalose-6-phosphate synthase activity (Operti et al., 1982). The *glcl* mutation reduces both trehalose and glycogen levels in the cell by maintaining the enzymes responsible for degradation (trehalase and glycogen phosphorylase) in a more active, phosphorylated form (Ortiz et al., 1983). In the yeast strain AP8, in which glycogen and trehalose metabolism are affected by activation of glycogen phosphorylase and of trehalase, we find no sporulation, as shown in Table 7-2. The results obtained for trehalase activity indicate that, in the diploid homozygous for the *glcl* mutation (AP8), trehalase is found in the cells already in its active form and is independent of further activation in vitro by ATP·Mg^{2+} and cAMP. The same reasoning could be applied to glycogen phosphorylase.

A defect in the regulatory mechanism of glycogen breakdown seems to

Table 7-1
Yeast strains employed: Genotypes and sources

1. Haploids

Strain	Genotype	Source
D311-3A	*a his1 trp2 lys2*	(1)
S288C	α wild type	(2)
1403-7A	*a try1 ura3 MAL4*[c]	(3)
1268-2B	α *ade6 trp2 trd*(TS ↑)	(1)
GS1-36	α *glc* 1-1	(2)
AM9-8B	*a bcy1*	(4)
Q6R2	α *lys2 MAL6 fdp*	(5)
VFP1-8C	*a cif trd*	(5)
AP28-1B	*a trd* (TS ↑)	(5)
AP28-3B	*a trd*	(5)

2. Diploids

Strain	a parent	α parent	Source of haploids
DE4	PG6-2C(*ade2-40 MAL2*[c])	PS-8C(*ade2-40 MAL2*[c])	(6)
AP8	BR15-14B(*glc1-1MAL4 his1*)	BR15-15D(*glc1-1MAL4 ura3 trp1*)	(5)
AP9	DE6-5D(*fdp lys2 MAL2*[c])	Q6R2(*fdp lys2 MAL6*)	(5)

Note: ↑ TS = high trehalose-6-phosphate synthase activity.
Sources: (1) J. R. Mattoon, University of Colorado, Colorado, U.S.A. (2) E. Cabib, National Institute of Health, Bethesda, Maryland, U.S.A. (3) Yeast Genetic Stock Center, Univ. California, Berkeley, California, U.S.A. (4) T. Ishikawa, Inst. Appl. Microbiology, Univ. of Tokyo, Japan. (5) Obtained in our laboratory. (6) F. Zimmermann, Technische Hochschule, Darmstadt, Germany.

exert a more significant effect upon sporulation capacity than the absolute amount of glycogen stored. The fact that diploid AP9, which has normal trehalase activity and normal glycogen metabolism, does show some sporulation suggests that glycogen is indeed a prerequisite for the formation of spores in *Saccharomyces*. However, trehalose-6-phosphate synthase activity shows a strong influence upon the preparatory steps for spore germination. In diploid AP9, homozygous for *fdp*, very reduced germinating capacity was observed, and no full tetrads were formed. This strain overproduces glycogen because G-6-P and UDPG cannot be converted into trehalose, yet germination is not enhanced. Similar results have already been reported by us for other diploid strains (Panek et al., 1981). These observations point towards a specific function for trehalose in triggering the germination process.

A family of genes first discovered as oncogenes in rat sarcoma viruses has recently been identified in *Saccharomyces cerevisiae*. A mutation in one of those genes (*RAS 2*) results in a premature sporulation response (in the

Table 7-2
Storage carbohydrates during sporulation and germination of some yeast strains

	Strain (genotype)		
	DE4 (*wt*)	AP8 (*glc 1*/*glc 1*)	AP9 (*fdp*/*fdp*)
Trehalose[a] content (mg/g dry wt)	10.4	16.2	2.2
Glycogen[b] content (mg/g dry wt)	74.6	44.6	96.4
Sporulation (%)	80.0	0.0	5.0
Germination (%)	87.0	—	10.0
4-spore tetrads (%)	60.0	—	0.0
Trehalose synthase[c]	271.0	356.0	27.5
Trehalase[d]	11.6 (51.0)	19.4 (22.7)	7.8 (18.3)

[a]Trehalose was extracted from cells after 72 hr in sporulation medium (sodium acetate 0.98%, glucose 0.05%, yeast extract 0.1%, agar 1.5%). Extraction and determination were as described by Oliveira et al., 1981.

[b]Glycogen was extracted after 72 hr in sporulation medium and determined as described by Becker, 1978.

[c]Trehalose-6-phosphate synthase activity was determined by a modification of the classical method as described by Operti et al., 1982. Activity is given in nmoles UDP $\times$ min^{-1} $\times$ mg $protein^{-1}$.

[d]Trehalase was determined as described by Ortiz et al., 1983. We consider basal activity to be the activity that is found in the cells at the point of harvest. The values in parentheses correspond to the activation obtained by incubating the enzyme preparation with protein kinase, cAMP and $ATP \cdot Mg^{+2}$. Activity is given in nmoles glucose $\times$ min^{-1} $\times$ mg $protein^{-1}$.

presence of glucose) and in a 100-fold increase in accumulated trehalose (Tatchell et al., 1985).

It is clear that trehalose breakdown provides energy for the germinating spore. Should we interpret this phenomenon exclusively as a result of an activation of trehalase (Thevelein et al., 1982) or also as a consequence of a modification of the membrane which would favour the interaction between substrate and enzyme? Küenzi and Fiechter (1969) claim that, even with a restricted supply of an exogenous substrate, trehalose is needed to carry out the budding process at a constant time.

The low levels of trehalose found in the wild-type diploid (10 mg/g) are noteworthy. The presence of normal levels of the enzymes involved in synthesis and mobilization is associated with normal sporulation and germination of the cells (Table 7-2).

Wild-type strains of *S. cerevisiae* when grown on glucose exhibit a biphasic growth curve. Trehalose normally accumulates only when cells become deprived of exogenous glucose, probably because of catabolite repression effects. Even so the levels attained during diauxie do not exceed 1-2% of

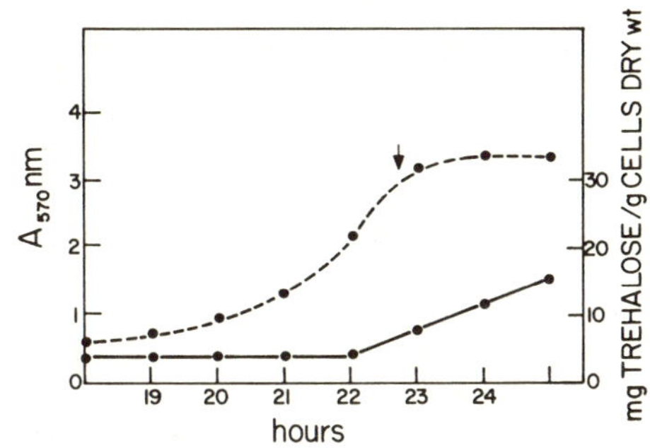

Figure 7-1. Trehalose accumulation during growth of a wild-type strain (D311-3A) on glucose. Cells were grown in YED medium at 28°C and 160 oscillations/min. Solid line corresponds to trehalose content as described by Oliveira et al. (1981). The dashed line represents growth of cells measured as absorbance at 570 nm. The arrow indicates glucose depletion.

the dry weight, as illustrated in Figure 7-1. Could the stress caused by glucose starvation deprive cells of the capacity of accumulating trehalose? When cells were collected at the onset of the transition phase and were transferred to nonproliferating conditions in buffered 20 mM glucose, a fourfold synthesis of trehalose was seen in 90 min, as shown in Figure 7-2. On the other hand, when cells were harvested during the first logarithmic phase and were incubated in resting conditions, no trehalose synthesis could be observed. It is noteworthy that the values obtained in vitro for trehalose synthase activity are the same whether cells are taken from the diauxic phase or from nonproliferating conditions (unpublished results). Therefore, glucose deprivation by itself is not the cause for restricted trehalose accumulation.

The next question raised was whether fully developed mitochondria were necessary to allow for trehalose accumulation. Cells pregrown in semianaerobic glucose medium when transferred to nonproliferating conditions were unable to accumulate trehalose (Figure 7-2). These results strongly suggest that trehalose accumulates only when a high energy charge is reached

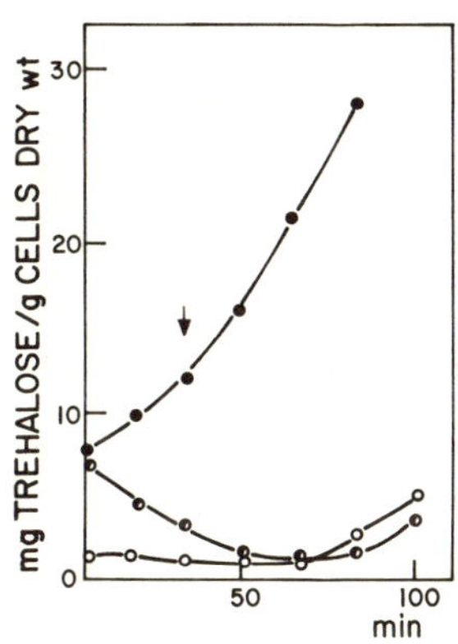

Figure 7-2. Trehalose accumulation in nonproliferating conditions. Strain D311-3A was incubated with 20 mM glucose in 0.05M sodium phosphate buffer, pH 6.0 at 28°C. (●) cells harvested from aerobic growth after glucose depletion; (○) cells harvested when glucose was still present in the medium; (◐) cells pregrown in semianaerobic conditions (flasks filled to the top) for 18 hr. The arrow indicates glucose depletion from the incubation buffer.

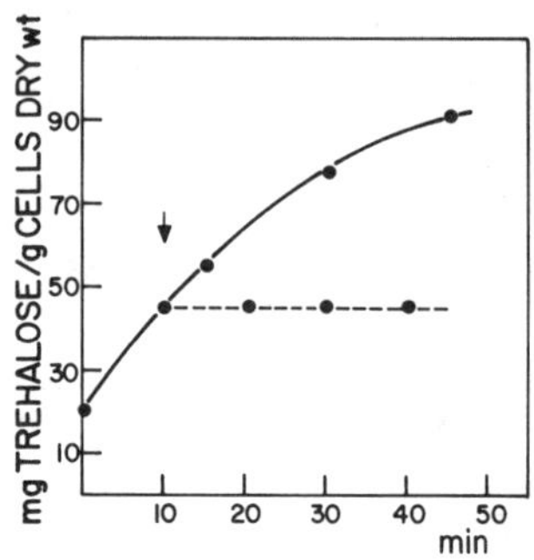

Figure 7-3. Effect of ammonium ions on trehalose accumulation in resting cells. Conditions as in Figure 7-2. The arrow indicates the moment of addition of 1 mM $(NH_4)_2SO_4$. The dashed line indicates that salt has been added.

by the cells. That, of course, is favoured by fully developed mitochondria. The possibility of an activator of trehalose-6-phosphate synthase formed in the mitochondria should also be considered.

Nonproliferating conditions, as opposed to growth conditions, deprive cells of a nitrogen source. Would trehalose accumulation under such a stress be a response to nitrogen starvation? Addition of 1 mM ammonium sulphate to cells actively accumulating trehalose, in the presence of glucose, causes an abrupt reduction of synthesis (Figure 7-3). This inhibitory effect had already been reported by Trevelyan and Harrison (1956). Ammonium ions activate phosphofructokinase, favouring glycolysis which competes with trehalose synthase for the same substrate: glucose-6-phosphate. Therefore, we may regard trehalose accumulation in buffered glucose as a response to the stress caused by nitrogen deprivation.

The next obvious step was to study the behaviour of trehalose in cells under deprivation of both glucose and nitrogen. In cells starved at 26°C in

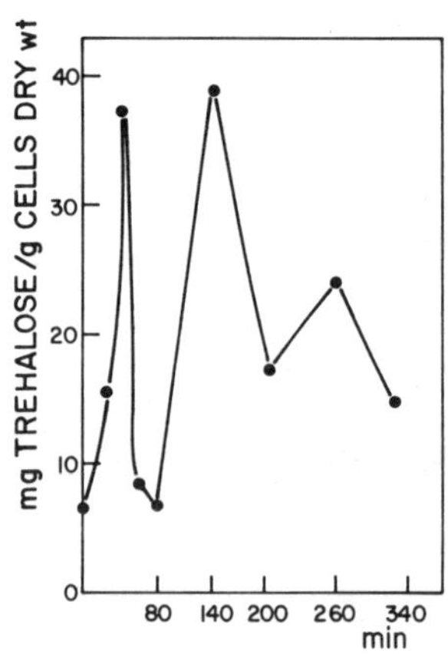

Figure 7-4. Trehalose profile during starvation of wild type strain. Cells were grown to early second log phase and were then transferred into 0.05 M sodium phosphate buffer, pH 6.0, and incubated without shaking. Trehalose was extracted and determined as in Oliveira et al., 1981.

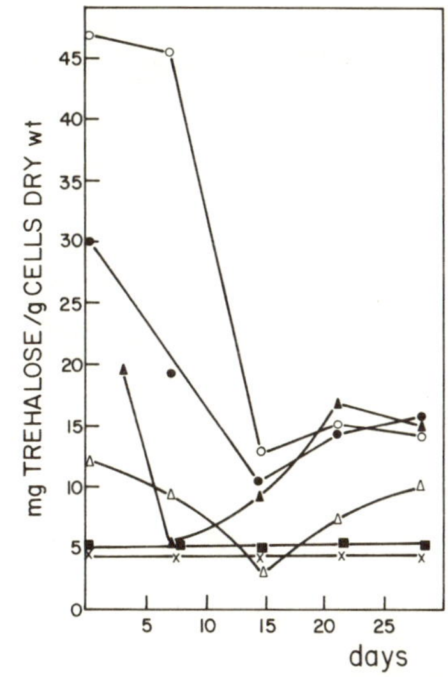

Figure 7-5. Trehalose profiles during long-term starvaton of wild-type and mutant strain. Conditions as in Figure 7-4. ▲ strain 1403-7A (*wt*); △ strain S288C (*wt*); ● strain 1268-2B (*trd*, TS ↑); ○ strain AP28-1B (*trd*, TS ↑); ■ strain Q6R2 (*fdp*); × strain VFP1-8C (*cif*, *trd*).

0.05 M sodium phosphate buffer, trehalose levels did not vary when samples were taken every hour for 3 hr (Panek, 1963). However, let us emphasize again that trehalose accumulation must be envisaged as resulting from the activities of the enzymes involved in its synthesis and degradation. These enzymes suffer an amplification in their regulation through the mechanism of interconversion of forms by a cyclic AMP-dependent phosphorylating-dephosphorylating cascade. Therefore, absolute values depend upon the concentration of effectors and the more or less active form of the enzyme at each sampling point. An analysis of Figures 7-4 and 7-5 shows that trehalose synthesis and breakdown occur in cycles in cells starved for short and long periods. The only strains that maintained a constant but extremely low trehalose value were those which harbour a lesion in trehalose-6-phosphate synthase activity (Table 7-3).

A study of viability under storage, desiccation, or any other stress should be undertaken with consideration for the capacity of the cell for synthesizing and mobilizing trehalose. The absolute values obtained for the disaccharide do not lead to conclusive results.

Let us now analyse the viability plots (Figure 7-6) in relation to the enzymatic values given in Table 7-3. Cells were incubated in standing flasks, in 0.05 M phosphate buffer at 26°C. In control strains the average value for trehalose-6-phosphate synthase is 170 nmoles UDP $\times$ min^{-1} $\times$ mg protein^{-1}. Basal levels of trehalase are very low, but the cells have the capacity of activating the enzyme at least 15- to 20-fold. In one of the control strains

Table 7-3
Trehalose-6-P-synthase and trehalase activities in various haploid strains of yeast

		Trehalase[b]	
Strain	T-6-P-Synthase[a]	Basal	$+ATP \cdot Mg^{2+}$ +cAMP
1403-7A	160.0	2.7	39.0
S288C	180.0	2.6	53.6
Q6R2	10.0	3.9	22.6
AM9-8B	25.0	13.0	13.0
GS1-36	230.0	24.0	42.0
1268-2B	350.0	41.2	52.3
AP28-1B	410.0	32.9	33.8
AP28-3B	170.0	72.8	72.3

Note: Cells were grown at 28°C in the following medium: 2% glucose, 1.3% yeast extract, 0.2% $(NH_4)SO_4$, 0.2% KH_2PO_4, pH 5.2. For trehalase determinations, cells were harvested at the onset of diauxie and 1 hr after depletion of glucose for trehalose synthase determinations.

[a]Activity is given in nmoles UDP $\times$ min^{-1} $\times$ mg protein^{-1} as described by Operti et al., 1982.

[b]Activity is given in nmoles glucose $\times$ min^{-1} $\times$ mg protein^{-1}. We consider basal activity that which is found in the cells at the point of harvest. The second column represents the activity obtained in vitro by incubating the homogenate with $ATP \cdot Mg^{2+}$ and cAMP as described by Ortiz et al., 1983.

(1403-7A), 40% viability is maintained over a period of 6 weeks. In strain S288C, which is considered by geneticists to be the prototype of a wild-type yeast strain, enzyme levels are normal, but only 20% viable cells were found after 2 weeks under our experimental conditions. These few cells, however, remained viable for over 4 weeks. Clearly, then, even when cells have normal trehalose metabolism, other genetic factors also influence viability under starvation.

Mutant strains with no significant trehalose-6-phosphate synthase activity do not survive over 2 weeks (Q6R2). On the other hand, when this enzymatic activity is normal, as in GS1-36 and AP28-3B, but breakdown of trehalose is favoured by the presence of an activated trehalase in its phosphorylated form, death of cells occurs after 3 weeks. In mutant strains in which both enzymes are very active (1268-2B and AP28 1B), we can observe that survival is greatly enhanced. Apparently, then, a good balance between synthesis and mobilization favours viability.

According to Lillie and Pringle (1980), glycogen and trehalose are degraded steadily as long periods of starvation proceed. Trehalose decline becomes more abrupt after 60 days, and the authors indicate a temporal correlation between carbohydrate depletion and the onset of cell death. Our

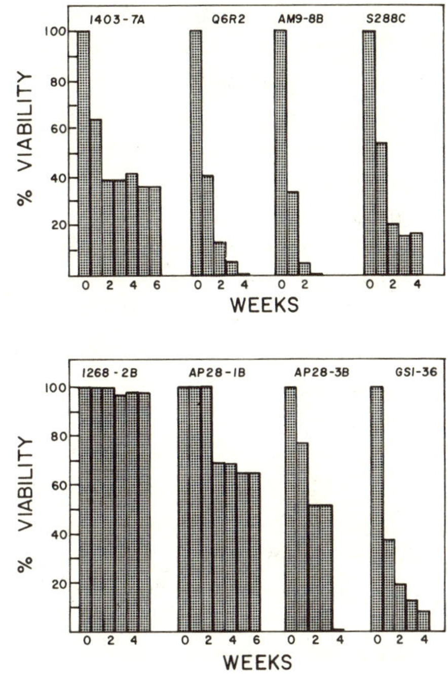

Figure 7-6. Survival plots of wild-type and mutant strains under nutrient deprivation. Cells were grown to early second log phase and then transferred to a stationary incubation in 0.05 M phosphate buffer pH 6.0 at 26°C. The percentage of viable cells was determined by the colony-forming efficiency on YPD plates.

results corroborate these findings and emphasize that strains with very low trehalose-6-phosphate synthase activity are far more fragile during starvation.

At this point we should add that an alternative pathway for trehalose synthesis, which does not depend upon the activity of trehalose-6-phosphate synthase, has been described (Panek et al., 1979). In contrast to the pattern we see in Figure 7-1 with a wild-type maltose negative strain (D311-3A), brewer's and baker's yeast grown on glucose accumulate significant amounts of trehalose during the transition phase from fermentative to oxidative metabolism. The common characteristic in these strains is the presence of a *MAL4* gene mutated to its constitutive form. We have demonstrated that all strains grown on maltose accumulate trehalose, even the *fdp* mutant, which lacks trehalose-6-phosphate synthase activity in vitro. The question we faced was whether maltose as such was the requisite for trehalose accumulation during growth. Using a constitutive *MAL2*^{c} strain and mutating it to its inducible (*MAL2*) and inactive (*mal2*) forms, we were able to demonstrate, by growing cells on different carbon sources, that the expression of a *MAL* gene constitutes the prerequisite for trehalose accumulation (Oliveira et al., 1981). Our working hypothesis for the moment is that the specific and unique relationship between maltose utilization and trehalose accumulation provides the cell with a mechanism which slows down the glycolytic flux and thus allows for mitochondrial derepression. This concept brings forth an additional role for trehalose in *Saccharomyces*. Only competent mitochondria capable of supporting normal oxygen-linked metabolism allow for efficient maltose utilization.

It became intriguing to verify whether the pool of trehalose linked to the enzymes responsible for maltose utilization would contribute to cell survival. We have named this pathway of trehalose synthesis system II, as opposed to system I, which requires trehalose-6-phosphate synthase activity.

An analysis of Table 7-4 shows that the high levels of trehalose accumulated during growth by *MAL* constitutive strains are depleted during the 72 hr in sporulation medium. Moreover, the presence of system II in these strains does not substitute for the deficiency in trehalose-6-phosphate synthase (AP9) and the enhanced activity of trehalase (AP8). Sporulation efficiency and consequently germination were not favoured by the alternative pathway expressed by the constitutive *MAL* gene.

Would resistance to nutrient deprivation be favoured by this additional trehalose accumulation system? We chose three strains which harbour the *MAL4* constitutive gene, and one of them which does not possess significant trehalose-6-phosphate synthase activity. As seen in Table 7-5, the trehalose accumulated during growth by system II was rapidly broken down when cells were maintained in buffer. Viability of the mutant with a lesion in system I was not prolonged (compare with strain Q6R2, which does not harbour a constitutive *MAL* gene). Furthermore, the high activity of trehalose-6-phos-

Table 7-4
Effect of trehalose synthesis system II upon sporulation efficiency

Strains	T-6-P-Synthase[a]	Trehalase[b]	Trehalose[c] 0[d]	Trehalose[c] 72h[e]	Sporulation (%)
DE4	271.0	11.6 (4.4)	127.5	10.4	80
AP8	365.0	19.4 (1.2)	56.6	16.2	0
AP9	27.5	7.8 (2.3)	77.6	2.2	5

Note: All diploids tested harbour dominant constitutive *MAL* genes.
[a]nmoles UDP × min^{-1} × mg protein^{-1}.
[b]nmoles glucose × min^{-1} × mg protein^{-1} (the values in parentheses are the activation ratios in vitro).
[c]mg/g cells dry weight.
[d]Cells grown on YED for 20 hr and sampled for trehalose extraction and determination according to Oliveira et al., 1981.
[e]Cells cultivated on sporulation medium (0.98% sodium acetate, 0.05% glucose, 0.1% yeast extract, and 1.5% agar).

phate synthase in strain AP28-1B seems to improve survival capacity of this strain when compared with the control strain (1403-7A).

The next aspect on which we should focus is the role of trehalose in conditions in which, specifically, the cytoplasmic membrane suffers alteration. Cells loaded with trehalose utilize 90% of the carbohydrate as soon as they are reinoculated into a fresh growth medium (Panek, 1963). Yet if maintained in buffer, trehalose is constantly being broken down and resynthesized (Panek, 1975). The main difference between these two conditions is that in a growth medium, preparation for cell division occurs during the "lag" phase. As claimed by Mazia (1961) "the energetic price of the division process itself is paid before division." Glycogen levels however, remain constant while trehalose is being degraded (Panek, 1963). Why has nature chosen this disaccharide to play a role during cell division and budding?

When yeast cells are loaded with trehalose by incubation in buffered glucose solution, 20% of the exogenous carbon source is stored as trehalose. The bulk of the disaccharide is found in the particulate fraction of lysed protoplasts. Mechanical disruption does not liberate trehalose, yet it can be easily extracted from intact cells with solvent mixtures such as ethanol-ether (1:1) or chloroform-methanol (3:1) or with ethanol at 100°C (Souza and Panek, 1968). Studying the yeastlike fungus *Pullularia pullulans*, Merdinger and Kohn (1967) have raised the hypothesis that the sugar is loosely bound to the lipids of the cytoplasmic membrane in intact cells and that this linkage is disrupted by organic solvents.

Table 7-5
Effect of trehalose synthesis system II upon viability of yeast cells

		Trehalose[a]					Viability (%)		
Strains	T-6-P-Synthase[b]	0[b]	3d	7d	14d	21d	7d	14d	21d
1403-7A	160.0	109.0	19.0	5.4	9.4	16.6	64	40	40
AP28-1B	410.0	90.0	23.0	47.0	13.0	15.0	100	100	69
VFP1-8C	20.0	60.0	18.0	4.5	4.2	4.0	41	19	6
Q6R2	10.0	7.1	—	6.7	4.2	4.6	40	14	6

Note: All strains tested harbour the constitutive *MAL4* gene except strain Q6R2, which has the inducible *MAL6* gene.

[a] mg/g cells dry weight.

[b] nmoles UDP × min^{-1} × mg $protein^{-1}$.

[c] Cells grown on YED for 20 hr and samples taken for extraction and determination of trehalose according to Oliveira et al., 1981. The cells were then washed and transferred to 0.05M sodium phosphate buffer, pH 6.0 and incubated for the times (days) indicated at 26°C. The percentage of viable cells was determined by the colony-forming efficiency on YPD plates.

We investigated further the hypothesis that there is a membrane-bound trehalose by using toluene in concentrations which affect permeability but do not lead to cell lysis. Although no alteration in trehalose levels was seen during starvation up to 3 hr (Panek, 1963), when toluene was added to the buffer, 85% of the accumulated trehalose was released from the cells in 2 hr (Souza and Panek, 1968). It is therefore reasonable to assume that changes in the membrane during budding may lead to the release of trehalose from its location, and it would therefore constitute the carbohydrate of choice for supplying energy during the "lag" phase of growth when no exogenous carbon source is yet being taken up. Although Keller et al. (1982) dispute our hypothesis that there is a membrane-bound trehalose, Küenzi and Fiechter (1969) give us some support by claiming that trehalose, the more mobile of the two storage carbohydrates, allows the cells to carry out the double-cell phase within a constant time.

Several lines of evidence suggest that trehalose is involved in sporulation and germination (Inoue and Shimoda, 1981; Panel et al., 1981; Thevelein et al., 1982). Each of these situations is a response to stress and definitely leads to alterations in the cell membrane.

It has been reported that cell viability after dehydration is 30% lower in ethanol-grown yeast than in cells grown on molasses. This observation correlates with the trehalose content, which is 25% lower in the first case. It is interesting to observe that the amount of trehalose is not altered in either case by the dehydration process (Beker et al., 1980).

When the diploids DE4, AP8, and AP9 (Table 7-2) were stored at 4°C in thin layers covered with 0.2 ml 0.05 M phosphate buffer pH 6.0, all strains remained totally viable for 3 weeks despite the lesions in trehalose metabolism present in diploids AP8 and AP9. When the control diploid (DE4) was filtered to dryness and then stored in a thin layer (10 mg of cells over a circle of 2 in. in diameter), total death occurred in 7 days. These cells contained only 2% of their dry weight in trehalose. In a following experiment we loaded the cells with trehalose by incubation in 20 mM buffered glucose for a period of 40 min after the carbon source had been consumed. This procedure allowed the cells to accumulate 7.5% of their dry weight in trehalose. When these cells were then filtered to dryness and stored at 4°C, 100% viability was maintained for 7 days, while at room temperature 50% of the cells died during the first week. After 14 days only 17% of the cells remained viable at both temperatures. Clearly, then, an increase in trehalose content favoured survival of drying.

Yeast cells possibly store trehalose during dehydration; what happens in the anhydrobiotic stage we do not really know, but trehalose may protect the membrane when cells come into contact with water and may allow them rapidly to resume their metabolic activities. This hypothesis is based on the binding of trehalose to the phosphate head groups of the phospholipid bilayer in the cytoplasmic membrane (Chapters 11, 12, 13, 17). By replacing water molecules, trehalose may contribute to the stability of yeast membranes.

It is becoming apparent from the literature that trehalose is effective as a cryoprotectant for the preservation of cells. The fact that a treatment of plant cells with trehalose is needed prior to freezing indicates that beneficial effects of the sugar may result from trehalose in the rehydration medium and also inside the cell (Bhandal et al., 1985). The capacity of yeast cells to synthesize trehalose would endow these cells with longevity and would attribute to trehalose an additional and specific role as a protector against autolysis.

Nature, we believe, is forever subtle but never malicious. Her secrets will be revealed if we follow the spark of an idea, asking and trying to explain exciting and mystifying questions of life. If wise nature has bestowed glycogen and trehalose upon yeast cells, and has provided them with sophisticated control mechanisms, we must be alert to the possible relevance of these compounds in the physiology of the yeast cell. Trehalose may play a role in preserving the integrity of the yeast cell membrane that could explain the coexistence of two storage carbohydrates in these cells.

REFERENCES

Becker, J.V. 1978. A method for glycogen determination in whole yeast cells. Analyt. Biochem. 86: 56-64.

Beker, M.J., Damberga, B.E., Krauze, I.J., Ventina, E.J. and Kontakevich, J.G. 1980. Factors of wine yeast resistance at their dehydration. *In* Current Developments in Yeast Research. Proc. Fifth Intern. Symp., London, Canada.

Bhandal, I.S., Hauptmann, R.M. and Widholm, J.M. 1985. Trehalose as cryoprotectant for the freeze preservation of carrot and tobacco cells. Plant Physiol. 78: 430-432.

Cabib, E. and Leloir, L.F. 1958. The biosynthesis of trehalose phosphate. J. Biol. Chem. 231: 259-275.

Charlab, R., Oliveira, D.E. and Panek, A.D. In press. Investigation of the relationship between sstl and fdp mutations in yeast and their effect on trehalose synthesis. Bras. J. Med. Biol. Res.

Clegg, J.S. 1965. The origin of trehalose and its significance during the formation of encysted dormant embryos of *Artemia salina*. Comp. Biochem. Physiol. 14: 135-143.

Crowe, J.H., Crowe, L.M. and Chapman, D. 1984. Preservation of membranes in anhydrobiotic organisms: the role of trehalose. Science 223: 701-703.

Edelmann, K., Stelwagen, P. and Oura, E. 1980. The influence of temperature and availability of oxygen on the carbohydrates of stored baker's yeast. *In* Proc. Sixth Intern. Ferment. Symp., Canada, pp. 51-56.

Elander, M. 1969. Trehalose-6-phosphasynthetase aus Bäckerhefe. Arkiv. für Kemi. 31: 17-30.

Elbein, A. 1974. The metabolism of α-α trehalose. Advances Carbohyd. Chem. and Biochem. 30: 227-256.

Ingram, L.O. 1976. Adaptation of membrane lipids to alcohols. J. Bacteriol. 125: 670-678.

Inoue, H. and Chimoda, C. 1981. Changes in trehalose content and trehalase activity during spore germination in the fission yeast *Schizosaccharomyces pombe*. Arch. Microbiol. 129: 19-22.

Kane, S.M. and Roth, R.J. 1974. Carbohydrate metabolism during ascospore development in yeast. J. Bacteriol. 118: 8-14.

Keller, F., Schellenberg, M. and Wiemken, A. 1982. Localization of trehalose in vacuoles and of trehalose in the cytosol of yeast. Arch. Microbiol. 131: 198-301.

Küenzi, M.T. and Fiechter, A. 1969. Changes in carbohydratecomposition and trehalase activity during the budding cycle of *Saccharomyces cerevisiae*. Arch. Microbiol. 64: 396-407.

Küenzi, M.T. and Fiechter, A. 1972. Regulation of carbohydrate composition of *Saccharomyces cerevisiae* under growth limitation. Arch. Microbiol. 84: 254-265.

Leloir, L.F. and Cabib, E. 1953. The enzymic synthesis of trehalose phosphate. Am. Chem. Soc. 75: 5445-5446.

Lillie, S.H. and Pringle, J.R. 1980. Reserve carbohydrate metabolism in *Saccharomyces cerevisiae*: responses to nutrient limitation. J. Bacteriol. 143: 1384-1394.

Londesborough, J. and Varimo, K. 1984. Characterization of two trehalases in baker's yeast. Biochem. J. 219: 511-518.

Matern, H. and Holzer, H. 1979. Endogenous proteolytic modulation of yeast enzymes. *In* Modulation of Protein Function, ed. Atkinson, D.E. and Fox, C.F., pp. 81-92. Academic Press, New York.

Matsumoto, K., Uno, I., Kato, K. and Ishikawa, T. 1985. Isolation and characterization of a phosphoprotein phosphatase deficient mutant in yeast. Yeast 1: 25-38.

Mazia, D. 1961. Biochemistry of the dividing cell. *In* Ann. Rev. Biochem, ed. Luck, J.M., Allan F.W. and Mackinney, G. Vol. 30, pp. 669-688. Academic Press, New York.

Merdinger, E. and Kohn, P. 1967. Isolation and identification of trehalose from *Pullularia pullulans*. Canadian J. Microbiol. 13: 1126-1128.

Navon, G., Shulman, R.G., Yamane, T., Ross Eccleshall, T., Lam, K.B., Baronofsky, J.J. and Marmur, J. 1979. Phosphorus-31 nuclear magnetic resonance studies of wild-type and glycolytic pathway mutants of *Saccharomyces cerevisiae*. Biochem. 18: 4487-4499.

Oestreicher, E.G. and Panek, A.D. 1982. Effect of Mg^{2+} on the activity of yeast trehalose-6-phosphate synthase. IRCS Med. Sci. 10: 418-419.

Oliveira, D.E., Rodrigues, E.G., Mattoon, J.R. and Panek, A.D. 1981. Relationships between trehalose metabolism and maltose utilization in *Saccharomyces cerevisiae*. Curr. Gen. 3: 235-242.

Operti, M.S., Oliveira, D.E., Freitas-Valle, A.B., Oestreicher, E.G., Mattoon, J.R. and Panek, A.D. 1982. Relationships between trehalose metabolism and maltose utilization in *Saccharomyces cerevisiae*. Curr. Genet. 5: 69-76.

Ortiz, C.H., Maia, J.C., Tenan, M.N., Braz-Padrão, G.R., Mattoon, J.R. and

Panek, A.D. 1983. Regulation of yeast trehalase by a monocyclic cyclic-AMP dependent phosphorylation-dephosphorylation cascade system. J. Bacteriol. 153: 644-651.

Panek, A.D. 1963. Function of trehalose in baker's yeast. Arch. Biochem. Biophys. 100: 422-425.

Panek, A.D. 1975. Trehalose synthesis during starvation of baker's yeast. Eur. J. Appl. Microbiol. 2: 39-46.

Panek, A.D. 1985. Trehalose metabolism and its role in *Saccharomyces cerevisiae*. J. Biotechnol. 3: 121-130.

Panek, A.D., Bernardes, E., Ortiz, C.H., Baker, S. and Mattoon, J.R. 1981. Biochemical genetics of trehalose metabolism in *Saccharomyces cerevisiae*. Ann. Acad. Bras. Ciências 53: 165-172.

Panek, A.D. and Mattoon, J.R. 1977. Regulation of energy metabolism in *Saccharomyces cerevisiae*. Arch. Biochem. Biophys. 183: 306-316.

Panek, A.D., Sampaio, A.L., Braz, G.R.C. and Mattoon, J.R. 1979. Genetic and metabolic control of trehalose and glycogen synthesis. New relationships between energy reserves, catabolite repression and maltose utilization. Cell. Mol. Biol. 23: 334-354.

Roth, R.J. 1970. Carbohydrate accumulation during the sporulation of yeast. J. Bacteriol. 101: 53-57.

Rothman-Denes, L.B. and Cabib, E. 1971. Glucose phosphate dependent and independent forms of yeast glycogen synthase: their properties and interconversion. Biochem. 10: 1236-1242.

Sá Correia, I. and van Uden, N. 1982. Effects of ethanol in thermal death and on the maximum temperature for growth of yeast *Kluyveromyces fragilis*. Biotechnol. Lett. 4: 805-808.

Souza, N.O. and Panek, A.D. 1968. Location of trehalase and trehalose in yeast cells. Arch. Biochem. & Biophys. 125: 22-28.

Suomalainen, H. and Pfaffli, S. 1961. Changes in the carbohydrate reserves of baker's yeast during growth and on standing. J. Inst. Brew. London 67: 249-254.

Tatchell, K., Robinson, L.C. and Breitenbach, M. 1985. RAS2 of *Saccharomyces cerevisiae*, is required for gluconeogenic growth and proper response to nutrient limitation. Proc. Natl. Acad. Sci. USA 82: 3785-3789.

Tenan, M.N., Ortiz, C.H., Dellamora-Ortiz, G., Mattoon, J.R. and Panek, A.D. 1985. Relationships between mutations affecting protein kinase and accumulation of energy reserves in *Saccharomyces cerevisiae*. FEMS Microbiol. Lett. 26: 217-220.

Thevelein, J.M. 1984. Regulation of trehalose mobilization in fungi. Microb. Rev. 48: 42-59.

Thevelein, J.M., den Hollander, J.A. and Schulman, R.G. 1982. Changes in activity and properties of trehalase during early germination of yeast ascospores: correlation with trehalose breakdown as studied by in vivo ^{13}C NMR. Proc. Natl. Acad. Sci. USA 79: 3503-3507.

Thomas, D.S., Hossack, J.A. and Rose, A.H. 1978. Plasma membrane lipid composition and ethanol tolerance in *Saccharomyces cerevisiae*. Arch. Microbiol. 117: 239-245.

Trevelyan, W.E. and Harrison, J.S. 1956. Studies of yeast metabolism. Biochem. J. 63: 23-33.

Uno, I., Matsumoto, K., Adachi, K. and Ishikawa, T. 1983. Genetic and biochemical evidence that trehalase is a substrate of cAMP- dependent protein kinase in yeast. J. Biol. Chem. 258: 10867-10872.

van de Poll, K.W. and Schamhart, D.H.J. 1977. Characterization of a regulatory mutant of fructose 1-6 biphosphatase in *Saccharomyces carlsbergensis*. Mol. Gen. Genet. 154: 61-67.

van der Plaat, J.B. and van Solingen, P. 1974. Cyclic 3′5′-adenosine monophosphate stimulates trehalose degradation in baker's yeast. Biochem. Biophys. Res. Commun. 56: 580-586.

van Solingen, P. and van der Plaat, J.B. 1975. Partial purification of the protein system controlling the breakdown of trehalose in baker's yeast. Biochem. Biophys. Res. Commun. 62: 533-560.

Wiemken, A. and Schellenberg, M. 1982. Does a cyclic AMP-dependent phosphorylation initiate the transfer of trehalose from the cytosol into the vacuoles in *Saccharomyces cerevisiae*? FEBS Lett. 150: 329-331.

8 Water and the Survival of Bacterial Spores

Grahame W. Gould

In contrast to the situation in large multicellular organisms, small and single-celled organisms that are free living must remain close to osmotic equilibrium with their surroundings. And yet their surroundings may often change enormously in osmolality over very short periods of time. There have therefore been strong evolutionary pressures operating to select efficient mechanisms that combat the effects of drying and those of low relative humidity, low water activity, high concentrations of salts or sugars, or freezing.

In general, it has been observed that the metabolically active vegetative forms of such microorganisms react to a rise in the concentration of environmental solutes by raising their internal solute levels so that they just exceed the external osmolality and thereby avoid excessive loss of water. They thus appear to operate a ''homeostatic'' mechanism with respect to cell water content. The internally accumulated solutes have been termed ''osmoregulatory'' or ''compatible'' solutes because they allow continued metabolic activity at concentrations well above those at which common solutes would cause inhibition. If the concentration of nonpenetrant solutes is so high that it exceeds the capacity of a cell to react in this way, then of course water is irreversibly lost. The partly dehydrated cell stops growing, becomes metabolically inert, or ''dormant,'' sometimes surviving in this state for long periods of time. The inert cell may increase greatly in resistance, for instance to heat, even by factors in excess of 1000-fold. In these respects, such cells superficially resemble bacterial endospores, which are by far the most dormant and resistant forms of life on earth.

Indeed, recently acquired data indicate that the central cytoplasmic contents of endospores, even when these cells are suspended in pure water, maintain a remarkable level of dehydration. This may be a major factor that accounts for their extreme dormancy and resistance, although the mechanisms by which this is achieved and the relative importance of additional factors are still conjectural.

Spore Dormancy and Resistance

Spores display extreme dormancy in that their endogenous metabolism is almost undetectable, even using sensitive radiotracer techniques. Their resistance to heat is very dependent upon strain and species. Generally spores formed by psychrophilic bacteria are much less heat tolerant than those of thermophilic bacteria. Spore heat resistance appears to have at least two distinct components. The first of these is an intrinsic property of the strain, which is related to the resistance of the vegetative cell. The second component is the *additional* heat resistance that is imposed on its contents by the spore. This is normally sufficient to raise the resistance of constituents of the spore between about 35 to 45°C (Warth, 1978).

Structure

The structure of the bacterial endospore differs fundamentally from that of the vegetative cell in that the spore is compartmentalised in a manner that is unique in prokaryotic cells (Figure 8-1). The compartmentalisation originates during spore formation when, within the mother cell, asymmetric septation occurs, causing a large and a small membrane-bounded compartment to form within the same rigid cell wall. The large compartment then engulfs the small one in such a way that the contents of the small compartment, or ''forespore,'' are now separated from the environment by three membranes, the middle one of which, topographically and functionally, has reversed polarity. Such a structure is so unusual in prokaryotic cells that, it has even been suggested, it may have evolved from the same pre-Cambrian event that led to the development of complex cells containing the first membrane-bounded organelles (Gould and Dring, 1979). In extant spore formers, development continues with the presumably osmotic shrinkage of the forespore compartment (Baillie et al., 1984; Algie, 1983) followed first by synthesis of peptidoglycan between the two (functionally ''outward-facing'') forespore membranes, then by deposition of a predominantly proteinaceous spore coat outside the outer forespore membrane. In most spore-forming bacteria, the mother cell then lyses to liberate the mature spore.

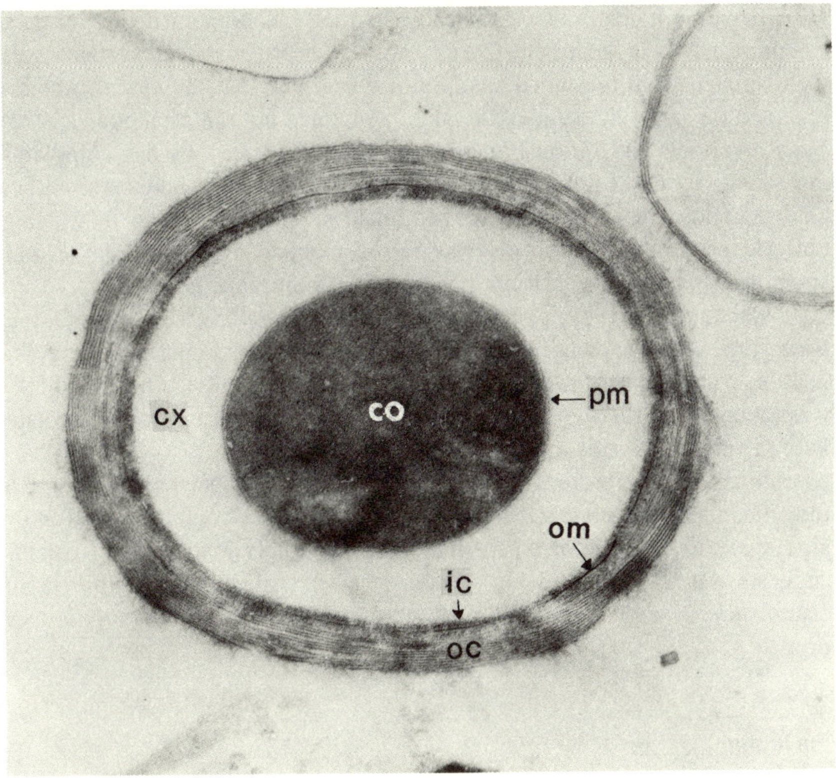

Figure 8-1. Endospore structure. Electron micrograph of a potassium permanganate-fixed spore of *Clostridium sporogenes*, showing the major compartments, the core (*co*) and cortex (*cx*), and the major structures: plasma membrane (*pm*), inner coat (*ic*), outer coat (*oc*), and the outer membrane (*om*). Bar indicates 0.1 μm.

Calcium and Dipicolinic Acid

Spores normally contain high levels of calcium and dipicolinic (DPA; pyridine 2,6-dicarboxylic) acid, typically about 2 and 10% of dry weight, respectively. There is good evidence from ion probe x-ray analysis for calcium (Johnstone et al., 1980) and from studies with radioactive DPA-labelled spores (Leanz and Gilvarg, 1973) that they are mostly present in the central cytoplasmic compartment, or "core," perhaps as some form of insolubilized calcium dipicolinate. (Calcium and DPA form a chelate with a stability constant of about 4.6 and normally precipitate at concentrations above about 40 mM.)

A role for DPA in resistance seemed likely following the isolation of

DPA-negative mutants of *Bacillus cereus* that also contained low levels of calcium and were heat sensitive. However, heat-resistant revertants were then isolated from these mutants, and it was found that, although some had regained the ability to synthesise DPA, some had not (Hanson et al., 1972). Such revertants had seriously defective dormancy (i.e. they had completely lost sensitivity to the normal germinants such as L-alanine and inosine and yet tended to lose resistance and to germinate simply on storage in water or buffers) particularly if the pH was raised much above 7 and if the ionic strength was raised also (Dring, personal communication).

Balance et al. (1979) isolated a mutant of *B. subtilis* the spores of which were DPA negative and produced heat-sensitive spores in normal culture but heat-resistant spores if DPA was added during sporulation. The added DPA was taken up by the developing spores, and the amount taken up correlated with the degree of resistance.

Altogether, therefore, the role of DPA has not yet been elucidated. The fact that the DPA-negative heat-resistant revertants contained no (or unanalysably low) levels of DPA, and correspondingly low levels of calcium suggests that a *direct* role in resistance is not likely. But a role in the establishment and/or *maintenance* of resistance and dormancy is very probable.

Ion Immobilisation

In addition to metals such as calcium (which may be immobilised largely as calcium dipicolinate) and magnesium, spores also contain levels of potassium and sodium that are not normally tightly sequestered in biological systems. However, dielectric studies have indicated that ions in the central core of the spore are somehow immobilised, a situation which contrasts with that in vegetative cells. Ions in the outer regions (coat and cortex) are mobile in some spores but immobilised in others (Carstensen et al., 1979).

It has been known since the pioneering work of Alderton and Snell (1963) that spores behave like cation exchangers so that, by suitable exposure to acid, followed by incubation in different salt solutions, a fraction of the spores' ion load may be replaced. When this is done, resistance is found to change, with "H-form" spores normally being most sensitive and Na-form, K-form, and Ca-form being most resistant. In some cases, such laboratory-created Na- and Ca-form spores were more resistant than the "native" spores from which they were derived.

More recently Marquis et al. (1981) have succeeded in extracting almost all the cations from spores of a particular strain of *Bacillus megaterium* without killing the cells. Such "stripped" spores lose some but not all heat resistance. These data have been interpreted as indicating that ion stasis contributes a *part* of the heat-resistance mechanism of the spore, with intrinsic (i.e. vegetative) resistance of the cell's constituents contributing another part and low water content in the core constituting the third component of the total resistance mechanism.

Basic Proteins

Spores contain a number of unique core-located basic polypeptides of characteristic composition, together with peptidases that specifically degrade them after the spore has been "triggered" and commenced germination. It has been suggested that their major function may be to provide a source of amino acids during germination. However, the fact that they constitute nearly half the core protein and bind to spore DNA (Setlow and Setlow, 1979) indicates that they may play a role in resistance as well, in particular to UV irradiation.

Peptidoglycan

The similarity of structure of peptidoglycan in all spores so far studied (Figure 8-2) implies an important structure-function relationship in the cortex. It is characterised by a low degree of peptide cross-linking of the glycan chains and a large net negative charge, so that it has been likened to a weak, loosely cross-linked cation-exchange polymer. It should undergo large volume decreases in the presence of high concentrations of salts similar to but greater than the decreases measured in more highly cross-linked vegetative peptidoglycans. Viscosity measurements of isolated spore peptidoglycan in the presence and absence of salts have confirmed that such changes do occur (Baillie and Murrell, 1974).

Much evidence implicates cortex peptidoglycan in resistance. For example, spores treated with reducing agents to make their coats permeable to lysozyme, which hydrolyses peptidoglycan, rapidly lose resistance when incubated with this enzyme (Gould and Hitchins, 1963). Some coatless mutants (Cassier and Ryter, 1971) or naturally occurring coat-defective spores are likewise sensitive to lysozyme; mutants with reduced levels of peptidoglycan in their cortexes have low resistance (Imae and Strominger, 1976); during germination, as heat resistance is lost, endogenous enzymes hydrolyse peptidoglycan in the cortex (Johnstone and Ellar, 1982). As the cortex region in the spore disappears during germination, so the core swells,

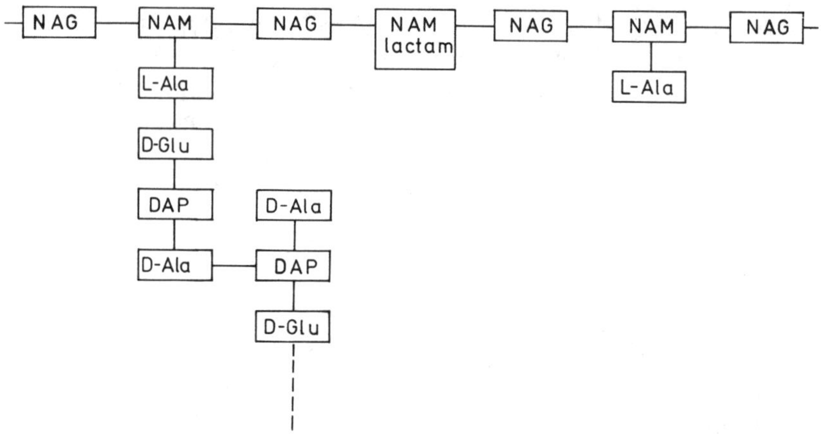

Figure 8-2. Spore peptidoglycan. Structure of the peptidoglycan polymer that has the same general structure in all endospores so far studied and exists in the cortex region. The polymer has a low degree of peptide cross-linking compared with most vegetative cell peptidoglycans, is electronegative, and contains characteristic muramyl lactam groups. (NAG, N-acetyl glucosamine; NAM, N-acetyl muramic acid; L-Ala, L-alanine; D-Glu, D-glutamic acid; DAP, diaminopimelic acid; D-Ala, D-alanine.)

within periods as short as 1 or 2 min in some species. The swelling time is too short to allow extensive synthesis, so that the swelling predominantly reflects uptake of water, or hydration of the core.

Water Status

The total water contents of spores suspended in water or near RH 100%, measured by a variety of techniques, ranges from about 67 to about 230 g per 100 g dry weight (Beaman et al., 1982). Those spores with the greatest heat resistances also have the lowest water contents and vice versa. However, these observations do not take into account the compartmentalisation within the spore. Indeed, an even better correlation has been seen between the ratio of core volume to core-plus-cortex volume and heat resistance; spores with a small ratio, and therefore an apparently highly condensed core, have higher resistances than spores with a large ratio (Bayliss et al., 1981; Beaman et al., 1982).

Direct measurements of the water content of the core cannot be made, but water content can be inferred in a number of ways. For example, measurements of the refractive index of spores, from about 1.4 to 1.53 (Ross and Billing, 1957; Leman, 1973), were compatible with water contents below about 25 g per 100 g dry weight for the spore or some region within it. Estimation of the mass of peptidoglycan in spores, and the volume it occupies, indicate that the cortex must be relatively dilute and the core relatively condensed. Determination of buoyant densities of spores in solutes of different molecular mass, which are therefore able to penetrate spores to different extents, was used by Gerhardt and his colleagues (Beaman et al., 1982; Tisa et al., 1982) to measure whole spore wet densities (1.17-1.27 g/ml) and dry densities (c. 1.44 g/ml). They deduced that core water contents were lower than whole spore water contents.

Beaman et al. (1984) estimated core water contents of *B. megaterium* spores within the range of 33 to 40 g per 100 g dry weight by measuring permeation of ^{3}H-labelled water and ^{14}C-labelled glucose into the various compartments within the spore. And these spores, with D_{90C} values of 1 to 6 min were not especially heat resistant.

A low core water content protects against ultraviolet radiation, which generates DNA photoproducts of the 5-thyminyl-5,6-dihydrothymine type in spores. These are typical products of the irradiation of dry DNA in the "A" configuration and contain fewer than six water molecules per nucleotide, in contrast to the cyclobutane-type thymine dimers characteristic of vegetative cells; they are also typical products of DNA in the high water-content "B" form (Stafford and Donnellan, 1968).

Further evidence of water status comes from the behaviour of spores during germination, when, as mentioned above, the core rapidly swells. This occurs at a time when it is actually losing dry matter (e.g. calcium dipicolinate) and therefore must be taking up water. Spores in this just-germinated state are heat sensitive if they are tested by heating in water, but if they are heated in concentrated solutions of solutes that do not readily penetrate the plasma membrane (and therefore remove water from the core by osmosis), the heat resistance typical of the ungerminated spore is fully regained (Dring and Gould, 1975; Figure 8-3). These observations strongly indicate that a low core water content is a major determinant of spore resistance and probably of dormancy too.

If the spores are heated in equiosmolal concentrations of penetrant solutes (e.g. glycerol), no regain of resistance is observed, suggesting that the osmotic dehydration of the core is responsible for returning the core to a sporelike state and reimposing resistance. Exactly the same phenomenon is seen, as mentioned earlier, when suspension of even nonsporing vegetative bacteria (Corry, 1974) and yeasts (Corry, 1976a) in strong solutions of nonpenetrant solutes can raise heat resistance by some thousandfold. It has been shown microscopically that the solutes causing osmotic loss of cytoplasmic water bring about a rise in the resistance of the cells (Corry, 1976b).

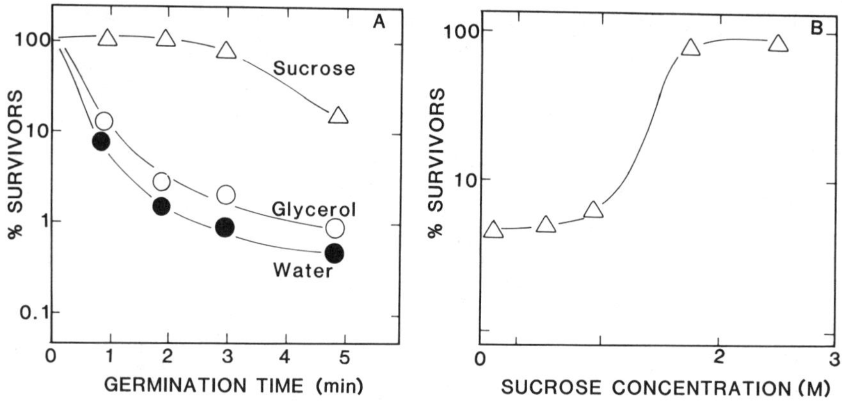

Figure 8-3. Reimposition of heat resistance on germinating spores by osmotic dehydration. Germination of spores of *Bacillus cereus* was initiated by the addition of L-alanine + inosine. At intervals, samples were removed and heated at 60°C for 10 min in water (●), 1.8 M glycerol (○), or 1.8 M sucrose (△). Figure 8-3a demonstrates the protection afforded by the nonpenetrant solute (sucrose) and the lack of protection afforded by the penetrant solute (glycerol) at the same molality. Figure 8-3b shows survivor levels after heating spores in a range of sucrose concentrations 2 min after the initiation of germination. Concentrations of sucrose between 1 and 2 M are necessary to reimpose resistance.

An estimate of the degree of dehydration of the core was obtained in a different way by Warth (1980). He compared the resistances of a number of enzymes when heated within intact spores with their resistances when heated in broken spore extracts and did so over a range of water activities which were obtained by control of the relative humidity during heating. The data obtained showed that the intact spore imposed an additional heat resistance of about 40°C on these enzymes. If this resulted solely from a reduction in water content in the core, then it would correspond to the effect of a reduction in the water activity within the core of a spore (i.e. at a_w 1.0), by about 0.25 to 0.3 units.

On the basis of our knowledge about the structure and composition of spores, what mechanism(s) could we envisage that could bring about, and then maintain, such a level of dehydration?

Mechanisms

Lewis et al. (1960) were the first to suggest that the core may be dehydrated by contraction of the cortex that surrounds it, for instance through the cross-linking of its excess of carboxyl groups by calcium. Later it was suggested that expansion of peptidoglycan in the cortex by electrostatic repulsion of the unshielded carboxyl groups was also a possible mechanism (Alderton and Snell, 1963). It was noted that expansion was more compatible with the known mass of peptidoglycan in spores and the large volume it occupies (Gould and Dring, 1975a). The cortex could act as an osmoregulatory organelle which would maintain the core in a low water state, even if the spore was suspended in pure water, as long as the level of low-molecular-mass components in the core in solution was low (Gould and Dring, 1975b). Warth (1980, 1985) reasoned that core dehydration equivalent to a reduction in a_w of about 0.25-0.3 was necessary to explain resistance on the basis of lowered water content alone. Thus simple osmoregulation by the cortex would be inadequate. It was therefore necessary to call upon cortically applied pressure, for instance through anisotropic expansion (Warth, 1985). Algie calculated the pressure generation required to maintain "effective a_w" of the core in spores near to 0.8, essentially by reverse osmosis, and reasoned that the calculated pressure was compatible with the likely tensile strength of peptidoglycans and related polymers (Algie et al., 1983).

From measured volumes, dry densities, and water sorption isotherms of spores and spore components, Algie (1984) calculated that the "effective a_w" of cores of spores of *B. stearothermophilus* was about 0.83 and that dehydration equivalent to an effective water activity of about 0.78 was the lowest that could be achieved by cortical pressure and reverse osmosis.

Without doubt, the more rigidly a macromolecule is held, the more resistant it will be to unfolding. Such increased resistance is commonly seen, for example, when enzymes are immobilised on solid supports (see Klibanov, 1979). Increases in resistance to heat by factors as great as 10^5-fold have been achieved by mechanical entrapment of enzymes, for instance of trypsin and chymotrypsin in polymethacrylate gels with water contents as high as 50% w/w, and presumably high water activities (Martinek et al., 1977). The effectiveness of such entrapment is, of course, increased at low water contents or water activities and decreased at high water content or high water activities. However, the *viscosity* of the solvent rather than water content or water activity per se is of key importance (Careri, 1982).

The broad NMR absorption pattern of spores indicated that water within them is more restricted, with respect to rotation and exchange, than water within vegetative cells held at the same RH. The data may be compatible with a highly viscous glassy state existing within the core and contributing to resistance and dormancy. Indeed, it has been suggested by Murrell and Wills (1977) that the dramatic destabilisation of spores that is caused by raised hydrostatic pressure (Clouston and Wills, 1969; Gould and Sale, 1970) results from a reduction in viscosity in the core. Like raised temperature and

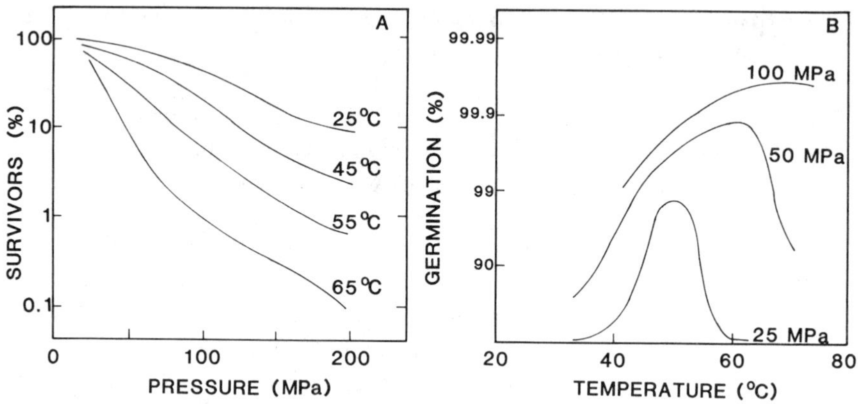

Figure 8-4. Destabilization of spores by raised hydrostatic pressure. Spores of *Bacillus coagulans* in 0.1 M pH 8.0 sodium phosphate were pressurised for 30 min, then heated at 70°C for 30 min to inactivate spores that had been destabilized ("germinated"). Figure 8-4a shows the large effect that temperature has on pressure-induced destabilization, which, in Figure 8-4b, is seen as an apparent increase in the optimum temperature for germination with increase in pressure.

increased hydration, raised pressure would tend to move such a system through its glass transition zone (see Appendix D), and pressure and temperature certainly interact strongly in their destabilising effects on spores (Figure 8-4).

Some data support alternative views. For instance, sorption and NMR studies by Bradbury et al. (1981) and Watt (1981) suggested that the water content of the core was as high as that of the cortex, and core water was even *more* mobile than water in the outer integuments, at least at relative humidities below 98%. These authors and Murrell therefore favoured some form of immobilisation within the core, for instance in a solid matrix of calcium dipicolinate, as a key factor in resistance and dormancy. Preliminary data have been reported (Murrell, 1981) indicating that "cocrystallisation" with calcium dipicolinate does raise the heat resistance of some, but not all, enzymes tested.

Conclusions

The current situation provides good evidence that a relatively low core water content is maintained within spores even when they are suspended in pure water and that this contributes in a major way to desiccation resistance and dormancy. Direct measurements of core water content cannot be made, but the best indirect estimations indicate a range from below about 25 to about 40 g per 100 g dry weight. Rapid hydration of the core occurs when germination is triggered and peptidoglycan in the cortex is hydrolysed. There is good evidence that this peptidoglycan plays a major role in the maintenance of a low core water content; at the molecular level the means by which this occurs is still unknown. The extreme stasis of ions in spores probably contributes additional resistance. Many of these observations are compatible with the core components' being in a metastable glass state, in which mobility of the water and other molecules that are present is so reduced that it essentially prevents metabolism and raises heat resistance via a large increase in viscosity and a consequent *kinetic* barrier to normal cell reaction and denaturation processes. However, the data that indicate a more mobile, higher core water status (Bradbury et al., 1981) must be considered. More direct measurement of water levels in the various spore compartments is urgently needed. Finally, the intrinsic resistance of the cells' proteins and other components represents the basic resistance on which the desiccation protection is built.

REFERENCES

Alderton, G. and Snell, N.S. 1963. Base exchange and heat resistance in bacterial spores. Biochem. Biophys. Res. Comm. 10: 139-143.

Algie, J.E. 1983. The heat resistance of bacterial spores and its relationship to the contraction of the forespore protoplasm during sporulation. Curr. Microbiol. 9: 173-176.

Algie, J.E. 1984. Effect of the internal water activity of bacterial spores on their heat resistance in water. Curr. Microbiol. 11: 293-296.

Algie, J.E., Lindsay, J.A. and Watt, I.C. 1983. Determination of the strength of the cell walls of vegetative cells of *Bacillus megaterium* and *Bacillus stearothermophilus*. Curr. Microbiol. 9: 13-17.

Algie, J.E. and Watt, I.C. 1984. Calculation of mass and water content between core, cortex and coat of *B. stearothermophilus*. Curr. Microbiol. 10: 249-254.

Baillie, E. and Murrell, W.G. 1974. Some hydrodynamic properties of spore peptidoglycan. Biochim. Biophys. Acta 372: 23-31.

Baillie, E.C., Germaine, G.R., Murrell, W.G. and Ohye, D.F. 1984. Photoreactivation, photoproduct formation and deoxyribonucleic acid state in ultraviolet-irradiated sporulating cultures of *Bacillus cereus*. J. Bacteriol. 120: 516-523.

Balance, G., Milhaud, P., Raulet, E., Silva, M.T. and Sousa, J.C.F. 1979. A *Bacillus subtilis* mutant requiring dipicolinic acid for the development of heat-resistant spores. J. Gen. Microbiol. 110: 365-379.

Bayliss, C.E., Waites, W.M. and King, N.R. 1981. Resistance and structure of spores of *Bacillus subtilis*. J. Appl. Bacteriol. 50: 379-390.

Beaman, T., Greenamyrem, J., Corner, T., Pankrantz, H. and Gerhardt, P. 1982. Bacterial spore heat resistance correlated with water content, wet density, and protoplast/sporoplast volume ratio. J. Bacteriol. 150: 870-877.

Beaman, T.C., Koshikawa, T., Pankratz, H.S. and Gerhardt, P. 1984. Dehydration partitioned within core protoplast accounts for heat resistance of bacterial spores. FEMS Microbiol. Lett. 24: 47-51.

Bradbury, J.H., Foster, J.R., Hammer, B., Lindsay, J., and Murrell, W.G. 1981. The source of the heat resistance of bacterial spores. Study of water in spores by NMR. Biochim. Biophys. Acta 678: 157-164.

Careri, G. 1982. Molecular hydration and its possible role in enzymes. *In* Biophysics of Water (F. Franks and S.F. Mathias, eds.), pp. 58-61. John Wiley, Chichester.

Carstensen, E.L., Marquis, R.E., Child, S.Z. and Bender, G.R. 1979. Dielectric properties of native and decoated spores of *Bacillus megaterium*. J. Bacteriol. 140: 917-928.

Cassier, M. and Ryter, A. 1971. Sur un mutant de *Clostridium perfringens* donnant des spores sans tuniques à germination lysozyme-dépendante. Ann. l'Inst. Pasteur, Paris, 121: 717-732.

Clouston, J.G. and Wills, P.A. 1969. Initiation of germination and inactivation of *Bacillus pumilis* spores by hydrostatic pressure. J. Bacteriol. 97: 684-690.

Clouston, J.G. and Wills, P.A. 1970. Kinetics of initiation of germination of *Bacillus pumilis* spores by hydrostatic pressure. J. Bacteriol. 103: 140-143.

Corry, J.E.L. 1974. The effect of sugars and polyols on the heat resistance of salmonellae. J. Appl. Bacteriol. 37: 31-43.

Corry, J.E.L. 1976a. The effect of sugars and polyols on the heat resistance and morphology of osmophilic yeasts. J. Appl. Bacteriol. 40: 269-276.

Corry, J.E.L. 1976b. Sugar and polyol permeability of *Salmonella* and osmophilic yeast cell membranes measured by turbidimetry, and its relationship to heat resistance. J. Appl. Bacteriol. 40: 277-284.

Danilatos, G.D., Denby, E.F., and Algie, J.E. 1984. The effect of relative humidity on the shape of *Bacillus apiarus* spores. Curr. Microbiol. 10: 313-316.

Dring, G.H. and Gould, G.W. 1975. Reimposition of heat-resistance on germinated spores of *Bacillus cereus* by osmotic manipulation. Biochem. Biophys. Res. Comm. 66: 202-208.

Gerhardt, P., Beaman, T.C., Corner, T.R., Greenamyre, J.T. and Tisa, L.S. 1982. Photometric immersion refractometry of bacterial spores. J. Bacteriol. 150: 643-648.

Gould, G.W. and Dring, G.J. 1975a. Role of an expanded cortex in the resistance of bacterial endospores. *In* Spores VI (P. Gerhardt, R.N. Costilow and H.L. Sadoff, eds.), pp. 541-546. Am. Soc. Microbiol., Washington, D.C.

Gould, G.W. and Dring, G.J. 1975b. Heat resistance of bacterial endospores and concept of an expanded osmoregulatory cortex. Nature, London, 258: 402-405.

Gould, G.W. and Dring, G.J. 1979. On a possible relationship between bacterial endospore formation and the origin of eukaryotic cells. J. Theor. Biol. 81: 47-53.

Gould, G.W. and Hitchins, A.D. 1963. Sensitization of bacterial spores to lysozyme and to hydrogen peroxide with agents which rupture disulphide bonds. J. Gen. Microbiol. 33: 413-423.

Gould, G.W. and Sale, A.J.H. 1970. Initiation of germination of bacterial spores by hydrostatic pressure. J. Gen. Microbiol. 60: 335-346.

Hanson, R.S., Curry, M.V., Garner, J.V. and Halvorson, H.O. 1972. Mutants of *Bacillus cereus* strain T that produce thermoresistant spores lacking dipicolinate and have low levels of calcium. Can. J. Microbiol. 18: 1139-1143.

Imae, Y. and Strominger, J.L. 1976. Relationship between cortex content and properties of *Bacillus sphaericus* spores. J. Bacteriol. 126: 907-913.

Johnstone, K. and Ellar, D.J. 1982. The role of cortex hydrolysis in the triggering of germination of *Bacillus megaterium* KM endospores. Biochim. Biophys. Acta 714: 185-191.

Johnstone, K., Ellar, D.J. and Appleton, T.C. 1980. Location of metal ions in *Bacillus megaterium* spores by high-resolution electron probe x-ray microanalysis. FEMS Microbiol. Lett. 7: 97-101.

Klibanov, A.M. 1979. Enzyme stabilization by immobilisation. Anal. Biochem. 93: 1-25.

Leanz, G. and Gilvarg, C. 1973. Dipicolinic acid location in intact spores of *Bacillus megaterium*. J. Bacteriol. 114: 455-456.

Leman, A. 1973. Interference microscopical determination of bacterial dry weight during germination and sporulation. Jena Review 5: 263-270.

Lewis, J.C., Snell, N.S. and Burr, H.K. 1960. Water permeability of bacterial spores and the concept of a contractile cortex. Science 132: 544-545.

Marquis, R.E., Carstensen, E.L., Child, S.Z., and Bender, G.R. 1981. Preparation and characterization of various salt forms of *Bacillus megaterium* spores. *In* Sporulation and Germination (H.S. Levinson, A.L. Sonenshein and D.J. Tipper, eds.), pp. 266-268. Am. Soc. Microbiol., Washington, D.C.

Martinek, K., Klibanov, A.M., Goldmacher, V.S., Tschernysheva, A.V., Moshaez, V.V., Berezin, I.V., and Glotov, B.O. 1977. The principles of enzyme stabilization II. Increase in the thermostability of enzymes as a result of multipoint non-covalent interaction with a polymeric support. Biochim. Biophys. Acta 485: 13-28.

Murrell, W.G. 1981. Biophysical studies on the molecular mechanisms of spore heat resistance and dormancy. *In* Sporulation and Germination (H.S. Levinson, A.L. Sonenshein and D.J. Tipper, eds.), pp. 64-77. Am. Soc. Microbiol., Washington, D.C.

Murrell, W.G. and Wills, P.P. 1977. Initiation of *Bacillus* spore germination by hydrostatic pressure: effect of temperature. J. Bacteriol. 129: 1272-1280.

Ross, K.F.A. and Billing, E. 1957. The water and solid content of living bacterial spores and vegetative cells as indicated by refractive index measurements. J. Gen. Microbiol. 17: 418-425.

Sale, A.J.H., Gould, G.W. and Hamilton, W.A. 1970. Inactivation of bacterial spores by hydrostatic pressure. J. Gen. Microbiol. 60: 323-334.

Setlow, B. and Setlow, P. 1979. Localisation of low-molecular weight basic proteins in *Bacillus megaterium* spores by cross-linking with ultraviolet light. J. Bacteriol. 139: 486-494.

Stafford, R.S. and Donnellan, J.E.S. 1968. Photochemical evidence for conformation changes in DNA during germination of bacterial spores. Proc. Natl. Acad. Sci. USA 59: 822-828.

Tisa, L.S., Koshikawa, T. and Gerhardt, P. 1982. Wet and dry spore densities determined by buoyant sedimentation. Appl. Environ. Microbiol. 43: 1307-1310.

Warth, A.D. 1978. Relationship between the heat resistance of spores and the optimum and maximum growth temperatures of *Bacillus* species. J. Bacteriol. 134: 699-705.

Warth, A.D. 1980. Heat stability of *Bacillus cereus* enzymes within spores and in extracts. J. Bacteriol. 143: 27-34.

Warth, A.D. 1985. Mechanisms of heat resistance. *In* Fundamental and Applied Aspects of Bacterial Spores (D.J. Dring, D.J. Ellar and G.W. Gould, eds.), pp. 209-225. Academic Press, London.

Watt, I.C. 1981. Water vapour absorption by *Bacillus stearothermophilus* endospores. *In* Sporulation and Germination (H.S. Levinson, A.L. Sonenshein and D.J. Tipper, eds.), pp. 253-255. Amer. Soc. Microbiol., Washington, D.C.

Wise, J., Swanson, A. and Halvorson, H.O. 1967. Dipicolinic acid-less mutants of *Bacillus cereus*. J. Bacteriol. 94: 2075-2076.

9 The Ecology of Dehydration in Soil Organisms

Diana W. Freckman

The objective of ecology is to understand the interactions between living organisms and their environment. The response of organisms to severe environmental stress is of special interest because survival is the ultimate criterion of success and also because the survival mechanisms and life history strategies involved have similarities among the different biota (Crawford, 1981; Wallwork, 1982), which could elucidate the patterns of nature.

In addition to the physiological mechanisms of survival, ecologists have been interested in the morphological or developmental form of survival and the physical and biological triggers that induce the survival response. Time is required for the organism to respond to the stress, and the response must be initiated by appropriate environmental cues which precede the harsh extreme. Organisms faced with severe environmental stress, defined here as any environmental factor producing physiological stress and decreasing the fitness of the organism, must adapt or die. "The strongest selective pressures exist for organisms that experience some form of environmental water stress, namely high or fluctuating salinity, desiccation, or freezing" (Yancey et al., 1982). Dehydration is such a stress, and the physiological and biochemical mechanism of response is the subject of this volume. In this chapter, emphasis will fall on the ecological response of organisms living in arid soil environments. We shall begin by discussing the habitat of the organisms and

The author thanks Walter G. Whitford for his contributions to the knowledge of desert soil ecology and for imparting his enthusiasm and love for the desert to those who interact with him. This work was sponsored by grants DEB 7823445-01 and BSR 8215357 of the National Science Foundation.

the soil, shall then examine survival mechanisms of selected biota to desiccation, and last shall consider how the interactions of biota with different responses to desiccation affect the functioning of an arid ecosystem.

The Soil Environment

The soil environment is a fascinating milieu with an enormous supply of ecological niches for its inhabitants. These niches are created by the combination of abiotic factors in the soil, such as fluctuating redox potentials, pH, matric and osmotic water potentials, and temperatures, interacting with the biotic factors of plant roots, microflora, and animals. Of these physical factors, none is more important to the biota than soil water. Water affects: (1) diffusion rates of solutes (e.g. nutrients to and from the biota); (2) gas exchange (soil aeration); and (3) motility.

Soil water, with dissolved ions and solutes, is held as a film in the pores and around soil particles by adhesion and capillarity. The matric potential is the free energy of water held by adhesion or capillary effects. This, plus the osmotic potential or the free energy of the solutes, comprises what is known as total soil water potential. Water potential is usually measured as negative bars (−bar), or megapascals (−0.1 MPa = −1 bar). Soil that is saturated, with all pores filled with water, has a water potential of zero, whereas soil that is fairly dry has a water potential of less than −1.0 MPa (or −10 bars). As the soil drains, water available for physiological processes and growth decreases. Harris (1981) has reviewed the effect of water potential on microflora and has grouped the microflora into four classes, depending on their ability to osmoregulate and develop at different water potentials.

As would be expected, the tolerance limits of organisms to decreasing soil moisture vary with genera and phyla, thus making any generalization on the effect of dehydration on soil organisms of doubtful value. This variation is illustrated by a comparison of four organisms and their response to decreasing water potential in soil (Figure 9-1): a gram-negative bacterium; two fungi, *Gaeumannomyces graminis* var. *tritici* and *Fusarium roseum* 'Culmorum' (Cook et al., 1972; Cook and Baker, 1983); and a fungivorous nematode species, *Aphelenchus avenae*. Maximum growth of the fungi was measured by hyphal length in soil and in agar plates, with adequate nutrients provided (Cook et al., 1972). Nematode motility and activity was measured as a change from an active vermiform to a coiled form (Demeure et al., 1979). Although maximum growth and motility for all organisms occurred at or above −0.1 MPa, *F. roseum* appeared to occupy a drier niche down to −7.0 to −8.0 MPa. Cook and Baker (1983) suggested that a limiting factor

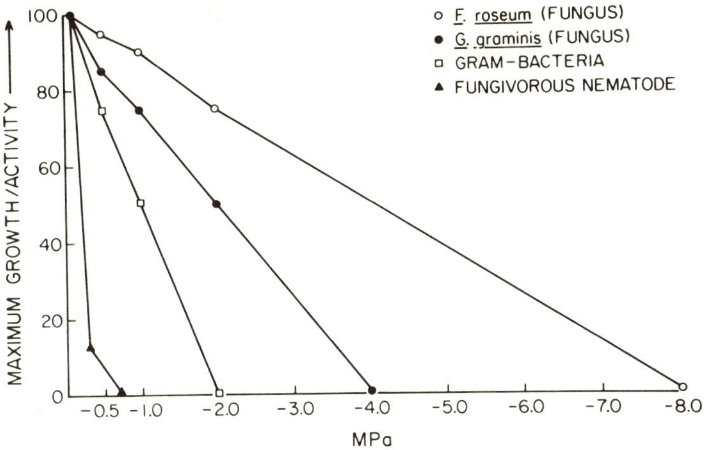

Figure 9-1. The effect of water potential on the growth and activity of 4 soil organisms: *Fusarium* (a fungus), *G. graminis* (a fungus; from Cook and Baker, 1983), a gram-negative bacteria, and a fungivorous nematode, *A. avenae*. (From Demeure et al., 1979, courtesy of *Journal of Nematology*.)

other than water potential must exist for *F. roseum* as soils wetter than −0.1 MPa. The activity of the nematode, *A. avenae*, appears to decrease long before its potential food sources, the fungi, are greatly affected by dehydration.

A group of soil organisms with similar responses to decreasing water potential is illustrated by studies of Demeure et al. (1979). Nematodes can, at any developmental stage, enter an ametabolic, anhydrobiotic state, which is reversible with addition of moisture. In 1969, Ellenby stated, "Very few species of nematodes are able to survive desiccation. . . . It is also not surprising that those which possess outstanding ability to survive desiccation . . . are forms which parasitize aerial parts of plants." Since that time many studies (see reviews in Demeure and Freckman, 1981; Womersley, 1981) have revealed that anhydrobiosis is a phenomenon experienced by many of the soil nematode species. In most soil nematodes examined (Glazer and Orion, 1983; Saeed and Rossner, 1984; Townshend, 1984), anhydrobiosis involves both the morphological response of a tight coiling of the vermiform worm and some physiological responses.

Demeure et al. (1979) examined three nematode species, representing three trophic or feeding levels, all isolated from habitats of different moisture regimes and found that all reached maximum coiling and entered anhydrobiosis at about −0.6 MPa (−6 bars in Figure 9-2). However, there were differences in the time needed to enter anhydrobiosis. The bacteriophage

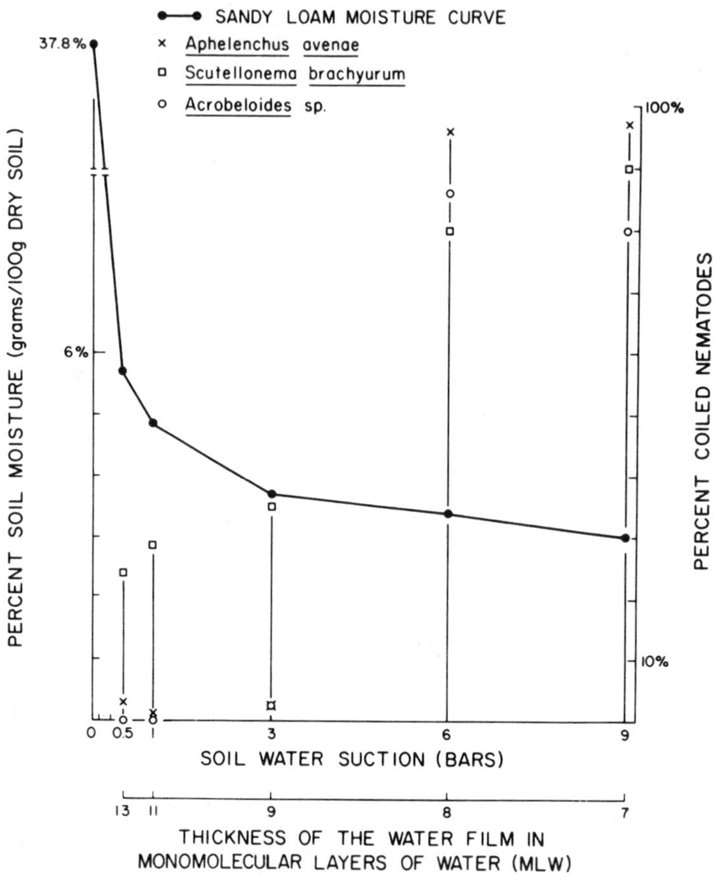

Figure 9-2. Soil moisture characteristics of a sandy loam (72.8% sand, 21.1% silt, and 6% clay) and percentage of coiling of *A. avenae* (×), *S. brachyurum* (□), and *Acrobeloides* sp. (○). (From Demeure et al., 1979, courtesy of *Journal of Nematology*.)

Acrobeloides sp. appeared to coil more slowly than the other two species, *Aphelenchus avenae* and *Scutellonema brachyurum*. It had been isolated from a desert soil, and its behaviour in this experiment indicated that it was adapted for maximum activity during the brief periods of moisture which occur infrequently in deserts. A plant parasite, *Scutellonema brachyurum*, isolated from a moist, semitropical planting, was less resistant to the decreasing soil moisture and began to coil at −0.01 MPa, but it is doubtful that the nematode entered anhydrobiosis at this moisture level.

Much of the research on the effect of water potential on soil biota has involved either a comparison of two moisture regimes or steadily decreasing moisture. The effects of fluctuating water potentials on survival of biota are more difficult to measure. Bryant et al. (1982), studying desiccation of protozoa, found no difference in the effect of either constantly wet or fluctuating wet-dry treatments on feeding activity or total density of amoebae in laboratory microcosms. Amoebae became encysted in dry soil and apparently resumed grazing on the bacteria during wet cycles. In another study in the field that focused on sclerotia, the fungus *Sclerotina minor* was found to have a higher survival rate in soils with water potentials varying weekly over the course of 1 year in a range from 0 to −4.30 MPa's (Abawi et al., 1985). The results of Bryant et al. were compatible with evidence indicating that the fungus is most abundant in the top few centimeters of field soil.

Pore size and texture of the soil are other factors influencing the amount of water available to the biota. Most soils have a wide range of particle sizes, from 2 mm coarse sand to <0.002 mm clay. Small soil pores exist between individual particles, and larger pores exist between aggregates. The percentage of sand and silt combined with the clay determines the pore size and influences the amount of water adhering to the soil particle. Many of the soil biota are dependent on the water in soil pores for motility. Even when clay is saturated, organisms in pores larger than 20 μm diameter will be immobilized. In pure clay, 20% of the pores are 20 μm or smaller, whereas 80% of pores in pure sand are >20 μm. Wallace (1959) demonstrated that movement of nematodes in soil was clearly a function of particle size and soil water. Movement was optimum when soil pores were half drained of free water. Demeure et al. (1979) extended this theory and suggested that the trigger mechanism for inducing anhydrobiosis involves the physical force acting on the thickness of the water film surrounding the nematode cuticle. The optimum thickness for the induction of anhydrobiosis was six to nine monomolecular layers of water molecules.

Crowe and Crowe (1982) noted that the removal of water from the nematode during slow dehydration causes the animals slowly to produce free glycerol and/or trehalose, which maintains the integrity of membranes while in anhydrobiosis. Robinson et al. (1984) found that an above-ground plant parasite, *Orrina phyllobia*, apparently desiccates and enters anhydrobiosis faster than other species tested and suggested that the reason was temperature-dependent variations in boundary-layer conductance around the nematode.

Different pore sizes also affect the life habits of other biota. Amoebae and bacteriophagous nematodes compete for bacteria, but amoebae are able to live in pores of a much smaller diameter than nematodes and thus can utilize bacteria which are unavailable to nematodes (Bamforth, 1980; Elliot et al., 1980).

Aeration is another critical factor for many of the organisms, particularly those whose life habit is aquatic and dependent on the water film around the soil particle. The concentration of oxygen in the soil atmosphere can vary from 21 to 0% over a short distance, and thus the soil profile has microsites of varying moisture and oxygen. All of these factors—water potential, osmotic potential, pore size, soil texture, and aeration—interact to alert an organism to approaching environmental stress. We shall now examine the mechanisms of adaptation to anhydrous conditions.

Survival Mechanisms of Soil Biota

The survival and life history of organisms in arid soil systems have been the subject of several recent books and reviews (Cloudsley-Thompson, 1975; Crawford, 1981; Louw and Seeley, 1982; Wallwork, 1982). Despite the apparent sparseness of vegetation, deserts are the domain of a diverse group of biota. A large and extremely diversified portion of the faunal biomass is invertebrates, which, like the abundant microflora, are easily overlooked because of small size and cryptic behaviour.

Surprisingly, dormancy is not a universal feature of desert soil biota, although many soil microflora and protozoa are inactive (Bamforth, 1984). Generally, the biota have responded to dehydration by escaping through one or a combination of behavioral, morphological, or physiological means.

Many organisms escape in space or avoid the stress by migrating or burrowing deeper into the soil; by being removed by wind; by producing eggs in different soil microsites ensuring survival of the species; and by a seasonal response such that reproduction coincides with the rainy season.

Another means of escaping dehydration is escaping in time. The biota using this strategy have developed mechanisms of desiccation tolerance and remain in the stressful environment. These organisms either maximize moisture gain from the environment or minimize water loss to the environment. Tolerance, or escaping in time, generally implies both a form of dormancy and use of stored food reserves for metabolic energy. The Namib desert snail *Sphincterochila zonata* is activated quickly from an inactive state in response to rainfall events. When the soil surface is sufficiently wet, the snail ingests up to 45 times as much water per day as it loses per day of hibernation (Shachak et al., 1976). *S. zonata*, by minimizing moisture loss and reducing its metabolic demands, survives for years without food (Schmidt-Nielsen et al., 1971). Other means of escaping in time include production of eggs or other structures which have a thick protective cover, serving to reduce moisture loss. The most extreme form of dormancy is anhydrobiosis, which, although common in animals such as rotifers, protozoa, tardigrades, and nematodes, has just recently been studied in desert systems. Collembola enter a delayed dormancy or diapause, which may be similar to anhydrobiosis (Greenslade and Greenslade, 1973; Poinsot-Balaquer, 1976; Verhoef and Witteveen, 1980).

These strategies, escape in time and space, are both used by many of the soil biota. For example, nematodes escape in space by migrating horizontally and vertically in the soil and escape in time by entering anhydrobiosis. The combinations of behavioural, morphological, and physiological changes developed in response to dehydration are not isolated capabilities but adaptations to stress which have served to increase the fitness of the organism.

Pilato (1979) examined adaptations, similarities in life histories, and other characteristics in soil rotifers, gastrotrichs, nematodes, and tardigrades and suggested that selection due to environmental factors was avoided or delayed, resulting in less genetic variability than is seen in other biota.

This "evolutionary end" has also been examined by Greenslade (1983) who advanced the new hypothesis of adversity, or *A-selection*. Biologists have defined two selection processes termed *r-selection*, for productivity in unpredictable habitats, and *k-selection*, for efficiency in predictable environments. On the basis of studies of invertebrates in extreme habitats which have similar characteristics, such as parthenogenesis and cryptobiosis, Greenslade suggested that adversity selection would favor conservation of adaptation in severe but predictable environments.

Biotic Interactions in a Desert Ecosystem

In desert ecosystems, soil fauna play important roles as regulators of rates of decomposition, nitrogen immobilization, and mineralization. Decomposition and mineralization are key processes that affect the productivity of ecosystems by regulating the rates at which mineral nutrients become available for plant growth. Recent work has elucidated the importance of soil fauna in regulating nutrient cycling (Freckman et al., in press; Whitford et al., in press).

Temperature extremes and water availability are thought to be the most important factors limiting production and activity of desert organisms (Whitford et al., 1981, 1983). Deserts are dry for long periods of time, with seasonal rainfalls. The "pulse-reserve hypothesis" (Noy-Meir, 1973) predicts a burst of activity with any moisture event. This hypothesis suggests that rainfall starts a pulse of energetic biotic activity which may be reduced rapidly to await the next rainfall pulse. Whitford et al. (1982) suggested that moisture may not be as limiting as originally thought and that the desert soil decomposer community has evolved adaptations that allow those organisms to process litter somewhat independently of rainfall (Mackay et al., in press).

In the Chihuahuan desert of New Mexico, USA, the decomposer community is composed primarily of bacteria, some fungi, protozoans, nematodes, microarthropods, and macroarthropods. Elimination of the soil animals has been shown to reduce the rate of decomposition (Santos et al., 1981). The functioning of this community is relatively simple, as is shown by the rate at which energy and matter enter and leave the system. Abundance, or numerical density per unit area per unit time, is usually inadequate to measure these transformations. Biomass, or the weight of a field population per unit time, holds more ecological significance for the transformations of energy and matter. Macroarthropods, specifically termites, are the most important in terms of abundance and biomass, serving to breakdown and move wood, feces, and organic plant material from above ground to below ground (Johnson and Whitford, 1975). The mites (Acari) are primarily prostigmatids and are the most abundant of the soil microarthropods, although Collembola and Psocoptera occasionally occur in litter. Mites feed on algae, bacteria, fungi, nematodes, and other mites (Peterson and Luxton, 1982). Nematodes are among the most numerous soil animals in desert soils, although in trophic structure they differ from their counterparts in mesic environments by being dominated by bacteriophagic, fungivorous, and omnivore-predaceous groups (Freckman et al., in press). In numbers and biomass protozoa in desert soils are at the minimal population level of mesic ecosystems (Bamforth, 1984). The primary decomposers in the desert are bacteria.

In the trophic chain, bacteria are food for small amoebae in small pore spaces (Bamforth, 1980, 1984); bacteriophagous and omnivorous nematodes feed on bacteria in larger pore spaces; other nematodes feed on fungi and plant roots, and these nematodes in turn are preyed upon by predaceous nematodes; tydeid mites appear to regulate the nematode community through predation. The amoebae, and bacteriophagous and omnivorous nematodes influence the rate of decomposition and mineralization by grazing on primary decomposers (Whitford et al., 1986) and by linking nutrients from the bacteria to higher trophic levels. Collectively their interactions create a complex trophic chain in a "simple" desert ecosystem.

To test the Noy-Meir pulse-reserve hypothesis, a year-long experiment was conducted in the Chihuahuan desert, near Las Cruces, New Mexico (Freckman et al., in press). Annual rainfall averages 211 ± 77 mm and occurs as late summer convectional storms. Summer temperatures reach 40°C and freezing temperatures occur from October through April. The response of the soil biota was measured monthly, before and after three levels of irrigation: control (natural rainfall), 6 mm/week, and 25 mm/month. Water was applied by overhead sprinklers. The rate of decomposition, the population densities, and the biomass of the soil biota remained unchanged with addition of water apart from increases in plant parasitic nematodes with the 6 mm/week treatment (Freckman et al., in press; Steinberger et al., 1984). The lack of effect with water was surprising, although studies on plant productivity and rainfall effects have shown that factors other than moisture, usually nitrogen, limit primary productivity in desert ecosystems.

Biota responded to the moisture cycles by escaping either in time (tolerating the stress) or in space (avoidance). Microflora, including bacteria, fungi, and protozoa, spend the greatest portion of their time in an inactive state, thus tolerating the stress. The lack of a response to the water regimes indicates that the microflora and protozoa may be carbon limited.

Of particular interest was the anhydrobiotic mechanism utilized by nematodes. Prior to irrigations, there were no differences in the percentages of anhydrobiotes on the three treatments. After irrigation, activity increased (anhydrobiosis decreased), particularly with the larger, monthly irrigation. A general moisture threshold appeared to exist at a soil moisture (gravimetric) of 4% (Figure 9-3), which corresponded to a matric potential of −0.4 MPa. If moisture was lower than 4.7%, more than 60% of the nematodes were anhydrobiotic and uncoupled from nutrient cycling. This water potential corresponds to that of Demeure et al. (1979) for nematodes in a similar soil

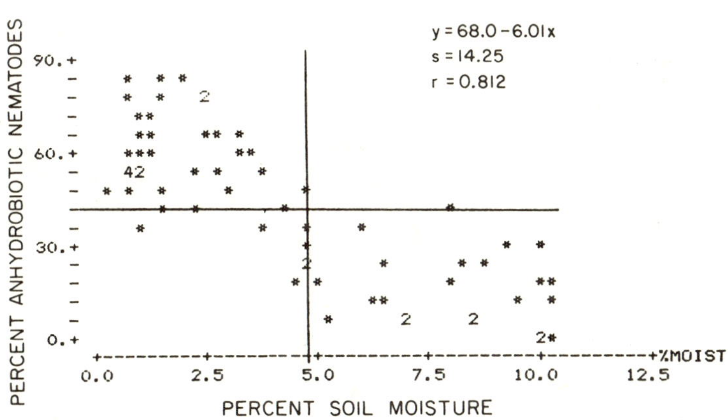

Figure 9-3. Monthly mean anhydrobiotic nematodes under three irrigation treatments for the year, before and after water was applied, plotted against gravimetric soil moisture. The + indicates the threshold of soil moisture, the point at which nematodes appear to become active or anhydrobiotic.

type (sandy loam). The nematodes were not 100% anhydrobiotic because of the varying moisture levels and heterogeneity of soil microsites.

Microarthropods survived dehydration by escaping in space. Orabatid and tydeid mites remained active by migrating diurnally from plant litter at the soil surface to the soil beneath litter. Litter moisture was 7% at 0800 hr but fell to less than 1% by midday.

The lack of a population response to watering pulses by the microarthropods was surprising. It appears that the active population of the soil biota participating in nutrient cycling in dry soils is a small subset of the total active population when the soils are wet. Whereas soil arthropods are active whenever temperatures are favorable, and thus uncouple decomposition and mineralization from abiotic constraints, nematodes and protozoa survive dry periods as cysts or anhydrobiotes and are active only during rainfalls or in moist microsites. The physiological, morphological, and behavioural responses to dehydration allow the organisms to process litter relatively independently of rainfall (Mackay et al., in press).

REFERENCES

Abawi, G.S., Grogan, R.G. and Duniway, J.M. 1985. Effect of water potential on survival of sclerotia of *Sclerotinia minor* in two California soils. Phytopathology 75: 217-221.

Bamforth, S.S. 1980. Terrestrial protozoa. J. Protozool. 27: 33-36.

Bamforth, S.S. 1984. Microbial distributions in Arizona, USA, deserts and woodlands. Soil Biol. Biochem. 16: 133-138.

Bryant, R.J., Woods, L.E., Coleman, D.C., Fairbanks, B.C., McClellan J.F. and Cole, C.V. 1982. Interactions of bacterial and amoebal populations in soil microcosms with fluctuating moisture content. Appl. Environ. Microbiol. 43: 747-752.

Cloudsley-Thompson, J.L. 1975. Adaptations of arthropods to arid environments. Ann. Rev. Ent. 20: 261-273.

Cook, R.J. and Baker, K.F. 1983. The Nature and Practice of Biological Control of Plant Pathogens. American Phytopath. Soc., St. Paul, Minn. 539 pp.

Cook, R.J., Papendick, R.I. and Griffin, D.M. 1972. Growth of two root-rot fungi as affected by osmotic and matric water potentials. Soil Sci. Amer. Proc. 36: 78-82.

Crawford, C.S. 1981. Biology of Desert Invertebrates. Springer- Verlag, Berlin, 314 pp.

Crowe, J.H. and Crowe, L.M. 1982. Induction of dehydration: membrane changes during drying. Cryobiol. 19: 317-328.

Demeure, Y. and Freckman, D.W. 1981. Recent advances in the study of anhydrobiosis in nematodes. *In* Plant Parasitic Nematodes (B.M. Zuckerman and R.A. Rohde, eds.), Vol. III, Academic Press, New York, pp. 204-225.

Demeure, Y., Freckman, D.W. and Van Gundy, S.D. 1979. Anhydrobiotic coiling of nematodes in soil. J. Nematol. 11: 189-195.

Ellenby, C. 1969. Dormancy and survival in nematodes. Symp. Soc. Exp. Biol. 23: 83-97.

Elliott, E.T., Anderson, R.V., Coleman, D.C. and Cole, C.V. 1980. Habitable pore space and microbial trophic interactions. Oikos 35: 327-335.

Freckman, D.W., Whitford, W.G. and Steinberger, Y. In press. Nematode population dynamics and activity in desert soils: effects of irrigation. Biol. Fert. Soils.

Glazer, I. and Orion, D. 1983. Studies on anhydrobiosis of *Pratylenchus thornei*. J. Nematol. 15: 333-338.

Greenslade, P.J.M. 1983. Adversity selection and the habitat templet. Amer. Natur. 122: 352-365.

Greenslade, P.J.M. and Greenslade, P. 1973. Epigaeic Collembola and their activity in a semi-arid locality in southern Australia during summer. Pedobiol. 13: 227-235.

Harris, R.F. 1981. Effect of water potential on microbial growth and activity. *In* Water Potential Relations in Soil Microbiology (J.F. Parr, W.R. Gardner and L.F. Elliott, eds.), pp. 23-95. Soil Sci. Soc. Amer., Spec. Publ. No. 9, Madison, Wisc.

Johnson, K.A. and Whitford, W.G. 1975. Foraging ecology and relative importance of subterranean termites in Chihuahuan desert ecosystems. Environ. Entom. 4: 66-70.

Louw, G.N. and Seeley, M.K. 1982. Ecology of Desert Organisms. Longman House, New York. 194 pp.

MacKay, W.P., Silva, S., Lightfoot, D.C., Pagani, M.I., and Whitford, W.G. In press. The effect of increased soil moisture and reduced soil temperature on a desert soil arthropod community. Amer. Mid. Nat.

Nicholas, W.L. 1984. The Biology of Free-Living Nematodes. Oxford Univ. Press, New York. 251 pp.

Noy-Meir, I. 1973. Desert ecosystems: environment and producers. Ann. Rev. Ecol. Syst. 4: 25-52.

Parker, L.W., Santos, P.F., Phillips, J. and Whitford, W.G. 1984. Carbon and nitrogen dynamics during the decomposition of litter and roots of a Chihuahuan desert annual, *Lepidium lasiocarpum*. Ecol. Monogr. 54: 339-360.

Petersen, H. and Luxton, M.A. 1982. A comparative analysis of soil fauna populations and their role in decomposition processes. Oikos 39: 287-388.

Pilato, G. 1979. Correlations between cryptobiosis and other biological characteristics in some soil animals. Boll. Zool. 46: 319-332.

Poinsot-Balaguer, N. 1976. Dynamique des communautés de Collemboles en milieu xérique méditerranéen. Pedobiol. 16: 1-17.

Robinson, A.F., Orr, C.C. and Heintz, C.E. 1984. Some factors affecting survival of desiccation by infective juveniles of *Orrina phyllobia*. J. Nematol. 16: 86-91.

Saeed, M. and Roessner, J. 1984. Anhydrobiosis in five species of plant associated nematodes. J. Nematol. 16: 119-124.

Santos, P.F., Phillips, J. and Whitford, W.G. 1981. The role of mites and nematodes in early stages of buried litter decomposition in a desert. Ecol. 62: 664-660.

Schmidt-Nielsen K., Taylor, C.R. and Shkolnik, A. 1971. Problems of heat, water and food. J. Exp. Biol. 55: 385-398.

Shachak, M., Chapman, E.A., and Orr, Y. 1976. Some aspects of the ecology of the desert snail *Sphincterochila* (*S.*) *zonata* (Bourguignat, 1853) (= S. Boisseri Charpentier, 1847). Argamon. Isr. J. Malacol. 5: 20-46.

Steinberger, Y., Freckman, D.W., Parker, L.W. and Whitford, W.G. 1984. Effects of simulated rainfall and litter quantities on desert soil biota: nematodes and microarthropods. Pedobiol. 26: 267-274.

Townshend, J.L. 1984. Anhydrobiosis in *Pratylenchus penetrans*. J. Nematol. 16: 282-289.

Verhoef, H.A. and Witteveen, J. 1980. Water balance in Collembola and its relation to habitat selection: cuticular water loss and water uptake. J. Insect Physiol. 26: 201-208.

Wallace, H.R. 1959. The movement of eelworms in water films. Ann. Appl. Biol. 47: 366-370.

Wallwork, J.A. 1982. Desert Soil Fauna. Praeger, New York. 296 pp.

Wallwork, J.A., Kamill, B.W. and Whitford, W.G. 1984. Life styles of desert litter-dwelling microarthropods: a reappraisal based on the reproductive behavioral of Cryptostigmatid mites. S. Afr. J. Sci. 80: 163-169.

Whitford, W.G., Freckman, D.W., Elkins, N.Z., Parker, L.W., Parmalee, R., Phillips, J. and Tucker, S. 1981. Diurnal migration and responses to simulated rainfall in desert soil microarthropods and nematodes. Soil Biol. Biochem. 13: 417-425.

Whitford, W.G., Freckman, D.W., Parker, L.W., Schaefer, D., Santos, P.F., and Steinberger, Y. 1983. The contributions of soil fauna to nutrient cycles in desert systems. *In* New Trends in Soil Biology (P. Lebrun, H.M. André, A. de Miedts, C. Gregoire-Wibo and G. Wauthy, eds.), Dieu-Brichart Publishers, Louvain-la-Neuve, pp. 49-61.

Whitford, W.G., Repass, R., Parker, L.W. and Elkins, N. 1982. Effects of initial litter accumulation and climate on litter disappearance in a desert ecosystem. Amer. Mid. Nat. 108: 105-110.

Whitford, W.G., Steinberger, Y., MacKay, W., Parker, L.W., Wallwork, J.A. and Weems, D. 1986. Rainfall and decomposition in the Chihuahuan desert. Oecologia 68: 512-515.

Womersley, C. 1981. Biochemical and physiological aspects of anhydrobiosis. Comp. Biochem. Physiol. 70B: 669-678.

Yancey, P.H., Clark, M.E., Hand, S.C., Bowlus, R.D., and Somero, G.N. 1982. Living with water stress: evolution of osmotic systems. Science 217: 1214-1222.

10 The Physical Properties and Metabolic Status of *Artemia* Cysts at Low Water Contents: The "Water Replacement Hypothesis"

James S. Clegg

Organisms that can lose practically all of their water in a completely reversible way (and there are many such cases; see Keilin, 1959; Crowe and Clegg, 1973, 1978; the present volume) provide some provocative and challenging questions. What are we to think of an organism that loses practically all of its cellular water at ordinary temperatures and pressures and ceases to metabolize but upon being rehydrated resumes all of the characteristics it previously exhibited? In the dried state it cannot easily be considered "alive," since we are told that the characteristics of living organisms are exhibited through the dynamics of their metabolism. On the other hand, we should be reluctant to call such an organism dead (unless we are willing to entertain the possibility of its resurrection), because it resumes an active life upon simple restoration of water. Thus, either one considers life to be a discontinuous process, or the reversibly dried organism presents to us a special level of biological organization, apart from any question of its being alive or dead.

An important phrase in the foregoing bears careful inspection: "ceases to metabolize." Do dried organisms actually stop metabolizing? I have addressed this important question previously and in some detail (Clegg, 1973,

I am grateful to Vivian Roe for speedy and professional manuscript preparation and to my colleagues in nuclear magnetic resonance, quasielastic neutron scattering, microwave dielectric, and differential scanning calorimetry [including C. F. Hazlewood (Baylor College of Medicine), H. E. Rorschach (Rice University), and their students and associates; R. M. Nicklow and N. Wakabayashi (Oak Ridge National Laboratory); E. H. Grant, R. J. Sheppard, S. Szwarnowski, V. E. R. McClean, and their associates (Queen Elizabeth College and King's College, University of London); and C. A. J. Hoeve (Texas A & M University)]. The research on which this chapter reports was supported by grants from the U.S. National Science Foundation (PCM 79-2509 and 82-17311).

1978a). Anticipating what follows in this chapter, I will take the position now that at or below water contents of about 0.1 g H_2O/g dry mass (g/g) cells cannot carry out a metabolism, if by *metabolism* we mean the operation of integrated enzyme-catalyzed chemical reactions that contribute to the maintenance of the cells in which they occur (see Chapter 3). Therefore, we must attempt to know whether or not such dried cells metabolize: if they *do*, however slowly, then much of the special quality is lost; if they do not, then we are dealing with a special sort of biology in which the organism exhibits two sets of properties and characteristics that differ *qualitatively* and in a *fundamentally* different way. This uncertainty of the metabolic status of dried but viable cells that has prompted previous students of the phenomenon to describe it as *cryptobiosis* (Keilin, 1959). Keilin's 1959 publication is an important review of work dating back to the time of Leeuwenhoek (about 1700) and shows clearly that the "metabolic vs. ametabolic" issue has lain at the core of our understanding of dried cells and is fundamental to our assessment of their biological significance, a point also emphasized by Hinton (1968). Because publications continue to appear in which claims are made for the existence of metabolism in severely dried cells (e.g. Chen, 1972, 1978), I will present further evidence here that rules out such a possibility, at least to my satisfaction.

A primary feature of the reversibly dried cell which we wish to understand concerns its ability to survive desiccation and rehydration. What mechanisms underlie this ability? In this paper I will consider what has come to be known as the *water replacement hypothesis* (WRH), a phrase presumably first coined by Crowe (1971). The initial hypothesis, based on the work of Webb (1965) and Warner (1962), was that polyhydroxyl compounds replace the structural water of cellular components as intracellular water levels are reduced, thereby preventing a variety of potentially lethal events from taking place (Clegg, 1974a). Figure 10-1 is a simplified picture of the WRH to which I will often refer. The hypothesis was formulated on the basis that many of the animals capable of reversible drying do, indeed, contain high concentrations of such polyhydroxyl compounds as glycerol, trehalose, and certain sugar alcohols (see Crowe and Clegg, 1973, 1978).

Support for the WRH was obtained from experiments on *Artemia* cysts and on glycerol protein mixtures studied by NMR (Clegg, Seitz, Seitz, and Hazlewood, 1982) and from work on nematodes (Crowe et al., 1978; Madin et al., 1978). But the strongest evidence has been compiled by John and Lois Crowe and their students and associates, who have clearly shown that trehalose does indeed replace the water adjacent to the surfaces of phospholipid vesicles and sarcoplasmic reticulum vesicles and, equally important, that trehalose preserves these systems upon severe drying (see Chapters 11-13).

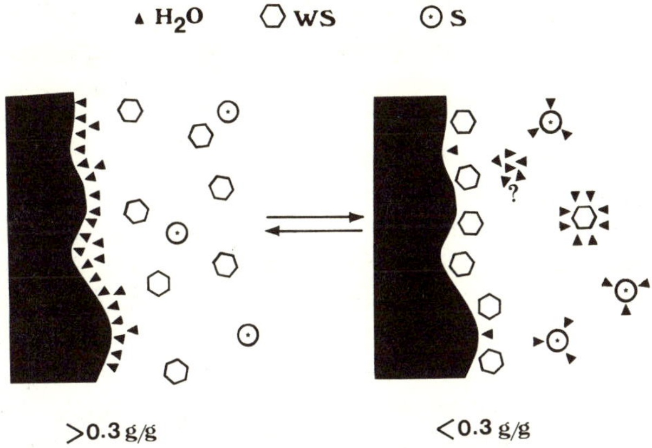

Figure 10-1. Simplified schematic representation of the "Water replacement hypothesis." The dark areas represent intracellular surfaces such as membranes or macromolecules. Filled triangles represent water molecules near these surfaces or hydrating mobile small solutes (S). Hydration water for small solutes, and "bulk-phase water," are considered present in the left diagram but are not depicted by symbols. No bulk-phase water is present in the diagram to the right. The hexagons represent polyhydroxyl solutes considered to be water substitutes (WS) at hydrations below about 0.3 g H_2O/g dry weight (g/g).

Glycerol, which I had previously regarded as a very strong candidate for replacing water (see Clegg, 1974a), is much less effective in these model systems and can even be deleterious.

The real test of the WRH must come from studies on *intact* cells or organisms. In this chapter I will summarize some previous recent work on this matter and will present some new, unpublished findings, which, I propose, can best be interpreted within the WRH framework. All these studies have been carried out on the encysted embryos of *Artemia*, an excellent model for such work.

The Cysts of *Artemia*

Artemia is an anostracon crustacean commonly called the brine shrimp, which inhabits saline lakes and solar salt operations worldwide. It is widely used in commerce, and its biology is reasonably well known (see Bagshaw and Warner, 1979; Persoone et al., 1980). A second international conference on *Artemia* was held in September 1985, and the published proceedings will soon appear.

Sexual varieties of the brine shrimp reproduce either viviparously or oviparously. In the latter case, encysted dormant embryos are released into the environment, which is usually brine. The normal fate of these cysts is desiccation, and they are capable of repeated hydration/desiccation cycles, which are part of their normal biology.

The cysts are available commercially in multikilogram quantities but can also be produced in gram quantities in large-scale laboratory cultures. The dried cysts can be stored for indefinite periods at low temperatures and pressures. They are among the most stable of all eucaryotic cells, surviving a wide variety of environmental extremes, and provide highly useful material for a variety of biological problems.

Several detailed accounts of the techniques used to wash, process, and experimentally manipulate these cysts have been published (Clegg, 1974b, 1978a, 1978b, 1979), which I will not duplicate here. All cysts used in our work are from the San Francisco Bay area. A few comments must, however, be made, and Figure 10-2 will be of use in that regard.

When fully hydrated, the cysts are spheres of diameter 210 ± 10 μm and contain close to 4000 cells in an inner cell mass whose diameter is 202 ± 8 μm. The shell is noncellular and complex, the outer 6 μm of which can be removed by hypochlorite treatment. That leaves behind only a 2 μm layer of

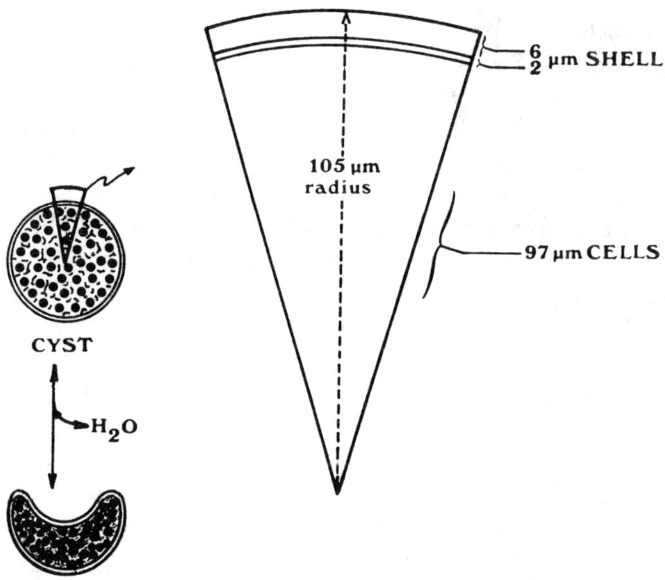

Figure 10-2. Diagram of fully hydrated and dehydrated *Artemia* cysts (lower left). The pie-shaped area gives average dimensions of the cyst and its two major subcompartments (shell and inner cellular mass) for the San Francisco Bay area strain.

a protein-chitin complex called the embryonic cuticle. The inner cell mass has no observable extracellular space.

When fully hydrated, the cysts contain close to 1.4 g/g of water, of which the *cells* contain nearly 1.7 g/g. We have not observed any desiccation treatment (short of lethal high temperatures above 104°C) that kills these cells. The use of gas-bombardment techniques at low temperatures under vacuum reduced the water content to <0.007 g/g without any measurable effect on cyst viability (Clegg et al., 1978). In summary, because of their relatively simple structure, their capacity to undergo repeated hydration-dehydration cycles over an enormous range of water contents, and their commercial availability inexpensively in large amounts, the *Artemia* cysts provide opportunities for experimental designs that are not commonly possible. Next, I summarize some of the results obtained with this system using a variety of techniques that probe the cysts' physical properties. We concentrate on low water contents here, but data have been obtained for the entire hydration range.

The Physical Properties of *Artemia* Cysts at Low Water Contents

Analysis of the changes in physical condition of cysts using six different types of probes are plotted in Figure 10-3. This compilation may appear bewildering at first, but there are advantages to be gained from this display that I will touch upon later. Also, individual data points are omitted for clarity, but in general they show very little scatter about the lines drawn in Figure 10-3. Finally, I will concentrate on interpreting these data within the context of the WRH and on the metabolic status of these cells as a function of water content.

Water Sorption Isotherms

The water sorption isotherms permit the calculation of the differential entropy ($\Delta\bar{S}$) and enthalpy ($\Delta\bar{H}$) changes accompanying the sorption of water from the vapor phase by dry cysts (<0.02 g/g). The data and their interpretation have been discussed at length (Clegg, 1978b), together with the experimental methods. I note here only those features of interest in the present context. First, both the entropy and enthalpy values level off at cyst hydrations above 0.1 g/g after undergoing marked changes at lower water contents. Such changes surely reflect substantial rearrangements in cyst components interacting with incoming water molecules. Negative $\Delta\bar{S}$ values are

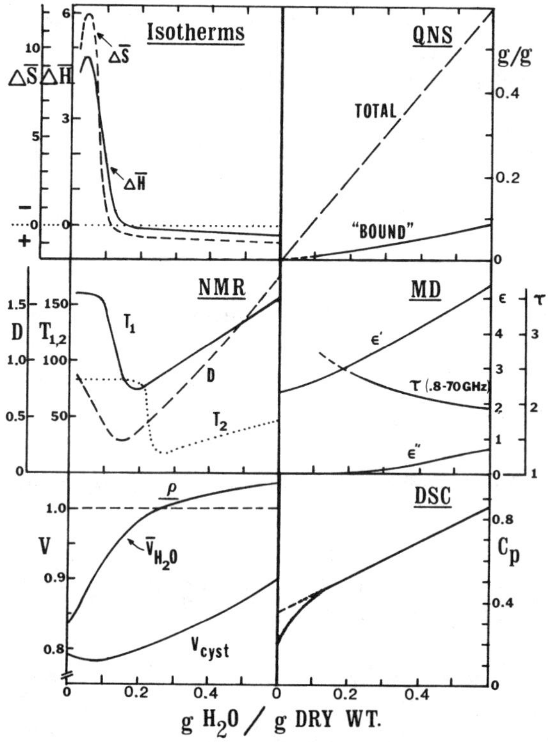

Figure 10-3. Summary of the physical properties of *Artemia* cysts at low water contents. Upper left: Differential values for the entropy (S) and enthalpy (H) changes accompanying the sorption of water vapor. $\Delta\bar{S}$ is in units of cal/mole degree, $\Delta\bar{H}$ is in units of kcal/mole. Middle left: The self-diffusion coefficient (D, in 10^{-6} cm^2/sec) and relaxation times (T_1, T_2, in msec) measured by nuclear magnetic resonance (NMR). Lower left: Cyst-specific volume (V, in cm^3/g) calculated from density (ρ) measurements. $\bar{V}_{H_2O}$ represents estimates for the specific volume of cyst water. The dashed line is the value for pure water. Upper right: The fraction of water considered to be "bound" (see text), compared with the total cyst water, deduced from analysis of quasi-elastic neutron scattering (QNS) spectra. Middle right: Microwave dielectric (MD) measurements of the permittivity (ϵ') and loss (ϵ'') are dimensionless, and τ in units of pico seconds. Lower right: Heat capacity (Cp) in units of cal/g/degree $\cdot$(1 + gH_2O/g dry weight) at 25°C. The dotted line is an extrapolation.

commonly taken to mean that the sorbed water molecules exhibit more order (fewer degrees of freedom) than those present in the bulk liquid (see Clegg, 1978b). Above cyst hydrations of about 0.15 g/g, both quantities change very little and are close to zero, suggesting that the energetics accompanying water sorption from the vapor phase have been completed and that the non-aqueous cyst components have achieved their primary hydration. Interpretations of isotherm data are hazardous (although the very small hysteresis of this system simplifies matters somewhat; Clegg, 1978b).

NMR Spectroscopy: T_1, T_2, and D

The two relaxation times (T_1 and T_2) and the self-diffusion coefficient (D) of water appear in the middle left panel of Figure 10-3. These data (Sietz et al., 1980, 1981), have been discussed at length in the context of the WRH (Clegg, Seitz, Seitz, and Hazlewood, 1982; Fullerton et al., 1983). The results reveal an unusual characteristic: instead of decreasing smoothly to low levels as water is removed, these parameters actually increase. The values of T_1 and T_2 undergo sharp changes, similar to those of a phase transition, at water contents of about 0.25 g/g (T_2) and 0.15 g/g (T_1). It should be recognized that, at such low water contents, these changes need not result *only* from changes in the rotation and translation of water molecules in the cells; however, at present we believe they do (Clegg, Seitz, Seitz, and Hazlewood, 1982). An example of a biological system which does exhibit a *smooth* decrease in T_1 and T_2 with decreasing water content is skeletal muscle (Cooke and Wein, 1973).

We should ask why the water molecules *increase* their motion at low *Artemia* cyst water contents, when precisely the opposite effect is to be expected? Also, what is the reason for their abrupt change in motion between cyst water contents of 0.15-0.25 g/g? We have proposed (Clegg, Seitz, Seitz, and Hazlewood, 1982) that such a result is fully consistent with the WRH. Water molecules that constitute the primary hydration layer of macromolecules (Figure 10-1, left) exhibit much slower motion than water molecules in a bulk aqueous phase. As that bulk aqueous phase is removed from the cells by dehydration, there approaches a point where most of the water exists only as primary hydration and T_1, T_2, and D reach their minima. At about that hydration, however, we suggest that polyhydroxyl compounds (hexagons in Figure 10-1) replace hydration water (black triangles), which is released and moves away from the surface, allowing that water an increase in its motions, thereby increasing T_1, T_2, and D. We have no direct evidence on the location of the "replaced" water but assume that it is involved with the hydration of small solutes (Figure 10-1, circles) rather than generating an aqueous phase (question mark, Figure 10-1). More on this point later.

We attempted to test these ideas by measuring relaxation times for bovine serum albumin—glycerol mixtures hydrated to different water levels (Clegg, Seitz, Seitz, and Hazlewood, 1982). These rather preliminary experiments did reveal abrupt increases in T_1 and T_2 (0.07 g/g) which did not occur in BSA powders without glycerol. At the time these studies were performed

(1979), it had not yet been suggested that trehalose was a much better candidate for the role of "water replacer" than glycerol. Yet glycerol probably plays several important roles in *Artemia* cysts (Clegg, 1978a). It is a remarkable substance, and I would be surprised if nature had not taken advantage of its remarkable properties in anhydrous biological systems.

If our present interpretation of the NMR data is correct, we should observe comparable behavior in cyst water as the critical hydration range is traversed, using other probes. It is gratifying that subsequent research on this system, using neutron scattering and microwave dielectric relaxation, provides some evidence for that expectation.

Quasi-Elastic Neutron Scattering

Using this powerful tool we have described in some detail the behavior of water in *Artemia* cysts at their maximum hydration (Trantham et al., 1984). As part of that study, a few experiments were done at lower water contents, and some of the resulting data are germane to the present account (Figure 10-3, top right). We observed that the amount of water assigned to the "bound" fraction was much less than expected at low water contents. That information was extracted from QNS spectra using a bound-free model in which the bound fraction could be identified as the Gaussian component of the scattering due to water (for full details, see Trantham et al., 1984). As Figure 10-3 shows, the estimated value of the "bound" fraction is far less than the total water present, even at *very* low water contents, certainly *not* an expected result. Table 10-1 shows the data from which the idealized curves were obtained. Thus, for cysts at a total water content of 0.1 g/g, only about 23% appear to exhibit the properties of "bound water." In the simplest case, one might expect 100% of the water to be bound at that hydration. The amount of bound water levels off as higher water contents are achieved and

Table 10-1
Estimation of "bound" water by quasi-elastic neutron scattering from *Artemia* cysts

Total cyst water	Bound water	% bound
0.101 ± 0.002	0.023 ± 0.016	23
0.313 ± 0.009	0.046 ± 0.047	15
0.780 ± 0.014	0.110 ± 0.050	14
1.206 ± 0.027	0.110 ± 0.040	9

Note: Table shows gH_2O/g dry weight ± SD.

remains constant, as expected, at about 0.11 g/g (Table 10-1). These results suggest that the sector of bound water decreases as cyst water content is lowered between about 0.8 and 0.1 g/g.

Microwave Dielectric Relaxation

This analysis also evaluates the motional properties of water. It has been applied to our system over its entire range of water contents (Clegg, Szwarnowski, McClean, Sheppard, and Grant, 1982, 1984). Figure 10-3 (middle right) shows the dielectric permittivity (ϵ') and loss (ϵ''), measured at a frequency of 2 GHz, for cysts as a function of hydration (22°C). The values for these parameters are higher than those given here because these data have not been corrected for the air space ($\epsilon' = 1$, $\epsilon'' = 0$) in the measuring dielectric cell. Unlike purified protein powders undergoing initial hydration, the cysts exhibit *no initial independence* of ϵ' on water content. For protein powders, that independence has been taken as a measure of bound water (i.e. water added to form the primary hydration layer). Because we do not observe such a "lag" in ϵ', we propose that this is a reflection of the paucity of such water in cysts undergoing initial hydration. Similar results have been observed at frequencies in the kilohertz and megahertz range (Clegg, 1978a, 1979). Those papers develop the interpretation in more detail. Of equal importance are the data obtained on the dielectric relaxation time (τ) over the frequency range of 0.8-70 GHz (Clegg et al., 1984). Analysis of these dispersions indicates that, as the water content is decreased, the apparent value of τ *increases*, quite contrary to expectation. Our data are not good enough to tell us precisely the water content at which the increase occurs, and although the experimental errors are very large (40% in τ) the outcome of these studies, on balance, is more consistent with the WRH than not.

In dielectric studies below cyst hydrations of 0.20 g/g, we see *no* significant response of cyst water to frequencies over the range 0.8-70 GHz (Clegg et al., 1984). This result proves that, if these cysts contained any water that exhibits the dielectric properties of water in aqueous solutions, it can be only a tiny fraction of the total amount present. Therefore we suppose that, when cyst water content falls appreciably below 0.3 g/g, few if any "water puddles" exist. Rather, we suggest that the water replaced from macromolecular surfaces during drying is involved with small, more rapidly moving solutes. That result would account nicely for our data (see also Chapter 16).

Density Measurements

All the foregoing studies probe the motional properties of water and, in some cases, other cyst components. We have also examined the mass-volume relationships of cysts as a function of water content (Clegg, 1984). Those data are valuable because they yield, in principle, information about structure. Figure 10-3 (lower left) summarizes a few results on our measurements of cyst density (ρ). The results are expressed as specific volumes ($V = \rho^{-1}$) and are for cysts whose outer shell has been removed. As water is added to dry cysts (≃0.02 g/g), the system exhibits a weak volume *contraction* until a water content of about 0.1 g/g is achieved. Above that level, volume increases smoothly with water content. Note the correspondence between the onset of an increase in volume and the point of stabilized entropy and enthalpy changes associated with water sorption. Our interpretation of this correspondence and of the near independence of cyst volume with water content (up to 0.1 g/g) is that a combination of electrostriction of water and rearrangements in cyst components effectively balances out (or nearly so) and results in the modest change in cyst volume we observe. The partial specific volume of cyst water ($\bar{V}$ H_2O, Figure 10-3) has been estimated by the method of slopes and intercepts (see Clegg, 1984). Below water contents of about 0.25 g/g, $\bar{V}$ H_2O is less than that of pure water (i.e. in density it is greater than the latter). Since our other data, notably isotherms and NMR, indicate that primary hydration events are completed at about 0.3 g/g, it is tempting to suggest that primary hydration water is more dense, on average, than pure water. Above 0.3 g/g we propose that increases in water content simply add to the volume of the bulk aqueous phase(s) in these cells. According to our reasoning, cysts above 0.3 g/g will have a water content whose specific volume is slightly greater (and whose density is less) than that of pure water. Unlike NMR and sorption isotherm data, we see no violent excursions in specific volume (density), only what appears to be a leveling off of this parameter at water contents above 0.3 g/g. It is admittedly difficult to interpret such data with any confidence for complex heterogeneous systems such as *Artemia*; the interpretation we have applied is by no means the only one.

The final studies to be described also bear on nonkinetic characteristics of cysts at low hydration, and they show some interesting parallels to the density measurements.

Differential Scanning Calorimetry

DSC studies have been done in conjunction with C. A. J. Hoeve. They provide direct information on the cyst heat capacity (*Cp*), which is a temperature derivative of the enthalpy (for macroscopic systems):

$$Cp = \frac{dH(T)p}{dT}, \qquad (10\text{-}1)$$

where H is the enthalpy and dT and T the temperature at constant pressure (p). If Cp is studied over some temperature range (T_1-T_2), both the enthalpy and entropy functions can be evaluated (see Privalov, 1980, for a useful account of biological studies of heat capacity). Thus, the DSC method is potentially of great value in our attempt to understand the properties of cysts at various water contents.

If dried cysts are hydrated to different "equilibrium" water contents, X (see Clegg, 1978b), and their Cp measured, then the partial heat capacity of the water (Cp_1) and nonaqueous components (Cp_2) can be calculated, given some assumptions, from the relationship:

$$Cp(1 + X) = \bar{C}p_2 + (\bar{C}p_1 \cdot X). \qquad (10\text{-}2)$$

Thus, a plot of $Cp(1 + X)$ versus X gives $\bar{C}p_2$ as the intercept, and $\bar{C}p_1$ can be obtained from the slope. This plot is shown in Figure 10-3 (lower right) where Cp is in units of $Cp(1 + X)$. First note that the heat capacity of the cyst water system changes linearly until water contents of less than 0.2 g/g are reached. Before interpretations of $\bar{C}p_1$ and $\bar{C}p_2$ are attempted, it should be noticed that these direct measurements indicate water-dependent changes that are also observed in the isotherm, NMR, and volume studies (Figure 10-3). I believe the initial changes in Cp up to about 0.1 g/g reflect primary hydration and structural rearrangements dependent on those events.

Figure 10-4 shows the effect of temperature on heat capacity for cysts over the hydration range of 0-0.18 g/g. Numbers to the right show the calculated partial heat capacity of cyst water over the temperature range of −140 to 24°C (values in the center are absolute temperatures). The insert shows the heat capacity of pure water (solid trace) and $\bar{C}p_1$ for cyst water as a function of temperature. The comparison reveals that cyst water has a greater heat capacity than pure water down to temperatures of about −116°C, at which it joins the curve for hexagonal ice (Ice-1, Eisenberg and Kauzmann, 1969). It appears that no abrupt transition in $\bar{C}p_1$ (cyst water) occurs between 24 and −56°C; thus, at cyst water contents below 0.2 g/g, none of the water in these cells freezes, and we may take this as evidence against the existence of a liquid phase with solvent properties. Because Cp/T is proportional to the entropy, S, we may propose that cyst water exhibits more degrees of freedom than Ice-1, at least down to −56°C. Also, it should be pointed out that the behavior of water in several proteins hydrated to about 0.1-0.2 g/g (Privalov, 1980) shows $\bar{C}p_1$ versus $T(K)$ relationships that are *very* similar to the one we have observed for *Artemia* cysts (insert, Figure 10-4) but *only* below 0°C: the hydration water of the proteins studied reaches that of pure water above 0°C, whereas $\bar{C}p_1$ (for cyst water) is always greater than Cp (pure water) under

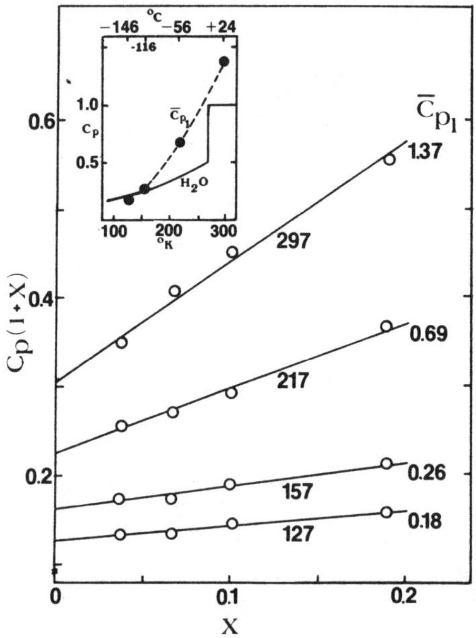

Figure 10-4. Heat capacity (cal/g/degree) as a function of cyst hydration (X, gH_2O/g dry weight) at various temperatures (°K). The partial heat capacities of water ($\bar{C}p_1$) at different temperatures (central numbers) are given to the right of each line. The insert shows the heat capacity of pure water as a function of temperature as well as $\bar{C}p_1$ for the cysts.

these conditions. It is tempting to suggest that the hydration water of cell surfaces (Figure 10-1, left) shows an increase in heat capacity when liberated, as the WRH predicts. That possibility would account for the difference between $\bar{C}p_1$ for cyst water and that for the water of protein hydration when both systems are hydrated to about 0.2 g/g or less. Thus, I deduce that the primary hydration water of proteins, in the *absence* of a suitable water replacement, remains at the protein surface and exhibits a heat capacity similar to that of pure water as observed for several proteins above 0°C (see Privalov, 1980). In contrast, that water when released exhibits more degrees of freedom, with an attendant increase in heat capacity.

Figure 10-4 also shows that the heat capacity of the *nonaqueous* cyst components ($\bar{C}p_2$)—the intercept—also changes with temperature at low water contents (0-0.18 g/g), and Figure 10-5 illustrates the relationship more clearly. These results on *Artemia* cysts are strikingly similar to those on four different proteins and DNA (Privalov, 1980). However, above cyst water contents of 0.2 g/g, the values $\bar{C}p_1$ and $\bar{C}p_2$ do not change with increasing water content (Figure 10-3, lower right). That result is in good agreement with the results from virtually all of the physical measurements we have carried out on these cysts: once the nonaqueous components achieve primary hydration, their properties and structure appear to be independent of further addition of water which, we deduce, simply adds to the volume of the bulk

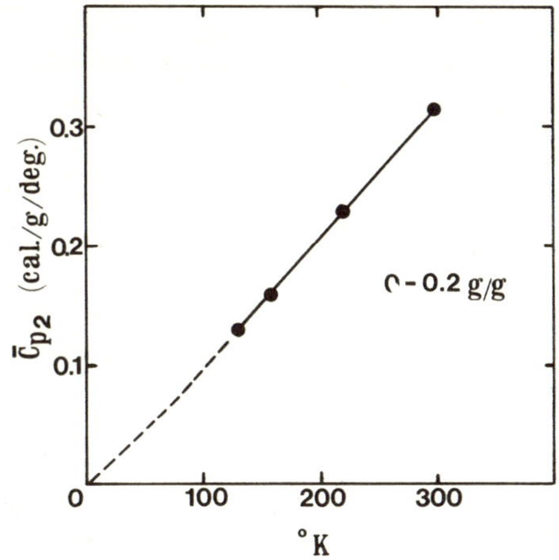

Figure 10-5. The partial heat capacity of nonaqueous cyst components ($\bar{C}p_2$) as a function of temperature (°K) over the hydration range given (g/g). Values of $\bar{C}p_2$ were obtained from intercepts (Figure 10-4).

aqueous phases in these cells. Those phases are, however, not simple aqueous solutions (Clegg, 1979).

In concluding this section of the chapter, a few comments on the expected viscosity of cysts at low water contents are in order. Although no measurements have been made of this property, some firm predictions can be made. At 0.2 g/g, the intracellular viscosity must be enormous. Consider only one cellular component, glycerol, present in the fully hydrated cyst at a concentration of about 10 weight percent with respect to water (Clegg, 1967). At 0.1 g/g, a reasonable guess would be an increase to at least 85 or 90 weight percent. The viscosity of such a solution is almost 1000 times that of water (Newman, 1968). This viscosity must retard the diffusional motion of all solutes in the cells, and the extent to which it does so is illustrated from the work of Mastro and Keith (1984). They measured the behavior of tempone, a spin label, in glycerol solutions by electron spin resonance spectroscopy. Both rotational and translational motion of the spin label were reduced 30-fold in 85 weight percent glycerol compared with pure water. We can demand that *all* metabolites in cysts at 0.2 g/g will experience a much greater retardation, since only glycerol has been considered: such cells are incredibly crowded, and we can anticipate profound effects on metabolism as a result.

The Metabolic Status of Cysts as a Function of Water Content

Figure 10-6 summarizes about 15 years of research on the metabolic dependence of these cells on cyst water content. We have no evidence for the

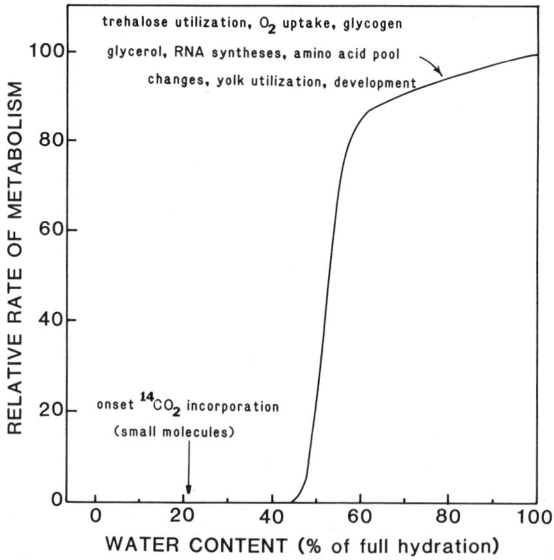

Figure 10-6. Summary of the hydration dependence of metabolic rates in *Artemia* cysts (see text). As the water content is lowered (toward the left), metabolic and synthetic processes are terminated. Initial water content (full hydration) was 1.4 g/g.

occurrence of enzymatic reactions below water contents of 0.3 g/g. Beginning at about that hydration, $^{14}CO_2$ is incorporated into a variety of small metabolites, and we have good evidence that those reactions are, indeed, metabolic (Clegg, 1978a). Further addition of water (to about 0.65 g/g) results in the initiation of a metabolism characteristic of fully hydrated cells; however, the *rates* of this metabolism depend strongly on water content until the cysts have achieved a hydration of close to 0.85 g/g, above which metabolic rate depends only weakly upon water content. These changes have been incorporated into a model (Clegg, 1978a, 1979) which describes in some detail the relationships we observe in Figure 10-6. Although I have made the case that these results on cyst metabolism have broad applications (see Clegg, 1979) and that the model is held to be valid for animal cells in general (Clegg, 1981; Mansell and Clegg, 1983), in the present chapter I concentrate only on metabolism at lower water contents (<0.1 g/g).

Can Biochemical Studies on Dried Cells Conclusively Show the Absence of Metabolism?

At the start of this chapter I promised I would defend the view that, below about 0.1 g/g water content, metabolism (*by any reasonable definition*) cannot occur. The question is: can that assertion be proven by experiment? I believe the answer is no: one can always object that "metabolism occurs but too slowly for your measurement to detect it." It does indeed seem difficult to prove the absence of metabolism by experiments that, themselves, attempt to sample it directly. Thus, if we examine available data on *Artemia*, we find no evidence for metabolic activity of any sort below water contents of 0.3 g/g. But lack of evidence does not disprove the presence of a metabolism even if we measure for periods of months or even years. Therefore, we arrive at the conclusion that the answer to our question must be no. If no other way to explore the metabolic status of dried cells existed, we would remain unable to describe their metabolic status and, therefore, to assess their biological significance. This uncertainty earlier workers had no way to avoid. In my opinion, there is indeed a solution, which I consider next.

Can Physical Studies of Dried Cells Conclusively Show the Absence of Metabolism?

Let us take as a starting point what we have learned about the properties of dried *Artemia* cysts and consider a cyst at 0.1 g/g. Figure 10-3 tells us some relevant facts. The heat capacity measurements tell us that there is no bulk aqueous phase with anything like the properties that such a phase would exhibit. The microwave dielectric and density measurements reinforce this conclusion. Below 0.1 g/g, the cyst components do not respond to applied electric fields over the frequency range 10 kHz to 70 GHz (Clegg, 1978a; Clegg et al., 1984). Thus, even metabolites, let alone enzymes, are essentially "frozen" at these water contents. The evidence on intracellular viscosity clearly accords with that interpretation. In view of our knowledge of what characterizes metabolism, as defined, it would seem impossible for such cells to do anything of the sort.

I can anticipate an objection to this interpretation: "How about the possibility that *localized pools* of an aqueous phase within the cells allow metabolism but are too small to be detected with physical probes?" This is an extension of the "objection to biochemical experiments argument." But we are not at the mercy of such reasoning when dealing with physical properties. Previously I have shown that an *Artemia* cyst at 0.1 g/g does not have sufficient water to hydrate all of its ribosomes, let alone all the other components that would require primary hydration in order to function. The work of Rupley and Careri (Chapters 15 and 16) and their associates on lysozyme hydration is informative in this regard: below 0.2 g/g, no enzyme activity is observed, and other examples of this sort on the hydration dependence of enzyme function can be given (see Clegg, 1973, 1978a). At 0.1 g/g, there simply is not enough water to go around, and what little water is present would be avidly sequestered by the binding sites of highest energy. This brings us to the possibility of localized "pools" of water. To defend their existence is to say that the water-water hydrogen bond is stronger than, for example, water-ion interactions. In other words, that liquid water would maintain its structure in the face of competition from other water binding sites, all under the conditions of severe water shortage. To me, at least, the "pool" reasoning is implausible, and I conclude that the answer is that they are prohibited from existing by simple physics.

An important point (Clegg, 1973, 1978a) must be made again here. A cyst at 0.1 g/g *will* have its chemistry, as will everything else at the pressure and temperature we consider here. But unless that chemistry consists of components of metabolism, then it is simply adventitious chemistry and physics. As I noted elsewhere (Clegg, 1973, 1978a), if one asserts that adventitious chemistry is metabolic, then one must also say that a rusting bridge, an Egyptian mummy, and potato chips must also metabolize. In that case, the cosmos "metabolizes"—an interesting notion but without much meaning in this context. One is well advised to realize that iron shavings consume molecular oxygen quite nicely. Only if it can be shown that oxygen uptake is coupled to a functioning electron transfer chain (or some other metabolic sequence) do these results have *metabolic* meaning.

How about the existence of a metabolism in these cells that is non-aqueous? One in which water is trivial, its functions being taken over by the water substitutes of the WRH? That truly interesting possibility has occurred to me. Yet there is no reason to believe it exists and some reason to discount it. The conditions known to *increase* viability of dried organisms (at about 0.1 g/g) are exactly those that suppress the metabolism of cells at ordinary water contents, notably molecular oxygen removal (anoxia) and low temperatures. One should also keep in mind a major biological reason for the existence of anhydrobiosis: a device serving to *conserve* reserves and *stabilize* the organism under adverse conditions until adequate water becomes available. It is worth pointing out that the longevity of the "dried cryptobiote" can be several hundreds of years (see Crowe and Clegg, 1973; Priestley and Posthumus, 1982).

Conclusions

I believe there is good reason to accept that the WRH is valid for *Artemia* cysts. It appears that the most likely candidate for the water substitute is trehalose, present in these cysts at about 15% of the dry weight (see Chapters

11-13). However, the participation of glycerol in cyst anhydrobiosis remains to be evaluated (Clegg, 1974a). I would not suggest that the trehalose system is the only mechanism involved in desiccation tolerance. On the contrary, a number of processes must apparently be involved in the adaptation of cells to reversible dehydration (see Chapter 1). What seems to be of inescapable importance is the need to avoid the damaging reactions that occur when structures are forced into contact as water coverage becomes inadequate or no longer exists. The WRH accounts for the way in which cells avoid such "proximity effects." At the same time, the removal of water eventually brings metabolism to a standstill. The water substitute(s) seems to play an inhibitory role here: while it preserves the structure, it *prevents the function*, an equally important role in stabilizing the dried cell, in my judgment, and one that has not received much attention previously.

I have gone to some length to explain why I think cells having water contents of less than 0.1 g/g are truly ametabolic and why that property is important to the dried cell. Having done so, we can consider such dried cells to be a unique biological state: their study can tell us a great deal about what we mean by the terms *life and living*, *death and dying*. To paraphrase Albert Szent-Györgyi (1971): biology has forgotten the dried cell—or has never discovered it.

REFERENCES

Bagshaw, J.C. and Warner, A.H. (eds.) 1979. Biochemistry of *Artemia* Development. University Microfilms International, Ann Arbor, MI, pp. 1-238.

Chen, S.S.C. 1972. Metabolic activities of dormant seeds during dry storage. Naturwissenschaften 59: 123.

Chen, S.S.C. 1978. Application of a volatile radioactive tracer to detect metabolic activities in dry wild oat seeds. *In* Dry Biological Systems (J.H. Crowe and J.S. Clegg, eds.), Academic Press, New York, pp. 175-184.

Clegg, J.S. 1967. Metabolic studies of cryptobiosis in encysted embryos of *Artemia salinov*. Comp. Biochem. Physiol. 20: 801-809.

Clegg, J.S. 1973. Do dried cryptobiotes have a metabolism? *In* Anhydrobiosis (J.H. Crowe and J.S. Clegg, eds.), Dowden, Hutchinson and Ross, Stroudsburg, PA, pp. 141-147.

Clegg, J.S. 1974a. Biochemical adaptations associated with the embryonic dormancy of *Artemia salina*. Trans. Amer. Micros. Soc. 93(4): 481-490.

Clegg, J.S. 1974b. Interrelationships between water and metabolism in *Artemia* cysts. I. Hydration-dehydration from liquid and vapor phases. J. Exp. Biol. 61: 291-308.

Clegg, J.S. 1978a. Hydration-dependent metabolic transitions and the state of water in *Artemia* cysts. *In* Dry Biological Systems (J.H. Crowe and J.S. Clegg, eds.), Academic Press, New York, pp. 117-154.

Clegg, J.S. 1978b. Interrelationships between water and cell metabolism in *Artemia* cysts. VIII. Sorption isotherms and derived thermodynamic quantities. J. Cell Physiol. 94: 123-138.
Clegg, J.S. 1979. Metabolism and the intracellular environment: the vicinal water network model. *In* Cell-Associated Water (W. Drost-Hansen and J.S. Clegg, eds.), Academic Press, New York, pp. 363-413.
Clegg, J.S. 1981. Interrelationships between water and cell metabolism in *Artemia* cysts. IX. Evidence for the organization of soluble enzymes. Cold Spring Harbor Symp. 46: 23-37.
Clegg, J.S. 1984. Interrelationships between water and cell metabolism in *Artemia* cysts. XI. Density measurements. Cell Biophysics 6: 153-169.
Clegg, J.S., McClean, V.E.R., Szwarnowski, S. and Sheppard, R.J. 1984. Microwave dielectric measurements on *Artemia* cysts at variable water content. Phys. Med. Biol. 29: 1409-1419.
Clegg, J.S., Seitz, P., Seitz, W. and Hazlewood, C.F. 1982. Cellular responses to extreme water loss: the water replacement hypothesis. Cryobiology 19: 306-316.
Clegg, J.S., Szwarnowski, S., McClean, V.E.R., Sheppard, R.J. and Grant, E.H. 1982. Interrelationships between water and cell metabolism in *Artemia* cysts. X. Microwave dielectrics. Biochim. Biophys. Acta 721: 458-468.
Clegg, J.S., Zettlemoyer, A.C. and Hsing, H.H. 1978. On the residual water content of dried but viable cells. Experientia 34: 734-735.
Cooke, R. and Wein, R. 1973. Nuclear magnetic resonance studies of intracellular water protons. Ann. N.Y. Acad. Sci. 204: 197-209.
Crowe, J.H. 1971. Anhydrobiosis: an unsolved problem. Amer. Naturalist 105: 563-573.
Crowe, J.H. and Clegg, J.S. (eds.) 1973. Anhydrobiosis. Dowden, Hutchinson and Ross, Stroudsburg, PA, 477 pp.
Crowe, J.H. and Clegg, J.S. (eds.) 1978. Dry Biological Systems, Academic Press, New York, 357 pp.
Eisenberg, D. and Kauzmann, W. 1969. The Structure and Properties of Water. Oxford University Press, New York, 99 pp.
Fullerton, G.D., Seitz, P.K. and Hazlewood, C.F. 1983. Application of the fast proton diffusion model to evaluation of water in *Artemia* cysts. Physiol. Chem. Phys. Med. NMR. 15: 489-499.
Hinton, H.E. 1968. Reversible suspension of metabolism and the origin of life. Proc. Roy. Soc. Lond. B. 171: 43-56.
Keilin, D. 1959. The problem of anabiosis or latent life: history and current concept. Proc. Roy. Soc. Lond. B. 150: 149-191.
Maden, K.A.C., Crowe, J.H. and Loomis, S.H. 1978. Metabolic transitions in a nematode during induction of and recovery from anhydrobiosis. *In* Dry Biological Systems (J.H. Crowe and J.S. Clegg, eds.), Academic Press, New York, pp. 155-174.
Mansell, J.L. and Clegg, J.S. 1983. Cellular and molecular consequences of reduced cell water content. Cryobiology 20: 591-612.
Mastro, A. and Keith, A.D. 1984. Diffusion in the aqueous compartment. J. Cell Biol. 99: 180s-187s.
Newman, A.A. 1968. Glycerol. Chemical Rubber Co., Cleveland, OH.
Persoone, G., Sorgeloos, P., Roels, O. and Jaspers, E. (eds.) 1980. *In* The Brine Shrimp *Artemia*, Vols. 1-3, Universa Press, Wettern, Belgium, 1024 pp.

Priestley, D.A. and Posthumus, M.A. 1982. Extreme longevity of lotus seeds from Pulantien. Nature, London, 299: 148-149.

Privalov, P.L. 1980. Heat capacity studies in biology. *In* Biological Microcalorimetry (A.E. Beezer, ed.), Academic Press, New York, pp. 413-451.

Seitz, P.K., Chang, D.C., Hazlewood, C.F., Rorschach, H.E. and Clegg, J.S. 1981. The self-diffusion of water in *Artemia* cysts. Arch. Biochem. Biophys. 210: 517-524.

Seitz, P.K., Hazlewood, C.F. and Clegg, J.S. 1980. Proton magnetic resonance studies on the physical state of water in *Artemia* cysts. *In* The Brine Shrimp *Artemia* (G. Persoone, P. Sorgeloos, O. Roels and E. Jaspers, eds.), Vol. 2, Universa Press, Wettern, Belgium, pp. 545-555.

Szent-Györgyi, A. 1971. Biology and the pathology of water. Persp. Biol. Med. 14: 239-249.

Trantham, E.C., Rorschach, H.E., Clegg, J.S., Hazlewood, C.F., Nicklow, R.M. and Wakabayashi, N. 1984. Diffusive properties of water in *Artemia* cysts as determined from quasi-elastic neutron scattering spectra. Biophys. J. 45: 927-938.

Warner, D.T. 1962. Some possible relationships of carbohydrates and other biological compounds with water structure at 37°C. Nature, London, 196: 1055-1058.

Webb, S.J. 1965. Bound Water in Biological Integrity. Charles C. Thomas, Springfield, IL, 215 pp.

11 Stabilization of Membranes in Anhydrobiotic Organisms

John H. Crowe
Lois M. Crowe

Despite the importance of anhydrobiotic organisms in everyday life and the importance of understanding the role of water in living systems (cf. Chapter 10), we understand surprisingly little of the mechanisms by which some organisms are capable of surviving dehydration. We are trying to elucidate some of these mechanisms and wish to describe in this review our progress over the past several years. We believe that studies in this area not only are of interest to the basic scientist but may also lead to improved methods for preserving biological materials.

We decided several years ago that, in view of the complexity of desiccation tolerance, rapid progress was not likely to be made towards understanding mechanisms of survival by studying intact organisms or even intact cells. Thus, we have adopted a decidedly reductionistic approach to the problem and have been studying effects of dehydration on subcellular components and molecules, particularly biological membranes, and the ways in which membranes might be stabilized in anhydrobiotic organisms. While we have made considerable progress, we recognize that much of what we have accomplished may be simplistic and that ultimately it will be necessary to reexamine in intact cells and organisms some of the ideas presented here.

Dehydration and Membranes

Dehydration of a biological membrane usually results in such massive upheavals that the structural and functional integrity of the membrane are irreversibly lost. While most scientists would accept this generalization, it is not necessarily clear why dehydration disrupts membrane structure. On the basis of our knowledge about the influence of water on membranes, one might make some predictions: a major factor influencing the associations of amphiphiles like membrane phospholipids in aqueous solution is the low solubility of the hydrocarbon chains in water—the well-known hydrophobic effect (Tanford, 1980). It follows that one might expect phospholipid bilayers to fall apart in the absence of water. However, numerous studies (e.g. Luzzati, 1968) indicate that many dry phospholipids remain in bilayers, although the physical properties of the lipids are altered by the dehydration. Thus, as bilayer structure may not be profoundly altered by dehydration, we are forced to look elsewhere for sources of damage due to dehydration.

We have adopted as a model system Ca-transporting vesicles isolated from lobster muscle. These membranes were chosen because they are relatively simple both structurally and functionally, possessing a Ca-ATPase (seen as intramembrane particles in freeze fracture; Figure 11-1a) restricted primarily to one fracture face of the membrane, and a physiological function which can be easily measured: Ca transport. Thus, the vesicles can be sequentially dried and rehydrated, and the morphologic consequences (as seen with freeze fracture) and physiology (studied by Ca transport) can be recorded.

When vesicles similar to those shown in Figure 11-1a were dehydrated, one consequence was seen immediately—the vesicles fused to form increasingly larger structures (Figure 11-1b). Dehydration-induced fusion is not unexpected, having been shown by other techniques in numerous membrane systems (e.g. Chapter 19, Appendix A). Since the literature had repeatedly suggested that fusion between biological membranes is accompanied by formation of hexagonal phase lipids near the point of fusion (Cullis and DeKruijff, 1979), we looked for hexagonal phase lipids in the dehydrating vesicles. Our evidence indicates their presence on the basis of freeze fracture (Crowe, L., and Crowe, 1982) and ^{31}P-nuclear magnetic resonance (Crowe, J., et al., 1983a, 1984b). Such data will be discussed in Chapter 12, but for the present suffice it to say that we initially accepted the suggestion from the literature (Cullis and DeKruijff, 1979) that these hexagonal-phase lipids were associated with fusion. We now believe that not to be the case, for the following reasons.

As membranes undergo dehydration, extensive lateral-phase separations of membrane phospholipids, cholesterol, and intramembrane particles are readily seen with freeze-fracture EM (Figure 11-1b). While the best evidence we have for these lateral phase separations is based on freeze fracture (Crowe, J., et al., 1983b), this is a subject of intense study at the moment using other methods. Since some phospholipids in pure form, most notably

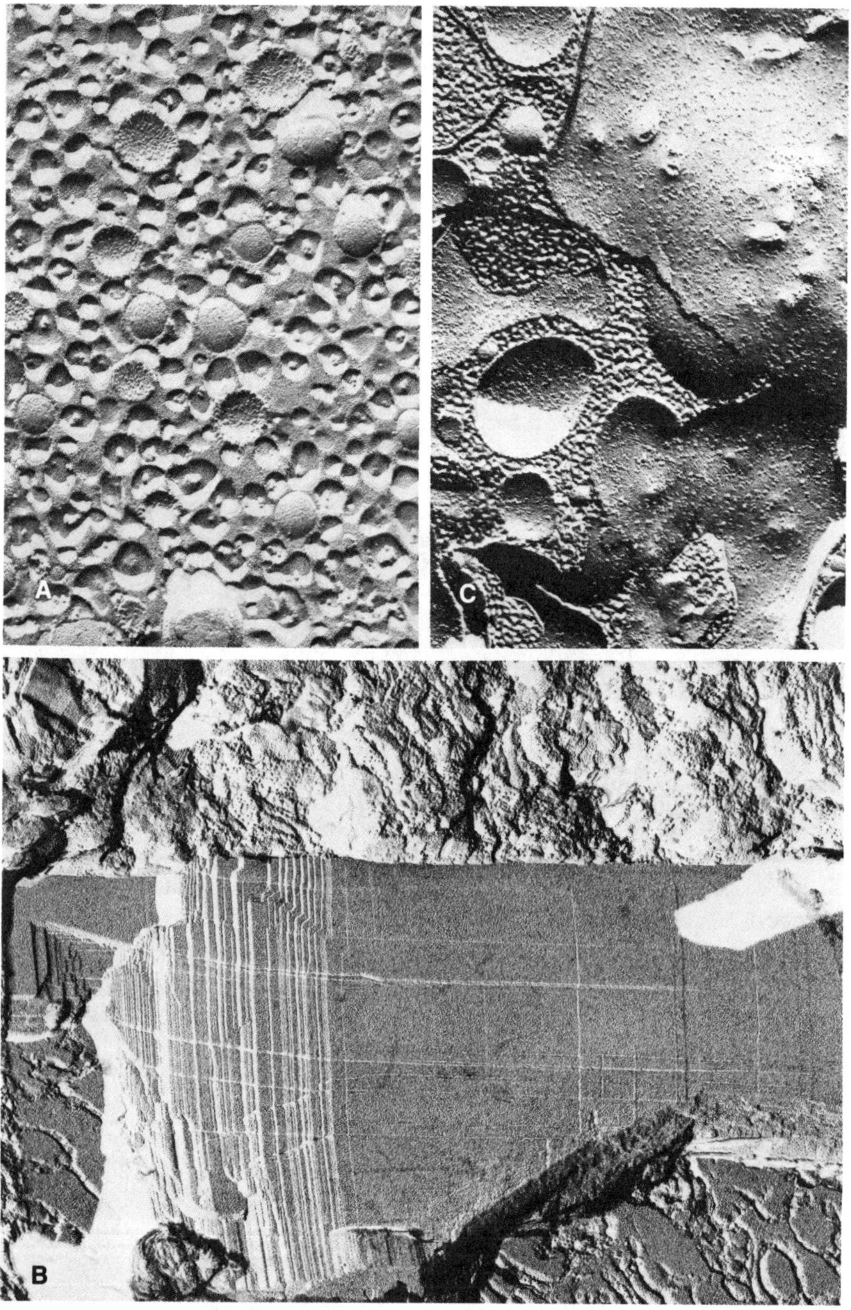

Figure 11-1. (a) Freeze-fracture image of Ca-transporting vesicles isolated from lobster muscle. Note the presence of intramembrane particles (IMPs), which represent the Ca-ATPase. (b) Ca-transporting vesicles that have been dehydrated. A large cholesterol crystal occupies the center of the field, phospholipid devoid of IMPs is at the lower right, and at the top of the field, phospholipid patches are phase separated from particle-rich areas. (c) Ca-transporting vesicles that had been previously dehydrated. Note the size of the vesicles in relation to panel (a), the distribution of IMPs on both fracture faces of the membrane, and the obvious aggregation of the IMPs.

phosphatidylethanolamine (PE), prefer not to exist in bilayers but instead form an inverted hexagonal phase above the phase transition temperature (e.g. Luzzati, 1968; Cullis and DeKruijff, 1979; Boni and Hui, 1983), we now suspect that the hexagonal phase lipids seen in freeze fracture represent phase-separated PE. Studies with ^{31}P-NMR support that viewpoint (Crowe, J., and Crowe, 1984). Upon rehydration, the vesicles which have experienced extensive phase separation show evidence of its damaging effects: intramembrane particles which are normally restricted to one face of the membrane before dehydration are found in equal numbers on both faces of the membrane. Moreover, the rehydrated vesicles show aggregations of particles (Figure 11-1c). In addition to the obvious morphological damage, these vesicles possess no ability to accumulate Ca^{++}. We conclude that the lateral-phase separations seen during dehydration are a key damaging event. The probable underlying causes will be discussed next.

Hydration State and Phase Behavior of Membrane Phospholipids

Polar head groups of phospholipids are hydrated to some extent, and the head groups are separated from each other by a ribbon of hydrogen-bonded water centered around the phosphate (e.g. Hauser et al., 1981). Recent computer modeling efforts of Gaber and his colleagues, discussed in Chapter 13, and other experimental work (Fringeli and Gunthard, 1976; Powers and Pershan, 1977, 1981; Wieslander et al., 1978; Schneider et al., 1979; Hauser et al., 1981; Volke et al., 1982; Smaby et al., 1983) have provided the best available evidence. While the arrangement of the water molecules is not well understood, it is known that water affects the physical properties of the phospholipids profoundly. Phosphatidylcholine has 8 to 12 water molecules hydrogen bonded around the head group (Chapman et al., 1967). When that water is removed, the physical properties of the phospholipids are altered; Chapman et al. (1967) found that, when dipalmitoylphosphatidylcholine (DPPC), which has a transition temperature (T_m) of about 41°C, is dehydrated, T_m rises 30-50°C, depending on the final water content. When water is added to the dry preparations, T_m falls until 8 to 12 water molecules/DPPC are added, after which the T_m reaches a constant value. The explanation for this finding, which we believe has profound importance to our understanding of dehydration damage to membranes, is as follows. One might expect the lateral spacing of the polar head groups to decrease with removal of water, thus increasing opportunities for van der Waals' interactions among the hydrocarbon chains. As a result, the dry phospholipids would be expected to enter gel phase at temperatures at which they would normally be in liquid crystalline phase. This phase transition is likely to play a key role in lateral-phase separations, as described in the following section.

Hypothesis for Dehydration-Induced Lateral Phase Separations

The existence of temperature-induced lateral-phase separations in membranes is well established (e.g. Wunderlich et al., 1978; Quinn, 1985), and at least qualitative models for the physical principles leading to such phase separations have been developed by Chapman et al. (1978) and have more recently been presented in a general way by Quinn (1985). The essence of these models is as follows. Membranes are heterogeneous, consisting of phospholipids, cholesterol, sphingolipids, proteins, and so forth. As a result of this heterogeneity, biological membranes often contain constituents with widely differing transition temperatures (T_m). Thus, as the membrane is cooled, the lipids with higher T_m pass from liquid crystalline to gel phase. This crystallization excludes the more fluid lipids from gel-phase domains, resulting in phase separation. With further cooling, T_m for other lipid classes is reached, with a repetition of the phase transition and exclusion of more fluid lipids from gel-phase domains. A parallel model for exclusion of proteins from lipid domains during cooling has been developed by Chapman et al. (1978).

Examination of phase properties of simple mixtures of phospholipids suggests that a similar phase separation is likely to occur during dehydration because of heterogeneity in T_m for the constituent phospholipids. We have very little information available concerning phase behavior of mixtures of phospholipids, but assuming that individual species in the mixture show behavior similar to that of pure forms, some predictions can be made about the consequences of dehydration. Figure 11-2 is a phase diagram for two commonly occurring phospholipids. In excess water and at temperatures >41°C, both lipids are in lamellar, liquid crystalline phase and form a heterogeneous bilayer mixture. But suppose the mixture is dehydrated. At a water content of six water molecules/phospholipid, one member of the pair (DPPC) will have undergone a transition to lamellar, gel phase, while the other remains in liquid crystalline phase. By analogy with cooling-induced phase separation, crystallization of the DPPC would be expected to exclude the more fluid PE. With further reductions in water content, the lamellar-to-hexagonal transition temperature falls, and the phase-separated PE would undergo a further transition to hexagonal phase. We suspect that phase sepa-

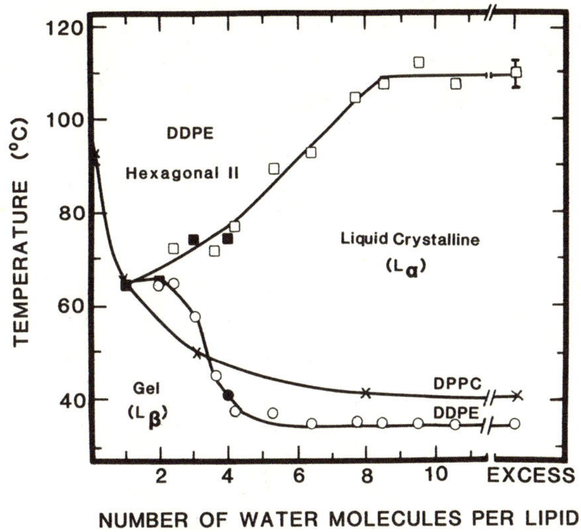

Figure 11-2. Hydration-dependent phase diagrams for dipalmitoylphosphatidylcholine and didodecylphosphatidylethanol- amine. DPPC data are from Chapman et al. (1967), and PE data are from Seddon et al. (1983).

rations may occur in intact membranes undergoing dehydration (see Figure 11-3). Therefore, phase behavior of membrane constituents during dehydration may play a key role in stabilizing membranes in anhydrobiotic organisms. With this background in mind, the time has come for us to consider the mechanisms by which stabilization might be accomplished.

Comparative Biochemistry of Anhydrobiotic Organisms

During the 1970s we studied in some detail the induction of anhydrobiosis in a soil-dwelling nematode (e.g. Madin and Crowe, 1975; Madin et al., 1985; Loomis et al., 1980). In order for these animals to survive dehydration, they must initially be dried slowly, during which time changes in their chemical composition occur, the most important of which appear to be synthesis of large quantities of trehalose and glycerol. Trehalose is an unusual molecule in that it is a nonreducing disaccharide of glucose, the monomers of which are joined in an α-1,1-glycosidic linkage (Figure 11-4). It is, in fact, the *only* nonreducing disaccharide of glucose—an important feature, as we shall see. In the anhydrobiotic nematodes we have studied, the dry weight may be as much as 20% trehalose and 5% glycerol. Survival of the animals in the dry state is correlated with synthesis of these compounds during induction of

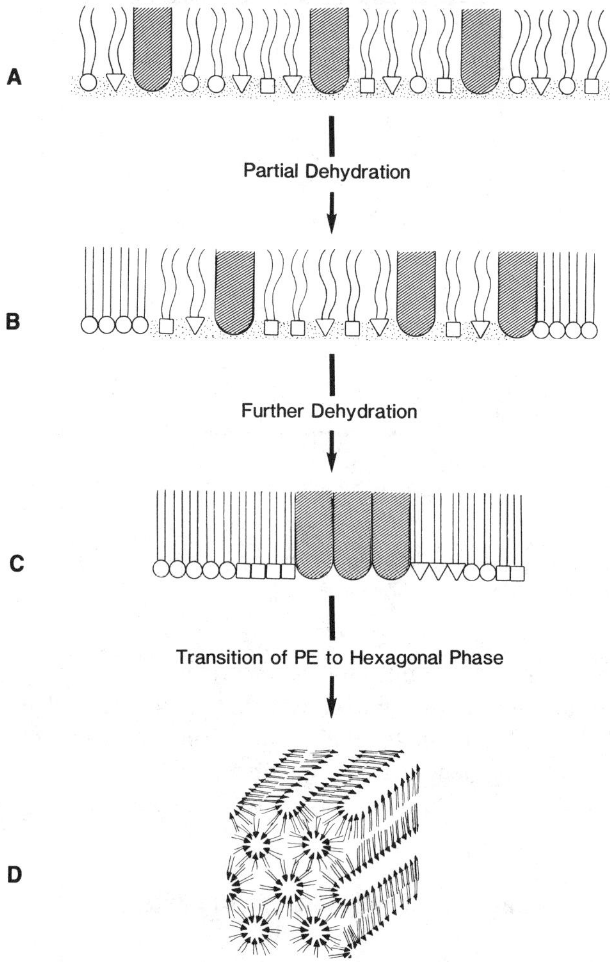

Figure 11-3. Diagrammatic summary of possible events leading to hydration-dependent lateral phase separations of components of a biological membrane. When the phospholipids in bulk water (a) are dehydrated, some components of the mixture might be expected to undergo a liquid-crystalline-to-gel-phase transition before others (according to the phase diagram). These gel-phase domains would exclude more liquid phospholipids, proteins, cholesterol, and so forth (b). With further dehydration, other phospholipids would enter gel phase, with another round of exclusion of other components from the gel-phase domains (c). As dehydration progresses, some phospholipids that prefer in pure form to be in hexagonal phase would undergo transition from lamellar to hexagonal phase (d).

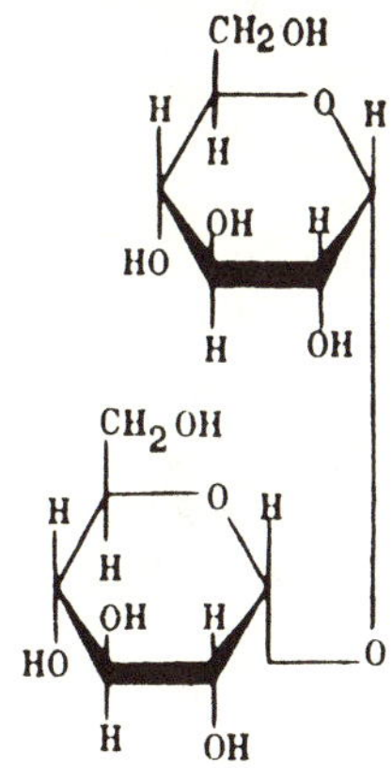

Figure 11-4. The chemical structure of trehalose.

anhydrobiosis (Madin and Crowe, 1975), and the loss of the ability to survive dehydration is correlated with degradation of trehalose and glycerol during rehydration (Crowe, J., et al., 1977). Regulation of synthesis of these molecules during induction of anhydrobiosis in nematodes is now reasonably well understood (Loomis et al., 1980; Madin et al., 1985).

Since trehalose and glycerol are synthesized in such large quantities in the nematodes we initially studied, we suspected that these molecules might be important in the mechanism of stabilizing the dry organisms. Because of our prejudices from reading the cryobiology literature, in which glycerol has been shown to be a potent cryoprotectant, we initially assumed that this might be the more important of the two molecules in stabilizing the dry organisms (Madin and Crowe, 1975). However, comparative biochemistry studies on a wide variety of anhydrobiotic organisms have not supported this hypothesis (Womersley, 1981). Trehalose is commonly found at high concentrations in a surprising array of anhydrobiotic organisms, including spores of certain fungi, macrocysts of the slime mold *Dictyostelium*, dry active baker's yeast, cysts of the brine shrimp, *Artemia*, and the desert resurrection plant (see Crowe, L., et al., 1984, for references). While glycerol has been shown to be present in some anhydrobiotic organisms, it is missing in many others (Womersley, 1981), suggesting that its role (if any) in anhydrobiosis is secondary to that of trehalose. We hasten to add, however, that other polyalcohols are commonly present that could take the place of glycerol.

Although acquisition of comparative biochemistry data is still far from complete, it may be argued, we conclude, that trehalose is somehow involved in stabilizing labile structures such as membranes in many anhydrobiotic organisms, and we have proceeded to examine that hypothesis.

Trehalose Stabilizes Dry Biological Membranes

Inhibition of fusion by trehalose. When the Ca-transporting model membranes described earlier are dried in the presence of trehalose at concentrations comparable to those found in anhydrobiotic organisms, the dry vesicles are seen with freeze fracture to exist in a trehalose matrix as collapsed cups (Figure 11-5a). No evidence of fusion between vesicles is seen during drying in the presence of trehalose, a finding that we have been able to quantify by measurements of freeze fracture images of the rehydrated vesicles (Figure 11-5b). More recently, we have studied inhibition of fusion during freezing (Rudolph and Crowe, 1985) and dehydration of pure phospholipid vesicles, using resonance energy transfer to monitor fusion. Those studies, to be discussed in Chapter 12, show that trehalose is capable of completely inhibiting fusion between phospholipid vesicles. It is also interesting that a trehalose-derivative cord factor has been shown to inhibit fusion between lysosomes and phagosomes containing intracellular parasitic bacteria (Goren et al., 1976). This inhibition of fusion may be responsible for the longevity of the bacteria in phagosomes.

Inhibition of lateral phase separations. In addition to the absence of fusion between Ca-transporting vesicles in the presence of trehalose, we have found no indications of the lateral phase separations seen in the absence of trehalose; for example, the cholesterol crystals and complex nonbilayer phases seen in the absence of trehalose are missing in vesicles dried in the presence of trehalose (Crowe, J., et al., 1983b). Furthermore, ^{31}P-nuclear magnetic resonance suggests the presence of a significant proportion of hexagonal-phase lipids in the vesicles dried without trehalose (Crowe, J., and Crowe, 1984). Similar studies of the vesicles dried in the presence of trehalose suggest that the hexagonal phase is missing. We are aware of the controversy about applying ^{31}P-NMR to study hexagonal-phase lipids (Thayer and Kohler, 1981; see also Appendix B) but offer this only as evidence that the bilayer structure of dry Ca-transporting vesicles is not seriously disrupted when they are dried in the presence of trehalose.

Structure and function in rehydrated vesicles. Vesicles dried in the presence of trehalose at concentrations greater than about 20% of the dry weight of the mixture upon rehydration are morphologically indistinguishable from freshly prepared ones (Crowe, J., et al., 1983a). The rehydrated vesicles are about the same size as the fresh ones, intramembrane particles are restricted predominantly to one face of the membrane, as in the fresh ones, and the intramembrane particles show none of the aggregation seen in rehydrated vesicles previously dried without trehalose (Figures 11-1, 11-5). Furthermore, the vesicles dried with high trehalose concentrations transport Ca

Figure 11-5. (a) Freeze fracture of Ca-transporting vesicles dried in the presence of trehalose. The vesicles appear as collapsed cups embedded in a matrix of trehalose. (b) Rehydrated Ca-transporting vesicles that had previously been dried in the presence of trehalose. Compare these vesicles with those shown in Figure 11-1.

normally and show normal Ca-dependent ATPase activity. In contrast with these results, when the vesicles are dried with progressively lower trehalose concentrations, morphological damage and loss of physiological function become increasingly evident (Figure 11-6).

We conclude that, at concentrations comparable to those found in many anhydrobiotic organisms, trehalose is capable of stabilizing these dry model

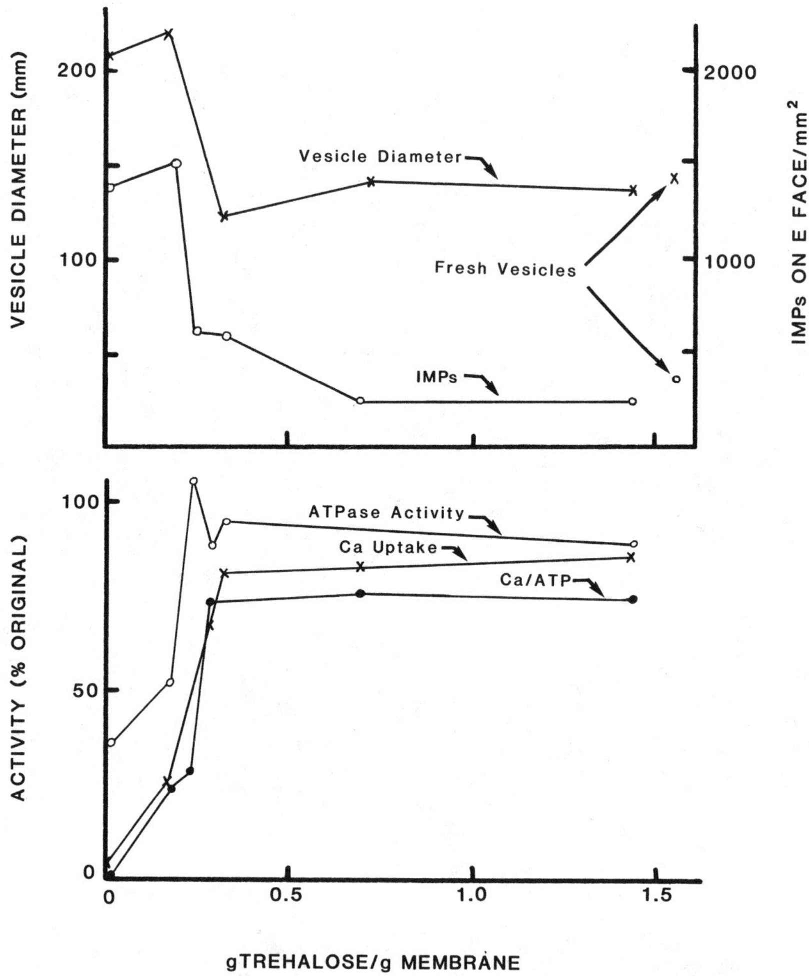

Figure 11-6. Effects of dehydration of Ca-transporting vesicles in various concentrations of trehalose. The top panel shows effects on size of the rehydrated vesicles and distribution of IMPs in the membrane. The bottom panel shows effects on ATPase activity, Ca uptake, and coupling between Ca uptake and ATP utilization. (Data from Crowe, J., et al., 1983a.)

membranes. We have not tested its effects on other membranes, so comparative data are still needed. We have used it to stabilize dry phospholipid vesicles (Crowe, L., et al., 1985). Before moving on to the next section, we want to emphasize a point that has escaped some workers who have been attempting to use trehalose as a stabilizing agent for other membrane systems and phospholipid vesicles (e.g. Madden et al., 1985). All the data presented here and elsewhere by us on this subject (Crowe, J., and Crowe, 1984; Crowe, J., et al., 1983b; Crowe, L., et al., 1985) have been expressed in terms of dry mass of trehalose per dry mass of membrane or lipid. In bulk solution, before the preparation is dried, the concentration of trehalose is often very low—usually in the millimolar range. Thus, it appears to be the dry mass ratio that is important, not the concentration of trehalose in bulk solution. We have recently found that at exceedingly high trehalose concentrations phospholipid bilayers may become destabilized, particularly if there is a large, asymmetric distribution of the sugar across the bilayer (see Chapter 12). The point we want to make is that some care must be taken to ensure that appropriate mass ratios of sugar to membrane are used.

Can Other Carbohydrates Stabilize Dry Membranes?

We have tested the abilities of a number of other sugars to stabilize dry Ca-transporting membranes and have established that trehalose is the most effective of those tested (Figure 11-7). Sugars other than trehalose fall into several categories.

Reducing sugars. Some reducing sugars such as glucose are initially effective at stabilizing the dry membranes, but with even moderately short storage times ATPase activity and Ca transport decline. We suspect that this effect is due to chemical reaction between the dry protein and glucose. This well-known reaction, known generally as the browning reaction (cf. Loomis et al., 1979), will be discussed further in the section on adventitious chemistry.

Other nonreducing sugars. Sucrose is capable of preserving membrane integrity in the dry state, but at least three times as much sucrose is required to achieve the same degree of stabilization as that achieved in the presence of trehalose.

Polyalcohols. Molecules such as glycerol are well known to have fusogenic properties that might be expected to be deleterious to membrane integrity. That appears to be the case; glycerol is the least effective of all the carbohydrates at stabilizing dry membranes and has destabilizing effects instead. This observation raises interesting questions concerning the com-

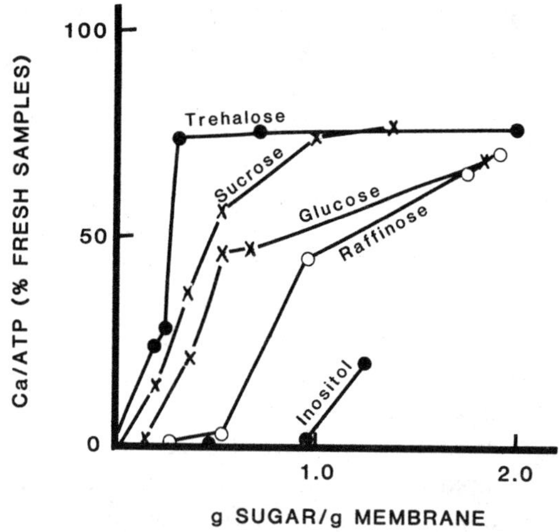

Figure 11-7. Effects of dehydration in various concentrations of the indicated sugars on Ca uptake by rehydrated vesicles. (Data from Crowe, L., et al., 1984.)

monly high concentrations of glycerol in anhydrobiotic organisms. While some understanding of the mechanism by which glycerol destabilizes phospholipid bilayers is being achieved (e.g. McDaniel et al., 1983), we have no experimental evidence concerning how those effects can be obviated.

We conclude that the ability to stabilize dry membranes is not a property that is general to all sugars. While some sugars other than trehalose possess this ability to a high degree, none of those tested is capable of preserving dry membranes at realistic physiological concentrations. We hasten to add two caveats: We have tested these carbohydrates with only one biological membrane, and we recognize that the effects reported here may vary from one membrane to the next. It is certainly possible that other sugars as effective as trehalose may exist. Nevertheless, we have suggested that there may be something about the trehalose molecule that imparts special effectiveness in stabilizing the dry membranes discussed here. We have begun experiments to identify what that property might be. We present results of some of those experiments in the following sections.

Effects of Sugars on Transition Temperature

If a key cause of damage to a membrane during dehydration is the liquid-crystalline-to-gel-phase transition, as we suggested in an earlier section, it seems likely that the mechanism by which sugars stabilize dry membranes would involve effects on this phase transition. We have begun to explore that possibility, using differential scanning calorimetry to measure transition temperature (T_c) for dry phospholipid-sugar preparations (Crowe, J., et al., 1984; Crowe, L., et al., 1985a, 1985b). In our initial experiments we used a model phospholipid whose phase properties are particularly well understood, dipalmitoylphosphatidylcholine (DPPC), dried with trehalose or other sugars from organic solvents to produce a homogeneous mixture. In subsequent experiments we have used dried vesicles of mixed phospholipids with trehalose and have produced similar but more complex results.

The results with homogeneous mixtures of DPPC and trehalose (Figure 11-8) show that, when DPPC is dehydrated, the melting temperature rises by about 30°C (Crowe, J., et al., 1984), in agreement with findings of previous

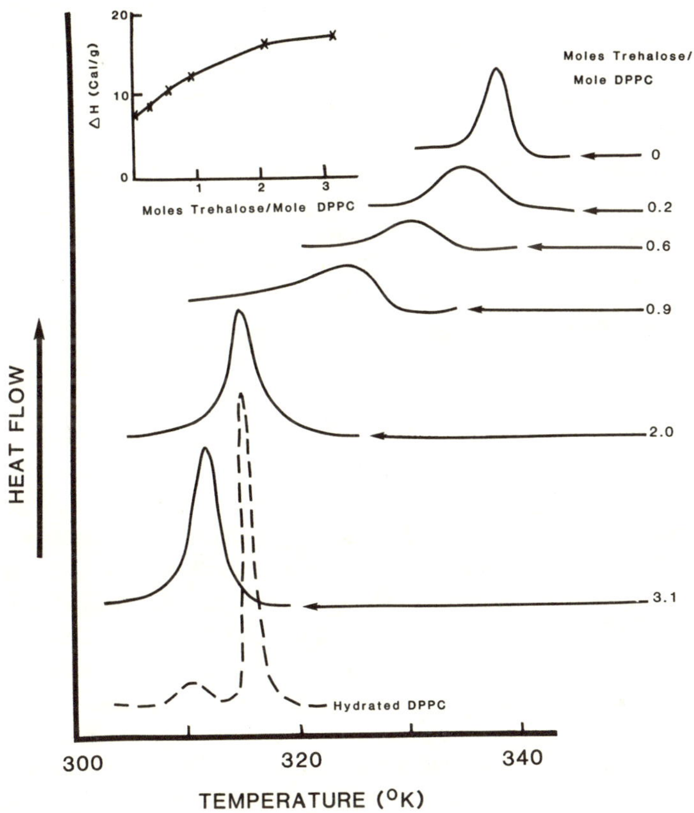

Figure 11-8. Melting endotherms for DPPC in excess water, dry, and in the presence of the indicated amounts of trehalose. (Data from Crowe, J., et al., 1984.)

workers (Chapman et al., 1967; Kodama et al., 1982). Thus, dry DPPC is in gel phase at temperatures at which it would normally be in liquid crystalline phase. But when even small amounts of trehalose were added to the DPPC before drying, T_c was depressed to a lower temperature and the melting endotherm was spread. With increasing amounts of trehalose T_c declined steadily, until at the highest concentrations of trehalose T_c was actually below that of the fully hydrated DPPC. In other words, large amounts of trehalose keep DPPC in the liquid crystalline phase at temperatures representative of the hydrated lipid. We suggest from this finding that trehalose could prevent damage to an intact membrane by depressing T_c and maintaining the constituent phospholipids in liquid crystalline phase.

We have carried out similar studies on effects of a number of other sugars on T_c of dry DPPC (Crowe, L., et al., 1985), finding that the various sugars studied all depress T_c by varying amounts, and they show the same order of effectiveness as in preserving dry membranes. For example, trehalose is the most effective at preserving dry membranes, and it is the most effective at depressing T_c of dry DPPC. By contrast, inositol is the least effective at preserving dry membranes and is also the least effective at depressing T_c. The other sugars fall in between these two extremes.

We interpret these results to be consistent with the hypothesis that depression of T_c and maintenance of dry membrane phospholipids in liquid crystalline phase is a key event in stabilizing dry membranes.

Mechanism of Interaction between Sugars and Dry Phospholipids

Solution properties of trehalose. All of the sugars and polyalcohols that have stabilizing effects on membranes and phospholipid vesicles interact strongly with water, and it has been postulated that these carbohydrates participate in the water lattice of bulk water by hydrogen bonding to water molecules (Franks, 1975). There is some thermodynamic evidence, in fact, that hydrogen bonding between at least some sugars and water is stronger than hydrogen bonding between the water molecules themselves. The ability of the sugars to hydrogen bond to water seems to be related to the spacing of the OH groups on the hydrocarbon backbone of the sugar; equatorial OH groups on pyranose and cyclitol rings could fit precisely into the water lattice in bulk water. It has also been suggested that, since axial OH groups do not possess the same spacing, the fit into the water lattice should not be as good (see Crowe, L., et al., 1984 for a more complete discussion of this matter).

On the basis of the above discussion, it follows that examination of solution properties of trehalose and other sugars might be relevant to a determination of the mechanism by which the sugars stabilize dry phospholipids. We are suggesting that the effective carbohydrates might be the ones that fit best into the water structure around polar head groups of phospholipids. The lines of available evidence do not support this proposition: (1) a partial phase diagram for trehalose does not differ significantly from that for similar sugars (e.g. sucrose), but it should be quite different if solution properties are to explain effects of this molecule on membrane phospholipids. (2) Examination of structural characteristics of various sugars (Crowe, L., et al., 1984) shows that there is no clear relationship between the number of equatorial OH groups available for hydrogen bonding and the relative effectiveness of the sugar at stabilizing dry phospholipid bilayers. For example, trehalose and cellobiose have the same number of OH groups available for intermolecular bonding, the same number of OH groups/unit volume, and all the OH groups are equatorial, yet trehalose is twice as effective as cellobiose at stabilizing dry membranes. Similarly, myoinositol, with five of its six OH groups in an equatorial position (more equatorial OH groups/unit volume than trehalose), and which forms strong hydrogen bonds with water, is one of the least effective sugars tested.

We conclude from the available data that solution properties of sugars and availability of equatorial OH groups to fit into the water lattice of bulk water do not provide an explanation of their relative effectiveness at stabilizing dry membranes. Therefore, an explanation for the mechanism of interaction must be sought elsewhere, and we do so in the following section.

Infrared spectroscopy. Infrared spectroscopy has in recent years become a useful method for studying group interactions in membrane components (e.g. Mantsch et al., 1982), and it has been particularly valuable in the present context. We have measured infrared spectra for the dry homogeneous mixtures of DPPC and sugars used in Figure 11-6. The most striking effects of trehalose on DPPC were in bands that have been assigned to the phosphate head group; the band at 1246 cm^{-1} was broadened, depressed in intensity, and shifted to 1240 cm^{-1} in the presence of trehalose. These effects can be titrated by adding various amounts of trehalose before the mixture is dried. Concurrent with the changes in the phosphate vibrations, alterations in bands assigned to trehalose are changed. Bands between 1400 and 1300 cm^{-1}, assigned to OH vibrations, were completely missing in the presence of DPPC. We interpret these data as suggesting that there is an interaction between OH groups on trehalose and the phosphate of the polar head group of the phospholipid. One further point: the shift in vibrational frequency for the phosphate is in the same direction and of similar magnitude as when water was added to the dry DPPC without trehalose, a finding which seems consistent with the suggestion that trehalose affects physical properties of membrane phospholipids in much the same way as hydrating them. In more recent studies we have shown similar interactions between trehalose and phospholipids in an intact biological membrane (Crowe, J., et al., 1984b).

We have carried out similar studies with other sugars, using the integrated intensity of the stretching vibration of the phosphate in the vicinity of 1240 cm^{-1} as a measure of the strength of interaction. The results (Crowe, L., et al., 1985) show that the various sugars tested have widely differing effects on this parameter and that their relative effectiveness at altering the infrared spectrum is correlated with their ability to preserve dry membranes. For example, trehalose is the most effective at preserving dry membranes and has the most pronounced effect on the infrared spectra. Inositol, by contrast, is the least effective.

We do not yet know what property of trehalose allows it to interact so strongly with phospholipids, but this problem is being examined through computer modeling studies on sugar-phospholipid interactions (Chapter 13).

Long-Term Stability of Dry Membranes: Adventitious Chemistry

There are many unanswered questions in the above discussion and we hesitate to add even more to the list. Nevertheless, we feel obliged to do so because adventitious, deleterious chemical reactions can proceed at an alarming rate in the membrane-carbohydrate preparations we have been studying in vitro—rates that are seemingly inconsistent with known longevities of anhydrobiotic organisms. How the sorts of adventitious reactions to be discussed briefly here are obviated in the intact organisms is not yet understood (see Chapter 5).

Oxidative effects. As one might expect, oxidation of lipid components of dry membranes rapidly destroys biological function. When Ca-transporting vesicles were dried in the presence of trehalose and stored for varying lengths of the time before they were rehydrated (Mouradian et al., 1984), the ability to accumulate Ca declined rapidly, falling to negligible levels within a month of storage (Figure 11-9). In a second paper on this subject (Mouradian et al., 1985), we found that this decline in ability to accumulate Ca coincided with lipid oxidation. Although Ca uptake was rapidly uncoupled from ATP utilization because of the oxidation, ATPase activity remained high until prolonged storage in air, when the protein apparently becomes directly affected by products of oxidation. By contrast, when the same membranes were stored under vacuum, little change in biological activity was seen for at least 100 days.

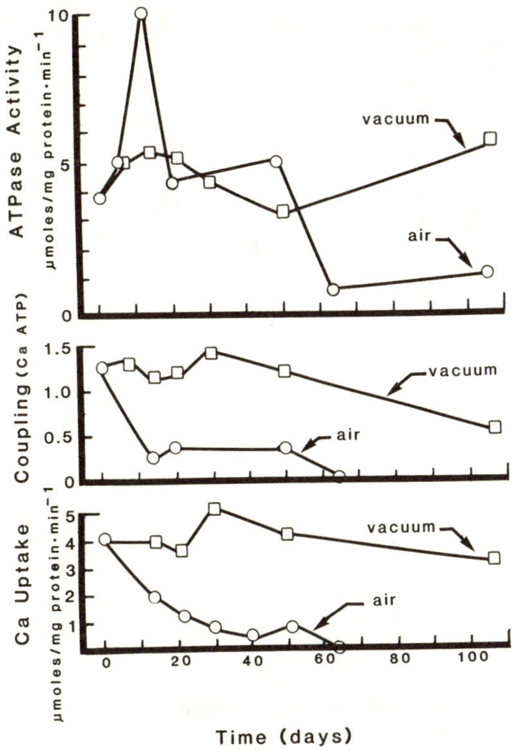

Figure 11-9. Effects of storage in air and in vacuum on stability of Ca-transporting vesicles dried with 0.5 g trehalose/g membrane, stored for the indicated intervals, and rehydrated, after which the measurements were made. (Data from Mouradian et al., 1985.)

We conclude from these experiments that, although trehalose can stabilize the dry membranes for short periods in air, it is incapable of doing so for extended periods. Therefore, it is likely that anhydrobiotic organisms possess mechanisms to obviate these oxidative effects. The nature of those mechanisms is not yet clear.

Browning reactions. Browning between reducing sugars and free amino groups of dry proteins yields an insoluble, discolored product (Feeny et al., 1975). All nonreducing sugars studied to date, by contrast, including trehalose, show no such browning reaction (Loomis et al., 1979; Figure 11-10). There is some evidence that the presence of trehalose inhibits the usual browning of proteins by reducing sugars; the data shown in the lower panel of Figure 11-10 represent changes in rates of browning of a dry protein as the availability of free amino groups for further reaction declines. The

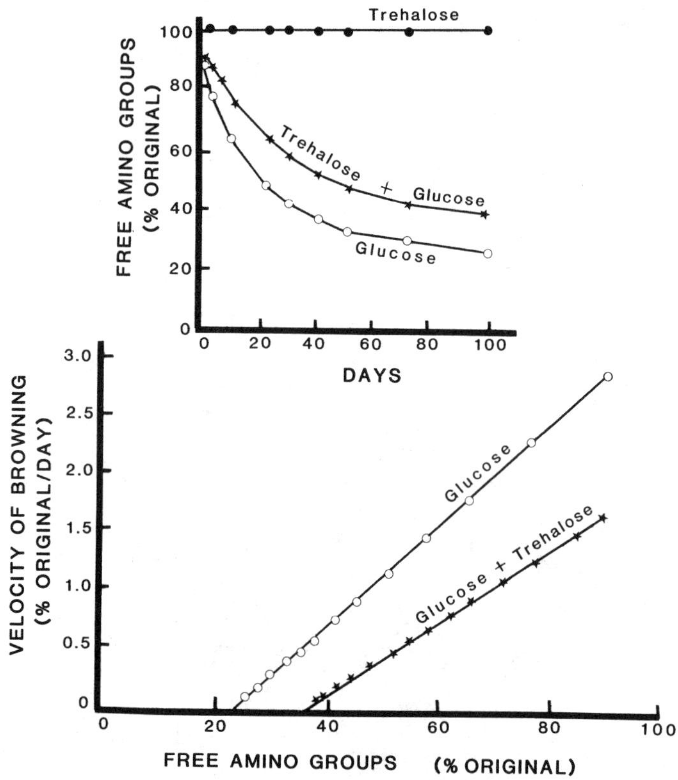

Figure 11-10. Effects of trehalose and glucose on free amino groups in a dry protein, bovine serum albumin. Upper panel: Free amino groups remaining in the protein with time of storage at 60% relative humidity. Lower panel: Velocity of disappearance of free amino groups as a function of concentration of free amino groups remaining accessible for reaction. (Data from Loomis et al., 1979.)

slopes of the lines are clearly different, indicating that, in the presence of trehalose, browning proceeds at a slower rate. Furthermore, both lines have positive intercepts on the abscissa, which may represent the portion of free amino groups on the protein that are unavailable for reaction. The fact that this fraction is larger in the samples with trehalose suggests that trehalose may actually alter the conformational state of the protein. More recent evidence from infrared spectroscopy is consistent with this notion. When spectra for dry and hydrated proteins were compared, a large shift in vibrational frequency was seen for the amide I band, which suggests that the peptide chains are disordered by the dehydration. But when the protein was dried in the presence of trehalose, no such shifts were seen; it appears that the dry protein was maintained in its native state in the presence of trehalose. We have no comparative data of this sort on interactions of other carbohydrates with proteins as yet, so we do not know their relative effects on protein structure or stability in the dry state. This is a topic of active research at the moment.

Conclusions

Results of studies over the past few years have brought us closer to understanding on a molecular level how many organisms are capable of surviving dehydration. This problem has fascinated scientists and philosophers for centuries, not only because of its inherent intellectual interest, but also because many of the organisms capable of entering an anhydrobiotic state are of great importance in human welfare. We admit to a philosophical fascination with these organisms and their extreme longevity; this fascination has driven much of our work on this subject. We are also aware that the results of these studies may have direct applications. For example, if we can reach an understanding of the mechanism by which trehalose imparts its stabilizing effects on dry membranes and proteins, it may be possible to design new stabilizing agents for use in preserving biological materials that are refractory to such treatment. It is important to understand the molecular basis for variability in the effectiveness of various carbohydrates at stabilizing dry membranes and proteins. We are striving for this understanding.

REFERENCES

Boni, L.T. and Hui, S.W. 1983. Polymorphic phase behaviour of dilinoleoylphosphatidylethanolamine and palmitoyloleoyl-phosphatidylcholine mixtures. Biochim. Biophys. Acta 731: 177.

Chapman, D., Peel, W.E. and Quinn, P.J. 1978. The modulation of bilayer fluidity by polypeptides and homogeneous catalysts. Ann. N.Y. Acad. Sci. 67: 67-83.

Chapman, D., Williams, R.M. and Ladbrooke, B.D. 1967. Physical studies of phospholipids. VI. Thermotropic and lyotropic mesomorphism of some 1,2-diacyl-phosphatidylcholines (lecithins). Chem. Phys. Lipids 1: 445.

Crowe, J.H. and Crowe, L.M. 1984. Effects of dehydration on membranes and membrane stabilization at low water activity. *In* Biological Membranes, Vol. 5 (D. Chapman, ed.). Academic Press, London. pp. 57-103.

Crowe, J.H., Crowe, L.M. and Chapman, D. 1984a. Preservation of membranes in anhydrobiotic organisms: the role of trehalose. Science 223: 701.

Crowe, J.H., Crowe, L.M., and Chapman, D. 1984b. Infrared spectroscopic studies on interactions of water and carbohydrates with a biological membrane. Arch. Biochem. Biophys. 232: 400.

Crowe, J.H., Crowe, L.M. and Jackson, S.A. 1983a. Preservation of structural and functional activity in lyophilized sarcoplasmic reticulum. Arch. Biochem. Biophys. 220: 477.

Crowe, J.H., Crowe, L.M. and Mouradian, R. 1983b. Stabilization of biological membranes at low water activities. Cryobiology 20: 346.

Crowe, J.H., Madin, K.A.C. and Loomis, S.H. 1977. Anhydrobiosis in nematodes: metabolism during resumption of activity. J. Exp. Zool. 102: 57.

Crowe, L.M. and Crowe, J.H. 1982. Hydration dependent hexagonal phase in a biological membrane. Arch. Biochem. Biophys. 217: 582.

Crowe, L.M., Crowe, J.H. and Chapman, D. 1985. Interaction of carbohydrates with dry dipalmitoylphosphatidylcholine. Arch. Biochem. Biophys. 236: 289.

Crowe, L.M., Crowe, J.H., Rudolph, A.S., Womersley, C. and Appel, L. 1985. Preservation of freeze-dried liposomes by trehalose. Arch. Biochem. Biophys. 242: 240.

Crowe, L.M., Mouradian, R., Crowe, J.H., Jackson, S.A. and Womersley, C. 1984. Effects of carbohydrates on membrane stability at low water activities. Biochim. Biophys. Acta 769: 141.

Cullis, P.R. and DeKruijff, B. 1979. Lipid polymorphism and their functional roles in biological membranes. Biochim. Biophys. Acta 559: 399.

Feeney, R.E., Blankenhorn, G. and Dixon, H.F.B. 1975. Carbonyl-amine reactions in protein chemistry. Adv. Prot. Chem. 29: 135.

Franks, F. 1975. Water, ice, and solutions of simple molecules. *In* Water Relations of Foods (R. B. Duckworth, ed.). Academic Press, London. pp. 3-22.

Fringeli, U.P. and Gunthard, H.H. 1976. Hydration sites of egg phosphatidylcholine determined by means of modulated excitation infrared spectroscopy. Biochim. Biophys. Acta 450: 101.

Goren, M.B., Hart, P.D., Young, M.R. and Armstrong, J.A. 1976. Prevention of phagosome-lysosome fusion in cultured macrophages by sulfatides of *Mycobacterium tuberculosis*. Proc. Natl. Acad. Sci. USA 73: 2510.

Hauser, H., Pascher, I., Pearson, R.H. and Sundell, S. 1981. Preferred conformation and molecular packing of phosphatidylethanolamine and phosphatidylcholine. Biochim. Biophys. Acta 650: 21.

Kodama, M., Kuwabara, M., and Seki, S. 1982. Successive phase-transition phenomena and phase diagram of the phosphatidylcholine-water system as revealed by differential scanning calorimetry. Biochim. Biophys. Acta 689: 567.

Loomis, S.H., O'Dell, S.J., and Crowe, J.H. 1979. Anhydrobiosis in nematodes: inhibition of the browning reaction of reducing sugars with dry proteins. J. Exp. Zool. 208: 355.

Loomis, S.H., O'Dell, S.J., and Crowe, J.H. 1980. Anhydrobiosis in nematodes: Control of the synthesis of trehalose during induction. J. Exp. Zool. 211: 321.

Luzzati, V. 1968. X-ray diffraction studies of lipid-water systems. *In* Biological Membranes (D. Chapman, ed.), pp. 71-124. Academic Press, New York.

McDaniel, R.V., McIntosh, T.J., and Simon, S.A. 1983. Nonelectrolyte substitution for water in phosphatidylcholine bilayers. Biochim. Biophys. Acta 731: 97.

Madden, T.D., Bally, M.B., Hope, M.J., Cullis, P.R., Schieren, H.P., and Janoff, A.S. 1985. Protection of large unilamellar vesicles by trehalose during lyophilization: retention of vesicle contents. Biochim. Biophys. Acta 817: 67.

Madin, K.A.C. and Crowe, J.H. 1975. Anhydrobiosis in nematodes: carbohydrate and lipid metabolism during dehydration. J. Exp. Zool. 193: 335.

Madin, K.A.C., Loomis, S.H., and Crowe, J.H. 1985. Anhydrobiosis in nematodes: control of carbon flow through the glyoxylate cycle. J. Exp. Zool. 234: 341.
Mantsch, H.H., Cameron, D.G., Tremblay, P.A., and Kates, M. 1982. An infrared spectroscopic characterization of the polymorphic phase behavior. Biochim. Biophys. Acta 689: 63.
Mouradian, R., Womersley, C., Crowe, L.M. and Crowe, J.H. 1984. Preservation of functional integrity during long term storage of a biological membrane. Biochim. Biophys. Acta 778: 615.
Mouradian, R., Womersley, C., Crowe, L.M. and Crowe, J.H. 1985. Degradation of functional integrity during long-term storage of a freeze-dried biological membrane. Cryobiology 22: 119.
Powers, L. and Pershan, P.S. 1977. Monodomain samples of dipalmitoyl phosphatidylcholine with varying concentrations of water and other ingredients. Biophys. J. 20: 137.
Quinn, P.J. 1985. A lipid-phase separation model of low-temperature damage to biological membranes. Cryobiology 22: 128.
Rudolph, A.S. and Crowe, J.H. 1985. Membrane stabilization during freezing: the role of two natural cryoprotectants, trehalose and proline. Cryobiology 22: 367.
Schneider, A.S., Middaugh, C.R., and Oldewurtel, M.D. 1979. Role of bound water in biological membrane structure: fluorescence and infrared studies. J. Supramol. Structure 10: 265.
Seddon, J.M., Cevc, G., and Marsh, D. 1983. Calorimetric studies of the gel-fluid and lamellar-inverted hexagonal phase transitions in dialkyl- and diacylphosphatidylethanolamines. Biochemistry 22: 1280.
Smaby, J.M., Hermetter, A., Schmid, P.C., Paltauf, F., and Brockman, H.L. 1983. Packing of ether and ester phospholipids in monolayers: evidence for hydrogen-bonded water at the sn-1 acyl group of phosphatidylcholines. Biochemistry 22: 5808.
Tanford, C. 1980. The Hydrophobic Effect. Wiley, New York.
Thayer, A.M. and Kohler, S.J. 1981. Phosphorus-31 nuclear magnetic resonance spectra characteristic of hexagonal and isotropic phospholipid phases generated from phosphatidylethanolamine in the bilayer phase. Biochemistry 20: 6831.
Volke, F., Arnold, K., and Garwich, K. 1982. The effect of hydration on the mobility of phospholipids in the gel state. A proton nuclear magnetic resonance spin echo study. Chem. Phys. Lipids 31: 179.
Wieslander, A., Ulmius, J., Lindblom, G., and Fontell, K. 1978. Water binding and phase structures for different *Acholeplasma laidlawii* membrane lipids studied by deuteron nuclear magnetic resonance and x-ray diffraction. Biochim. Biophys. Acta 512: 241.
Womersley, C. 1981. Biochemical and physiological aspects of anhydrobiosis. Comp. Biochem. Physiol. 70B: 669.
Wunderlich, F., Kreutz, W., Mahler, P., Ronal, A., and Heppeler, G. 1978. Thermotropic fluid ordered "discontinuous" phase separation in microsomal lipids of Tetrahymena. An x-ray diffraction study. Biochemistry 17: 2005.

12 Hydration-Dependent Phase Transitions and Permeability Properties of Biological Membranes

Lois M. Crowe
John H. Crowe

When dry seeds or isolated seed embryos are placed in water, they imbibe water rapidly in the first few minutes of rehydration. There follows a longer, slow phase until they are fully rehydrated (Simon and Raja Harun, 1972; Leopold, 1980; Murphy and Noland, 1982). As water is taken up, solutes such as ions, sugars, and proteins leak from the seeds or embryos, very rapidly at first and then much more slowly. That this rapid initial leakage is not due to loss of extracellular material from the surface of the seeds is shown by successive dryings and rehydrations which show the same rapid initial leakage. However, if seeds are slowly imbibed (for example on moist filter paper) before being put into bulk water, leakage of solutes is greatly restricted (Simon and Raja Harun, 1972).

Very similar results have been obtained from dry nematodes. When nematodes (*Aphelenchus avenae*) are dried slowly, they undergo extensive metabolic changes which enable them to survive almost complete desiccation, a condition which is called anhydrobiosis (Madin and Crowe, 1975). If, instead, the wet worms are quickly dried by placing them in dry air, they cannot revive upon rehydration. When anhydrobiotic or quick-dried worms are placed in water, they, like seeds, leak rapidly for a short while, then more slowly. The leakage of the quick-dried worms, however, is much higher than that of the anhydrobiotic nematodes. Heat-killed seed embryos or cotyledons also show significantly higher leakage than viable seed parts (Leopold, 1980; Murphy and Noland, 1982).

In the case of seed embryos placed directly in water, leakage slowed if the embryos had been rehydrated to an initial water content of 20-30% (Simon, 1978). Both quick-dried and anhydrobiotic nematodes show a very rapid decrease in the rate of leakage when they are hydrated to more than 20% water content at high humidity, although again the leakage from anhydrobiotic nematodes is significantly less than from quick-dried nematodes (Figure 12-1). Thus, though there are real differences between "living" and "dead" dry organisms, membrane leakage is not an uncontrolled process in either case. Experiments such as these led Simon (1974, 1978) to hypothesize that membrane structure in dry organisms is disorganized, with phospholipids in an inverted hexagonal (H_{II}) phase and integral membrane proteins displaced from their environment in the bilayer.

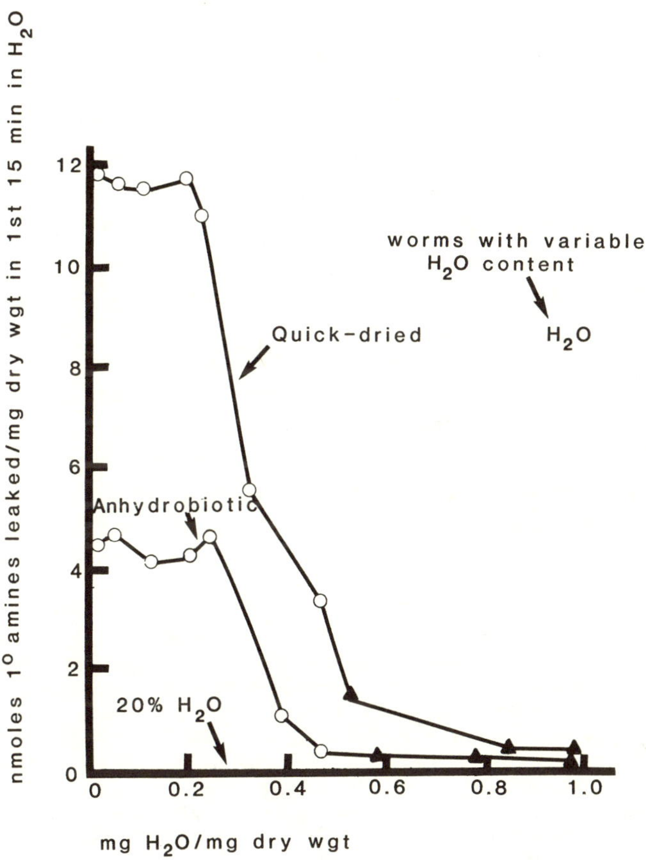

Figure 12-1. Leakage of primary amines from quick-dried and anhydrobiotic worms in the first 15 min after placement in water. The worms were first rehydrated to various degrees (mg H_2O/mg dry weight of worm) by exposing them to high humidities (○) or moistened filter paper (▲). (Redrawn from Crowe et al., 1979, courtesy of *Journal of Experimental Zoology*.)

Work by Chapman et al. (1967) had shown that phosphatidylcholines decrease their gel to liquid crystalline transition temperature as they go from the dry to the hydrated state, reaching a minimum at about 20-25% water content. About 20-25% water must be present before free water can be detected by calorimetry, suggesting that about 10 to 20 water molecules are bound in the region of each phospholipid head group. Luzzati and Husson (1962) had also shown that isolated brain phospholipid undergoes a lamellar-to-H_{II} phase change below about 20% water. One might expect, then, that membranes at less than 20% water content would be disturbed, even to the extent of forming nonbilayer phases such as H_{II}.

The suggestion by Simon led many workers to test his hypothesis with several different systems and techniques. McKersie and Stinson (1980) and Seewaldt et al. (1981) have used x-ray diffraction to study both extracted phospholipids and seed particles and have found only lamellar phases at all water contents. Phosphorus NMR of both pollen and its isolated phospholipids showed the presence of a bilayer down to at least 11% moisture content (Priestley and de Kruijff, 1982). In our laboratory we have looked at quick-dried and anhydrobiotic nematodes with freeze fracture, and while there are some membrane differences between the two preparations, all the membranes appear to be in a bilayer state (Crowe et al., 1978). Dry, dormant cysts of *Artemia salina* do not show any hexagonal-phase phospholipid when they are freeze fractured (L. Crowe, unpublished).

The absence of H_{II} phospholipid in freeze-fracture preparations is not conclusive evidence of its absence; however, if present it must be there in very small quantity. If isolated sarcoplasmic reticulum (SR) membranes from lobster muscle are air dried and freeze fractured, there are very extensive areas of phase-separated lipids, including patterns attributable to H_{II} (Crowe and Crowe, 1982). When lyophilized SR is examined by ^{31}P NMR at a

Figure 12-2. Freeze-fracture replicas of lobster remoter muscle. This muscle is characterized by an unusually high proportion of membranous components; about 75% of the volume of the muscle is occupied by the T system and SR (Rosenbluth, 1969). Fixation and infiltration were carried out on the whole muscle. (a) Longitudinal fracture of hydrated muscle, fixed briefly in glutaraldehyde and infiltrated with glycerol. Intramembrane particles are found mainly on the P fracture face of the membranes. The T system (T) in this muscle is in the form of large sheets or sacs of membrane; the sarcoplasmic reticulum (SR) appears as fenestrated sheets and tubules. Myofibrils (MF) are also shown. (b) Oblique fracture of muscle which has been briefly fixed in glutaraldehyde, infiltrated with trehalose, and air dried. Myofibrils are much more closely packed in these preparations, and the SR appears highly condensed. Intramembrane particles are still associated with the membranes. (c) Oblique fracture of muscle which has been briefly fixed in glutaraldehyde and air dried. The SR is no longer distinguishable as individual membranes, and large patches of H_{II} lipid have separated out in the areas between the myofibrils. The bars equal 0.1 μm.

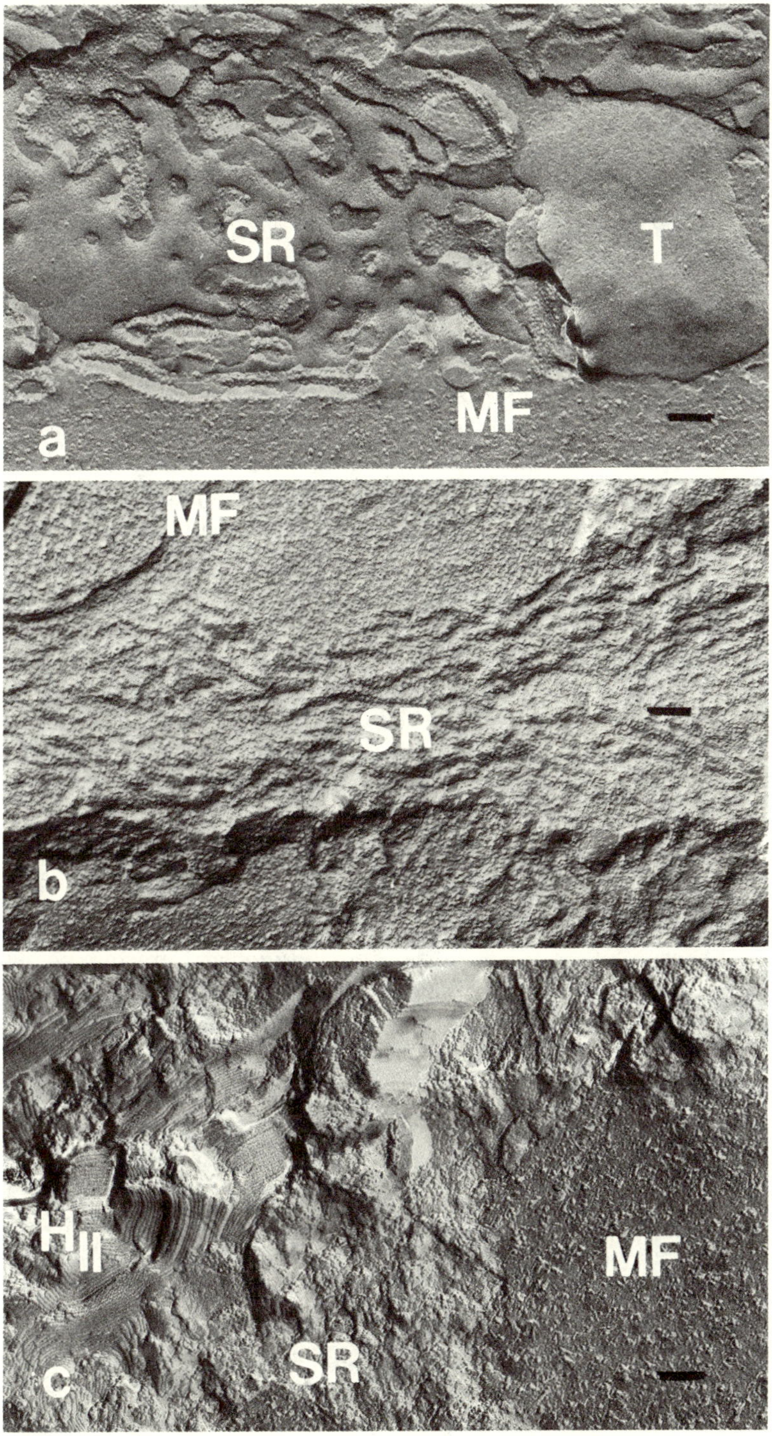
SR
T
MF
a
MF
SR
b
H
II
MF
SR
c

moisture content of about 9%, a spectrum typical of hexagonal-phase phospholipid is seen (Crowe et al., 1983). This formation of H_{II} phase is not confined to isolated membranes: if whole lobster muscle is air dried, extensive hexagonal-phase lipid can be found (Figure 12-2c). If the isolated membranes or whole intact muscle fibers are first placed in trehalose solution before drying, the phase separations are largely eliminated (Figure 12-2b). It would appear, then, that membranes in organisms which do not normally dry out will undergo changes to nonbilayer phases. The addition of trehalose to the latter membranes will inhibit such changes (Crowe et al., 1982, 1983).

In view of the fact that the presence of hexagonal phase phospholipid has not been demonstrated in organisms which normally survive desiccation, we would like to suggest an alternative explanation for the leakage of dry organisms. Although phospholipids may still form an intact bilayer in dry systems, this bilayer is composed of a large array of different phospholipids, some of which may be in gel phase when dry at ambient temperatures. As these gel-phase lipids are rehydrated, they can undergo a gel-to-liquid-crystalline-phase transition. Many studies have shown that liposomes show increases in permeability to water as well as to ions, carbohydrates, and other solutes while undergoing the gel-to-liquid-crystalline transition. Initial leakage from desiccation-tolerant systems placed in water may occur because of hydration-induced phase transitions in the bilayers.

Liposomes as Model Systems

Studies on multicellular systems such as seeds and nematodes are difficult to interpret because of the complexities of composition and structure involved. Biological membranes contain proteins which can become displaced or denatured as the membrane is dried. Changes in permeability may come about because of protein structural changes as well as because of phospholipid phase changes.

In order to sort out these effects, many investigators have begun to use simpler liposome systems consisting of one or two phospholipids, sometimes with the addition of integral membrane proteins. One such study shows that the incorporation of membrane proteins increases water permeability by an order of magnitude (Carruthers and Melchior, 1983). Some binary mixtures of phospholipids appear to show rapid flux through what are assumed to be boundary regions between gel-phase and liquid-crystalline-phase lipids. Cholesterol decreases the water permeability of dimyristoyl-phosphatidylcholine (DMPC) when present between 10 and 40 mol % (Carruthers and Melchior, 1983). Both dipalmitoylphosphatidylcholine (DPPC) and dipalmitoylphosphatidylglycerol (DPPG) liposomes exhibit a maximum in the self-diffusion rate of ^{22}Na at their respective gel-to-liquid-crystalline transition temperatures (Papahadjopoulos et al., 1973). Cations and anions leak into liposomes of saturated phosphatidylcholines in a manner dependent upon acyl chain length and hydrated ion size, with increased permeability at the transition temperature of the phosphatidylcholine (Blok et al., 1975). It was suggested that ions diffused through statistical pores, with the probability of pore formation increasing at discontinuities or defects at the boundary of liquid and solid domains in the bilayer (Blok et al., 1975). Liposomes of DMPC have a high glucose permeability when equilibrated at their transition temperature but not below or above it, while DPPC liposomes leak glucose as their phase equilibrium is perturbed by temperature cycling through the transition temperature (Blok et al., 1976). Theoretical treatments of some of these results (Papahadjopoulos et al., 1973; Marsh et al., 1976) correlate permeability characteristics of liposomes with the proportion of interfacial lipid at the boundary between gel and liquid regions.

Although these experiments were all done with hydrated liposomes at various temperatures, they support the idea that leakage can occur from rehydrating liposomes or membranes as the phospholipids hydrate and undergo a phase transition. Such leakage may even be enhanced in natural membranes composed of a variety of lipids, perhaps phase separated during drying as some or all go into gel phase when they dehydrate at ambient temperatures. As the membranes rehydrate, different lipids may undergo phase transitions at a particular environmental temperature, while others remain in gel state, leading to an increase in boundaries between phases.

Dehydration and Liposome Leakage

Earlier experiments with isolated lobster SR had shown us that when it was air dried or lyophilized, lipid-phase separations occurred including the formation of H_{II} phase, multilamellar structures, cholesterol crystals, and the exclusion of intramembrane particles from lipid domains (Crowe and Crowe, 1982; also see Chapter 11). When these membranes were rehydrated, they were found to be unable to accumulate calcium; intramembrane particles had been displaced from their normal orientation in the bilayer and aggregated, and the membranes had undergone extensive fusion (Crowe et al., 1982). When about 0.3 g trehalose/g dry membrane (about 20-25% of the dry weight) was added to the SR before lyophilization, membrane structure and function were preserved. Other carbohydrates will also preserve the SR membranes, but considerably more mass of carbohydrate is required (Crowe, Mouradian, Crowe, Jackson, and Womersley, 1984).

Subsequent studies of the interactions between carbohydrates and DPPC showed that sugars were able to lower the gel-to-liquid-crystalline-phase transition temperature of dry DPPC to that of hydrated DPPC, with trehalose being effective at much lower concentrations than other carbohydrates. Infrared spectroscopic studies led us to suggest that trehalose and other carbohydrates exert their effects by hydrogen bonding to the phosphate of the lipid head group and increasing the head group spacing and therefore the fluidity of the DPPC (Crowe et al., 1984b). Since some carbohydrates have considerable effect on the physical properties of DPPC, and are also able to preserve structure and function in SR membrane vesicles, we decided to test the effects of carbohydrates on vesicles of pure phospholipids during dehydration.

Unilamellar vesicles, about 50 nm in diameter, were made from aqueous solution by a rotary drying method (Shew and Deamer, 1985). When trehalose was included in the solution, almost the entire mass of lipid was converted to unilamellar vesicles, with some trehalose trapped inside. Markers can also be trapped at the same time. We used isocitrate as our marker, since there is a sensitive fluorometric assay available for it (Plaut, 1962). Unilamellar vesicles without trehalose inside were made by freezing, thawing, and sonicating a suspension of very small unilamellar vesicles (Pick, 1981). Retention following lyophilization was determined by assaying the isocitrate outside the vesicles, then lysing the vesicles with Triton to release trapped isocitrate. We found that it was necessary to have trehalose both inside and outside the vesicles in order to maximize retention of isocitrate. With no trehalose present or with trehalose present on only one side of the membrane, retention was very low. When it was present on both sides of the membrane, retention increased dramatically until it reached 100% at about 1 g trehalose/g phospholipid. Variable amounts of trehalose outside were obtained by dialyzing the unilamellar vesicles or passing them over a Sephadex column and then adding back desired amounts of trehalose. The percentage of retention was determined by comparing the lyophilized samples with fresh samples from the same preparation. Results from a typical experiment are shown in Figure 12-3. Maximum retention of isocitrate by these vesicles occurs when the mass ratio of trehalose to phospholipid is about 1:1; about 5-10% of the trehalose is trapped inside the vesicles and the rest outside.

Fusion and Liposome Leakage

Studies using hydrated liposomes show that they leak during fusion (Wilschut and Papahadjopoulos, 1979; Hammoudah et al., 1981). This characteristic is not surprising, since fusion is thought to come about through

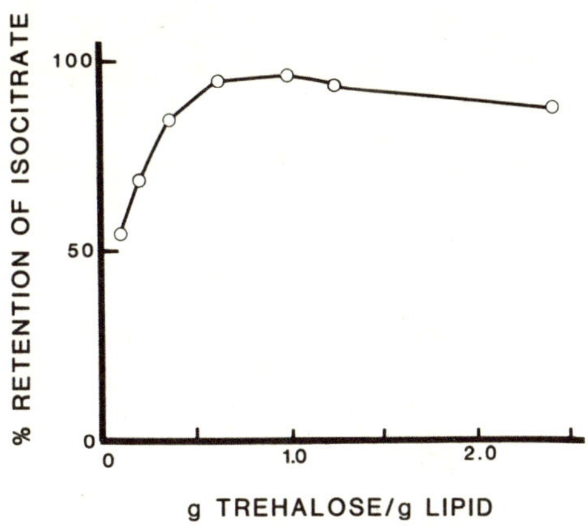

Figure 12-3. Isocitrate retention by liposomes lyophilized at various trehalose: lipid mass ratios. Retention is expressed as a percentage of the isocitrate trapped by the freshly-made liposomes.

defects in the bilayer and perhaps through nonbilayer structures during the fusion event (Cullis and Hope, 1978; Hui et al., 1981). This leakage suggested to us that leakage of isocitrate might result from fusion between liposomes during either the freezing or drying of lyophilization. Trehalose does inhibit fusion of liposomes when they are frozen and thawed in aqueous solutions of trehalose but at much higher concentrations than we use in any preparations which we are going to lyophilize (Rudolph and Crowe, 1985). During drying, trehalose could be acting simply as a spacer to keep liposomes separated, or it could affect liposome physical properties. Freeze fractures of a lyophilized dry sample at a high trehalose:phospholipid mass ratio show small individual vesicles embedded in a matrix of trehalose, while liposomes which have no added external trehalose have undergone extensive fusion (Figure 12-4). When a sample such as the one in Figure 12-4a is rehydrated, the resulting unilamellar vesicles have the same average diameter as the original freshly made liposomes. Vesicles resulting from the rehydration of the fused sample shown in Figure 12-4b are largely multilamellar, and the average diameter of the unilamellar vesicles is greatly increased.

We obtained a more quantitative estimate of fusion and its inhibition by trehalose and other carbohydrates using resonance energy transfer between fluorescent probes (lipid analogues) incorporated into the bilayer (Uster and Deamer, 1981; see also Appendix A). This method involves preparing two populations of liposomes, one with a "donor" probe and one with an "acceptor" probe incorporated into the bilayer. The two populations of vesicles are mixed together at the beginning of the experiment. Upon excitation, energy can be transferred from the donor to the acceptor probe if they are in close proximity, since the donor emission and acceptor excitation wavelengths overlap. The result is a quenching of the donor fluorescence. Mock-

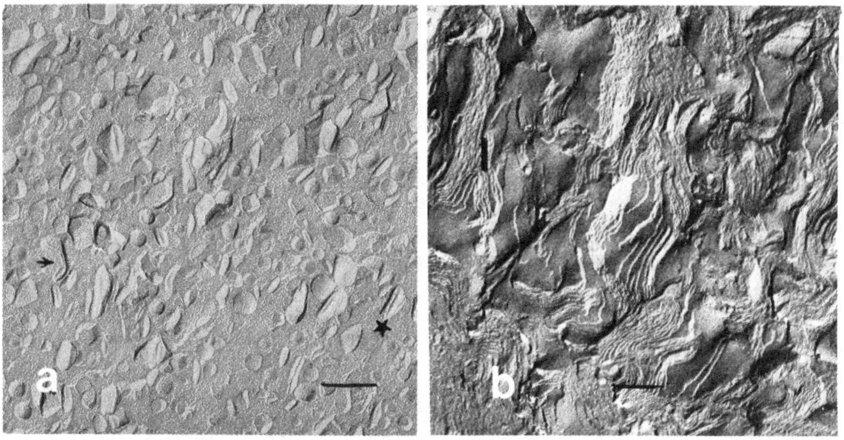

Figure 12-4. Freeze-fracture micrographs of lyophilized liposomes. (a) Made by rotary drying, these liposomes have about 0.1 g trehalose/g total phospholipid inside them and 2 g trehalose/g phospholipid outside. Individual liposomes are embedded in a matrix of trehalose and have collapsed into a cup shape (arrow). The majority of the fracture images are spherical, but some liposomes are cross-fractured near the edge of the cup (star). (b) Liposomes from the same preparation but with no trehalose added back after removal of external trehalose are shown. The extensive fusion has turned this into a multilamellar sample. The bars equal 0.1 μm.

fused vesicles are prepared in the same way that both donor and acceptor probes are added together to the chloroform solution of POPC and PS. Maximum donor fluorescence (0% probe intermixing) is measured by adding equal amounts of separate populations of unlyophilized donor and acceptor unilamellar liposomes to the fluorescence cell. Probe intermixing is measured as a proportion of the difference between 0 and 100% donor quenching.

The results of one such experiment are shown in Figure 12-5. Unilamellar liposomes which were dried with no trehalose present showed 90% probe intermixing; liposomes with about 0.1 g trehalose/g lipid showed about 70% probe intermixing and, with the addition of more trehalose outside the liposomes fusion during lyophilization, fell rapidly to a minimum value of about 15%. When these data are compared with those in Figure 12-3, it can be seen that about four to five times as much trehalose is required to maximize retention as is required to minimize fusion. Clearly, preventing fusion during drying is not the whole answer to preventing leakage from liposomes.

These resonance energy transfer experiments also showed us that liposomes that are frozen and then thawed had experienced four to five times more probe intermixing than those which had been frozen and dried at low trehalose concentrations. These differences suggest that much of the damage during freeze-thaw cycle comes from the thawing event.

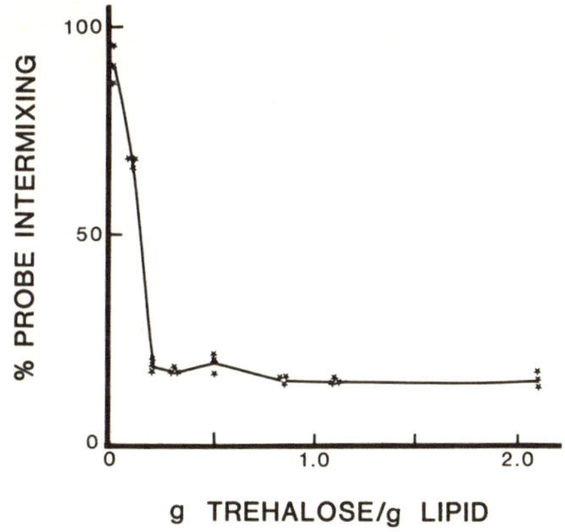

Figure 12-5. Estimate of POPC-PS liposome fusion by resonance energy transfer (described in text). Probe intermixing is minimized at a trehalose:lipid mass ratio of 0.2 g/g. This is less than the amount of trehalose required to maximize retention (see Figure 12-3).

Dehydration and Phase Transitions in Liposomes

Phase separation of mixtures of phospholipids in hydrated liposomes is dependent on the nature of the head group, the difference in length of acyl chains, differences in saturation of acyl chains, and the presence of molecules such as cholesterol (Luna and McConnell, 1978; Phillips et al., 1972; Cullis and de Kruijff, 1978; Cullis and Hope, 1980; Boni and Hui, 1983; Carruthers and Melchior, 1983). The effect of hydration on phase separations of phospholipid mixtures has not been studied as far as we know, but Quinn (1985) has combined data from papers by Chapman et al. (1967) and Seddon et al. (1983) which show the effects of dehydration on DPPC and didodecylphosphatidylethanolamine (DDPE; see Chapter 11). If these two phospholipids behave independently in a mixture, it is clear that dehydration at a constant temperature would lead to phase separation of lamellar DPPC and of DDPE; at very low water concentrations it might lead to the formation of hexagonal phase from DDPE.

A convenient way of looking at phase separations in phospholipid mixtures is with differential scanning calorimetry. Although most mixtures of phospholipids show complete miscibility while in the liquid crystalline phase, many will tend to separate as they are cooled to the gel phase of the lipid with the higher transition temperature. As the mixture is cooled further, other lipids will gel at their characteristic transition temperatures, leading to further lateral separation of different lipids. This process could also occur as membrane lipids dry at constant temperature, as suggested by phase diagrams of different lipids (Chapman et al., 1967; Seddon et al., 1983). Mixtures of dry phospholipids which have undergone lateral phase separation will show more than one transition in calorimetric scans.

We have used DSC to look at the transition temperatures of a mixture of 90 mol % POPC and 10 mol % PS both as a multilamellar system colyophilized with trehalose and as lyophilized unilamellar liposomes like those used for the retention and fusion studies described above. The transition temperatures (T_m) shown in all cases are the peaks of the transitions. Many of the transitions were very broad, and in some cases it was difficult to identify peaks, although computer analysis minimized this problem.

The phosphatidylserine used in these preparations had a very low T_m (Figure 12-6) which was only slightly affected by the addition of trehalose. Pure POPC on the other hand, had a T_m well above room temperature which was strongly affected by the addition of trehalose, so that at a mass ratio of 1:1 the POPC transition was actually below that of pure PS, a drop of about 70°C. A mixture of POPC and PS, lyophilized with various mass ratios of trehalose:lipid shows two transitions at lower mass ratios of trehalose, a high one which is about 6-8° lower than pure POPC and a low one which is about 15° lower than pure PS. The lowering of one T_m suggests that phase separation is not complete. The significant fact of Figure 12-6, however, is that, between 0.3 and 1.0 g trehalose/g lipid, the high-temperature transition completely disappears and only a low-temperature transition is seen. In other words, when sufficient trehalose is present, phase separation of the POPC and PS does not occur. The enthalpy of the high-temperature transition falls and the enthalpy of the low temperature rises as the mass ratio of trehalose in the POPC-PS mixture increases, suggesting that, as trehalose increases, the proportion of lipid participating in the low-temperature transition rises.

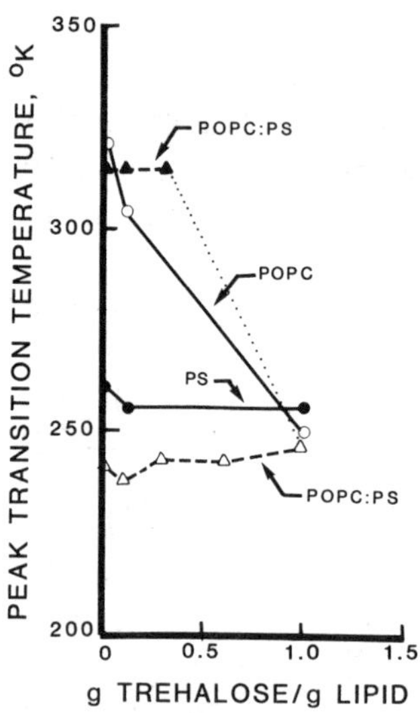

Figure 12-6. Differential scanning calorimetry of bulk POPC on PS. T_m of POPC, PS, and a 9:1 mixture of POPC:PS at various trehalose:lipid mass ratios. At low trehalose content, the mixture of POPC:PS shows two transitions (hence the two curves), but at high trehalose content (0.6 g/g) there is only a low-temperature transition. T_m of PS is very little affected, but the transition temperature of POPC is driven down by increasing trehalose. All of these samples were prepared by lyophilization from a mixture of trehalose, lipid, benzene, and methanol.

The results of DSC using lyophilized unilamellar liposomes are not so clear-cut. These liposomes, made by the rotary drying method, had a constant fixed small amount of trehalose trapped inside them (about 0.05-0.1 g/g). After they were put through a Sephadex column to remove external trehalose, various amounts of trehalose were added back to the outside of the liposomes and they were lyophilized. Hydrated liposomes made in this way are all in the same size range: 85% of the vesicles are between 25 and 70 nm, the other 15% are all below 150 nm. The method of preparation gave the liposomes various degrees of asymmetry in the amount of trehalose on each side of the bilayer. At all concentrations of trehalose we detected two transitions in the dry liposomes (Figure 12-7). When no trehalose was present inside or outside, the high transition was at the same temperature as the POPC:PS mixture which was lyophilized from benzene-methanol—not surprising, since these unilamellar vesicles had certainly fused during drying (see Figure 12-4b). When small amounts of trehalose were present inside and outside the bilayer, there was a slight drop in the temperature of the high transition: these liposomes were also fused and resembled multilayers with a small amount of trehalose between some of the bilayers (Figure 12-4b). At higher trehalose mass ratios, the temperature of the high transition increased by about 14-20°. At the trehalose mass ratios used in these preparations it is unlikely that vesicle fusion has occurred (see Figures 12-4a, 12-5), and we suggest that the presence of a high transition is due to the presence of an inner half bilayer that is not seeing any increase in trehalose concentration. The

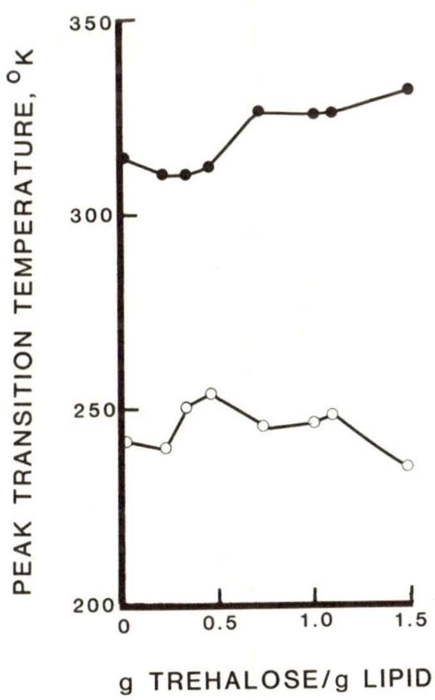

Figure 12-7. DSC of POPC:PS unilamellar liposomes made by the rotary drying method and lyophilized at various trehalose:lipid mass ratios. At all trehalose: lipid mass ratios, there is both a high-and low-temperature transition, T_m, in contrast to curves for the multilamellar preparations plotted in Figure 12-6. At the lowest trehalose concentrations, the liposomes have fused into multilamellar preparations (cf. Figures 12-4, 12-5). At higher trehalose content, the preparations consist of unilamellar vesicles.

enthalpy change of this high transition is fairly constant with the addition of increasing amounts of trehalose and is approximately one-third to one-half the enthalpy change of the transition of POPC:PS multilayers, again suggesting that it comes from the inner half of the bilayer.

If the constant enthalpy change of the high transition in the dry liposomes comes from the inner half of the bilayer which has a constant amount of trehalose interacting with it, then increasing enthalpies from the low transition should reflect changes associated with increasing amounts of trehalose interacting with the POPC:PS of the outer half of the bilayer. The enthalpy change of the low transition of the dry liposomes increases with increasing amounts of trehalose (Figure 12-8), as in the multilayer preparations, and when these enthalpies are compared with the mass ratio of trehalose outside the liposomes:phospholipid of the outer bilayer, the enthalpy changes are similar. If the enthalpy of the low-temperature transition is compared to the percent retention of trapped solute in lyophilized unilamellar vesicles, the curves can be seen to be similar, with both reaching a maximum between 0.6 and 0.8 g trehalose per g phospholipid.

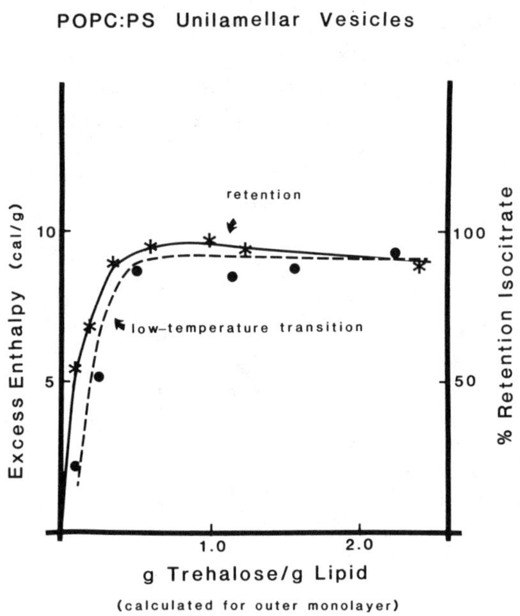

Figure 12-8. Excess enthalpies of the low DSC transitions of lyophilized liposomes of POPC:PS and multilayers of POPC:PS lyophilized from a benzene/methanol mixture. The trehalose:lipid mass ratios of the liposomes have been calculated from the trehalose *outside* the liposomes and the phospholipid of the *outer* half of the bilayer. The excess enthalpies of both types of preparations are similar.

This suggestion of phase separation of the inner from the outer monolayer of the lyophilized liposomes has some basis in experimental data. A number of laboratories have shown that small unilamellar vesicles of DPPC have a broadened phase transition and lower T_m than multilamellar liposomes (Suurkuusk et al., 1976; van Dijck et al., 1978). This effect is thought to be due to strain introduced into the bilayer by the high radius of curvature of small unilamellar liposomes. Hydrodynamic and NMR data, combined with geometric analysis, suggest that there is about twice as much lipid in the outer half bilayer as in the inner half of small unilamellar vesicles, and the two monolayers have different acyl chain packing (Mason and Huang, 1978). High-resolution ^1H NMR saturation transfer experiments provide evidence that there are differences in packing and mobility of the acyl chains in the two monolayers, with the outer acyl chains being more restricted in motion. Hydrated vesicles appear to show a single calorimetric transition, but dry monolayers of unilamellar vesicles are not necessarily as tightly coupled. Upon drying, the original spherical vesicles collapse into a cup shape with a very small internal volume (Figure 12-4a). Most of the fracture images from such a shape will be concave or convex spheres, while a few vesicles will be fractured to show the cup shape, and some which are cross-fractured near the edge of the cup will show the internal space. The geometry of such a shape is very complex, and we have not yet attempted to predict the disposition or spacing of the phospholipid molecules in each monolayer. The increase in the T_m of the high-temperature transition may reflect a tighter packing of the head groups of the inner monolayer.

A number of small molecules such as glycerol and methanol (McDaniel et al., 1983; O'Leary and Levin, 1984), ethanol (Simon and McIntosh, 1984), chlorpromazine, tetracaine, and phenyl and benzyl alcohols (McIntosh et al., 1983) have been shown to cause interdigitation of acyl chains from each half of bilayers of multilamellar DPPC. These effects have been noted in hydrated systems and in pure glycerol. Preliminary data (M. Caffrey, personal communication) indicates that, when DPPC is lyophilized from a benzene-methanol solution, it forms multilayers. If trehalose is added to the mixture before lyophilization in equal mass ratios with DPPC, the dry samples in the gel phase have a lamellar repeat of 38 Å, and the phospholipids are in the regular hexagonal packing, L_β with the chains in an all-*trans*, fully extended conformation. These data suggest that, in the presence of trehalose at a 1:1 mass ratio with multilayers of DPPC, the acyl chains are interdigitated to about 50% of their length. From these data on DPPC multilayers, we can infer that the two halves of the bilayer in lyophilized liposomes are not interdigitated, since it is hard to envision how interdigitated phospholipids could have uncoupled transitions.

The low-temperature transition showed some variability, with an increase and then a decrease in transition temperature with the addition of trehalose outside the liposomes. The temperature of the low transition at a mass ratio of trehalose:lipid of 1:1 was the same whether the preparation was unilamellar liposomes or multilayers lyophilized from benzene-methanol.

From our interpretation of the DSC results with lyophilized unilamellar vesicles and with homogeneous multilayer:trehalose preparations, it appears that dry POPC:PS vesicles do not leak upon rehydration when (1) the dif-

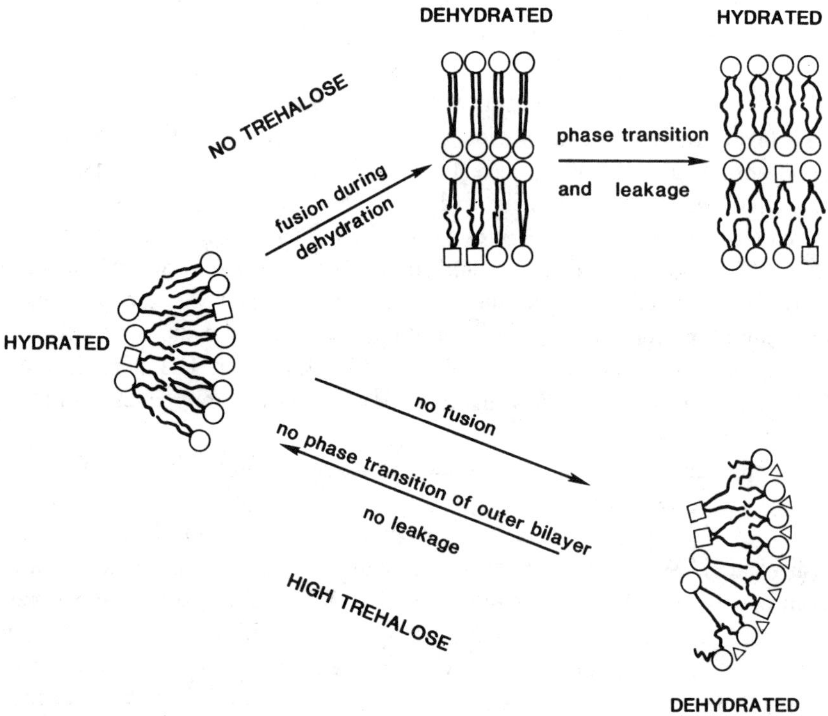

Figure 12-9. Diagram of events which may occur during lyophilization and rehydration of unilamellar liposomes at no trehalose (or at low trehalose concentrations) and at high trehalose concentrations. At room temperature, the hydrated POPC and PS are in liquid crystalline phase. When lyophilization is complete, the lipids warm to room temperature. If little trehalose is present (upper sequence), the lipids phase separate and upon rehydration the POPC undergoes a phase transition and the liposomes leak. When the trehalose concentration is high (lower sequence), fusion is inhibited, and the POPC and PS form a single phase when dry. The POPC and PS of the outer half of the bilayer are in liquid crystalline phase when dry at room temperature. They do not undergo a phase transition during rehydration and the liposomes do not leak. POPC, (○); PS, (□); trehalose, (△).

ferent phospholipids of the outer half of the bilayer are not phase separated and (2) when the outer half of the bilayer does not undergo a phase transition upon rehydration. These occur when there is sufficient trehalose present outside the liposomes to bring the transition temperature of POPC down so that it forms a single phase with PS; as a consequence, that part of the phospholipid is in liquid crystalline phase when rehydrated at room temperature. The difference in trehalose concentration needed to prevent fusion and that needed to maximize retention is most probably related to the amount needed to prevent phase separation and phase transitions when rehydrating. A diagrammatic representation of the effect of trehalose on solute leakage when dry POPC:PS liposomes are rehydrated at room temperature is given in Figure 12-9. This diagram is specific for this lipid mixture; however, the generalization can be made that dry liposomes (and presumably membranes in whole organisms) will show minimal leakage when they are (1) prevented from fusing, (2) prevented from undergoing phase separation during drying, and (3) prevented from undergoing phase transitions during rehydration.

Effects of Other Carbohydrates

As discussed in Chapter 11, trehalose is found widely in anhydrobiotic organisms, so it is satisfying to see that it has such dramatic effects on preserving both biological membranes and synthetic liposomes. We have shown that trehalose is effective at much lower concentrations than other carbohydrates both in preserving structure and function in sarcoplasmic reticulum (Crowe et al., 1982) and in affecting physical properties of DPPC (Crowe, Crowe, and Chapman, 1985). We were also interested in finding out its comparative effectiveness in preventing fusion and leakage in dry liposomes. Isocitrate retention following lyophilization was followed for a number of carbohydrates, with the results shown in Figure 12-10. Only one other carbohydrate, the disaccharide maltose, is able to prevent leakage as well as trehalose. For practical reasons which are set forth in Chapter 11, maltose is not useful for preserving dry membranes or dry liposomes. The next best sugar, sucrose, never achieves the high degree of preservation that trehalose does. Inositol, as with membranes and DPPC, has almost no effect on liposomes. Preservation by the monosaccharides glucose and galactose was very low, never reaching more than 40% retention. Experiments with trapped radioactive glucose and trehalose showed that, while very little trehalose leaked from freshly made vesicles, glucose leaked very rapidly. Since the liposomes with glucose and galactose were dialyzed at least overnight, they had almost no carbohydrate inside them by the time they were lyophilized.

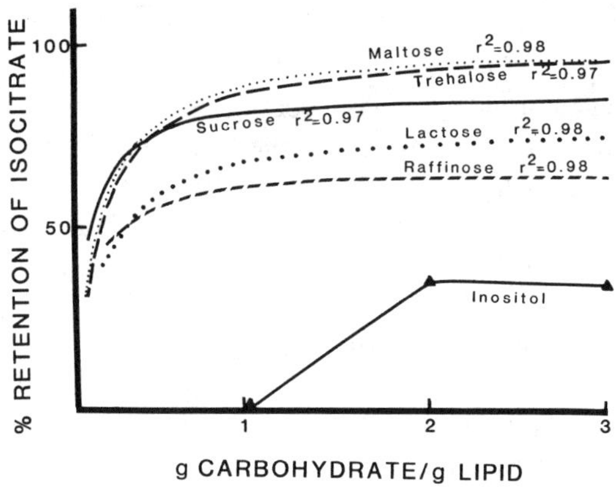

Figure 12-10. Retention of isocitrate by DPPC liposomes made by the rotary drying method and lyophilized with various amounts of carbohydrates. The lines, except for inositol, are fitted curves calculated using a BASIC curve-fitting program (Spain, 1982).

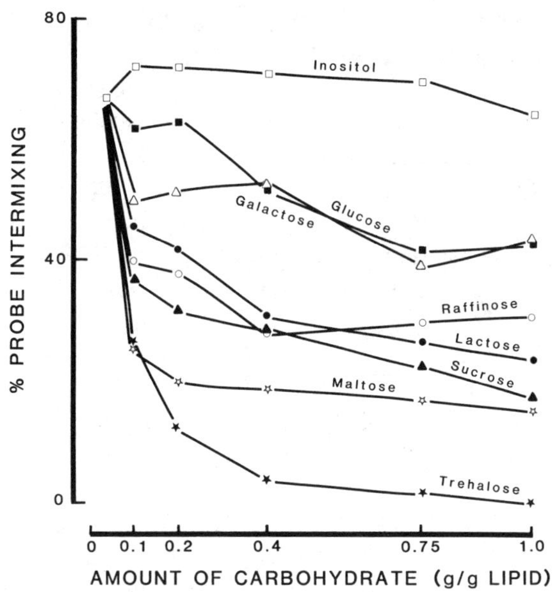

Figure 12-11. Fusion of DPPC liposomes (as estimated by resonance energy transfer) lyophilized with various mass ratios of several carbohydrates.

Resonance energy transfer experiments were used to estimate the amount of fusion of lyophilized liposomes in the presence of the various carbohydrates (Figure 12-11). The liposomes were made the same way as the ones for retention and fusion experiments described above. Except for glucose and galactose the amount of carbohydrate needed to minimize fusion was in the range of 0.2-0.4 g/g, much less than that needed to maximize retention. The high amount of fusion with glucose and galactose is probably related to the leakage of these carbohydrates from the liposomes. Inositol is ineffective in preventing fusion.

Conclusions

Many carbohydrates can stabilize liposomes against fusion and against leakage during freeze drying. Trehalose, which is found in high concentration in anhydrobiotic organisms, is most effective against fusion, and it and maltose are most effective in preventing leakage during dehydration. Trehalose has also been found to drive down the T_m of vesicles of POPC and prevent phase separation of POPC and PS in homogeneous dry multilamellar preparations of a mixture of these phospholipids. A tentative interpretation of calorimetric data from lyophilized liposomes suggests that the inner and outer monolayers of the vesicle wall are independent of each other with regard to phase changes when there is enough trehalose present outside the liposomes to prevent phase separation of POPC and PS in the outer monolayer. Trehalose may serve as a protectant against desiccation damage in anhydrobiotic organisms by suppressing phase separations and phase transitions in dry membranes. Such effects would lead to maintenance of integrity of membranes during rehydration.

REFERENCES

Blok, M.C., van Deenen, L.L.M., and de Gier, J. 1976. Effect of the gel to liquid crystalline phase transition on the osmotic behavior of phosphatidylcholine liposomes. Biochim. et Biophys. Acta 433: 1.

Blok, M.C., van der Neut-Kok, E.C.M., van Deenen, L.L.M., and de Gier, J. 1975. The effect of chain length and lipid phase transitions on the selective permeability properties of liposomes. Biochim. et Biophys. Acta 406: 187.

Boni, L.T., and Hui, S.W. 1983. Polymorphic phase behavior of dilinoleoylphosphatidylethanolamine and palmitoyoleoylphosphatidylcholine mixtures. Biochim. et Biophys. Acta 731: 177-185.

Carruthers, A., and Melchior, D.L. 1983. Studies of the relationship between bilayer water permeability and bilayer physical state. Biochemistry 22: 5797.

Chapman, D., Williams, R.M., and Ladbrooke, B.D. 1967. Physical studies of phospholipids. VI. Thermotropic and lyotropic mesomorphism of some 1,2-diacyl-phosphatidylcholines (lecithins). Chem. Phys. Lipids 1: 445.
Crowe, J.H., Crowe, L.M., and Chapman, D. 1984a. Preservation of membranes in anhydrobiotic organisms: the role of trehalose. Science 223: 701.
Crowe, J.H., Crowe, L.M., and Chapman, D. 1984b. Infrared spectroscopic studies on interactions of water and carbohydrates with a biological membrane. Arch. Biochem. Biophys. 232: 400.
Crowe, J.H., Crowe, L.M., and Jackson, S.A. 1982. Preservation of structural and functional activity in lyophilized sarcoplasmic reticulum. Arch. Biochem. Biophys. 220: 477.
Crowe, J.H., Crowe, L.M., and Mouradian, R. 1983. Stabilization of membranes at low water activities. Cryobiology 20: 346.
Crowe, J.H., Lambert, D.T., and Crowe, L.M. 1978. Ultrastructural and freeze fracture studies on anhydrobiotic nematodes. *In* Dry Biological Systems, J.H. Crowe and J.S. Clegg, eds. Academic Press, London, New York.
Crowe, J.H., O'Dell, S.J., and Armstrong, D.A. 1979. Anhydrobiosis in nematodes: permeability during rehydration. J. Exp. Zool. 207: 431.
Crowe, L.M. and Crowe, J.H. 1982. Hydration-dependent hexagonal phase lipid in a biological membrane. Arch. Biochem. Biophys. 217: 582.
Crowe, L.M., Crowe, J.H., and Chapman, D. 1985. Interaction of carbohydrates with dry dipalmitoylphosphatidylcholine. Arch. Biochem. Biophys. 236: 289.
Crowe, L.M., Crowe, J.H., Rudolph, A., Womersley, C. and Appel, L. 1985. Preservation of freeze-dried liposomes by trehalose. Arch. Biochem. Biophys. 242: 240.
Crowe, L.M., Mourdian, R., Crowe, J.H., Jackson, S.A., and Womersley, C. 1984. Effects of carbohydrates on membrane stability at low water activities. Biochim. Biophys. Acta 769: 141.
Cullis, P.R. and de Kruijff, B. 1978. Polymorphic phase behavior of lipid mixtures as detected by PNMR. Evidence that cholesterol may destabilize bilayer structure in membranes containing phosphatidylethanolamine. Biochim. Biophys. Acta 507: 207.
Cullis, P.R., and Hope, M.J. 1978. Effects of fusogenic agents on membrane structure and function of erythrocyte ghosts and the mechanism of membrane fusion. Nature 271: 672.
Cullis, P.R., and Hope, M.J. 1980. The bilayer stabilizing role of sphingomyelin in the presence of cholesterol. A PNMR study. Biochim. Biophys. Acta 597: 533.
Hammoudah, M.M., Nir, S., Bentz, J., Mayhew, E., Stewar, T.P., Hui, S.W., and Kurian, R.J. 1981. Interactions of La^{3+} with phosphatidylserine vesicles: binding, phase transition, leakage, PNMR and fusion. Biochim. Biophys. Acta 645: 102.
Hui, S.W., Stewart, T.P., Boni, L.T., and Yeagle, P.L. 1981. Membrane fusion through point defects in bilayers. Science 212: 921.
Leopold, A.C. 1980. Temperature effects on soybean imbibition and leakage. Plant Physiol. 65: 1096.
Luna, E.J., and McConnell, H.M. 1978. Multiple phase equilibria in binary mixtures of phospholipids. Biochim. Biophys. Acta 509: 462.
Luzzati, V., and Husson, F. 1962. The structure of the liquid-crystalline phases of lipid-water systems. J. Cell Biol. 12: 207.

McDaniel, R.V., McIntosh, T.J., and Simon, S.A. 1983. Nonelectrolyte substitution for water in phosphatidylcholine bilayers. Biochim. Biophys. Acta 731: 97.

McIntosh, T.J., McDaniel, R.V., and Simon, S.A. 1983. Induction of an interdigitated gel phase in fully hydrated phosphatidylcholine bilayers. Biochim. Biophys. Acta 731: 109.

McKersie, B.D. and Stinson, R.H. 1980. Effect of dehydration on leakage and membrane structure in *Lotus corniculatus* L. seeds. Plant Physiol. 66: 316.

Madin, K.A.C., and Crowe, J.H. 1975. Anhydrobiosis in nematodes: carbohydrate and lipid metabolism during dehydration. J. Exp. Zool. 193: 335.

Marsh, D., Watts, A., and Knowles, P.F. 1976. Evidence for phase boundary lipid: permeability of Tempo-choline into dimyristoylphosphatidylcholine vesicles at the phase transition. Biochemistry 15: 3570.

Mason, J.T., and Huang, C. 1978. Hydrodynamic analysis of egg phosphatidylcholine vesicles. *In* Liposomes and Their Uses in Biology and Medicine. D. Papahadjopoulos, ed. New York Acad. Sciences, New York, p. 29.

Murphy, J.B., and Noland, T.L. 1982. Temperature effects on seed imbibition and leakage mediated by viscosity and membranes. Plant Physiol. 69: 428.

O'Leary, T.J., and Levin, E.W. 1984. Raman spectroscopic study of an interdigitated lipid bilayer. Dipalmitoylphosphatidylcholine dispersed in glycerol. Biochim. Biophys. Acta 776: 185.

Papahadjopoulos, D., Jacobsen, K., Nir, S., and Isac, T. 1973. Phase transitions in phospholipid vesicles. Fluorescence polarization and permeability measurements concerning the effect of temperature and cholesterol. Biochim. Biophys. Acta 311: 330.

Phillips, M.C., Hauser, H., and Paltauf, F. 1972. The inter- and intra-molecular mixing of hydrocarbon chains in lecithin/water systems. Chem. Phys. Lipids 8: 127.

Pick, U. 1981. Liposomes with a large trapping capacity prepared by freezing and thawing of sonicated phospholipid mixtures. Arch. Biochem. Biophys. 212: 186.

Plaut, G.W.E. 1962. Isocitric dehydrogenase (TPN-linked) from pig heart. Methods Enzymol. 5: 645.

Priestley, D.A. and de Kruijff, B. 1982. Phospholipid motional characteristics in a dry biological system. Plant Physiol. 70: 1075.

Quinn, P.J. 1985. A lipid phase separation model of low-temperature damage to biological membranes. Cryobiology 22: 128.

Rosenbluth, J. 1969. Sarcoplasmic reticulum of an unusually fast- acting crustacean muscle. J. Cell Biol. 42: 534.

Rudolph, A.S., and Crowe, J.H. 1985. Membrane stabilization during freezing: the role of two natural cryoprotectants, trehalose and proline. Cryobiology 22: 367.

Schuh, J.R., Banerjee, U., Muller, L., and Chan, S.I. 1982. The phospholipid packing arrangement in small bilayer vesicles as revealed by proton magnetic resonance studies at 500 MHz. Biochim. Biophys. Acta 687: 219.

Seddon, J.M., Cevc, G., and Marsh, D. 1983. Calorimetric studies of the gel-fluid (L_β-L_α) and lamellar-inverted hexagonal (L_α-H_{II}) phase transitions in dialkyl- and diacylphosphatidylethanolamines. Biochemistry 23: 1280.

Seewaldt, V., Priestley, D.A., Leopold, A.C., Feigenson, G.W., and Goodsaid-Zalduondo, F. 1981. Membrane organization in soybean seeds during hydration. Planta 152: 19.

Shew, R., and Deamer, D. 1985. A novel method for encapsulation of macromolecules in liposomes. Biochim. Biophys. Acta 816: 1.
Simon, E.W. 1974. Phospholipids and plant membrane permeability. New Phytol. 73: 377.
Simon, E.W. 1978. Plant membranes under dry conditions. Pestic. Sci. 9: 169.
Simon, E.W., and Raja Harun, R.M. 1972. Leakage during seed imbibition. J. Exp. Bot. 23: 1076.
Simon, S.A., and McIntosh, T.J. 1984. Interdigitated hydrocarbon chain packing causes the biphasic transition behavior in lipid/alcohol suspensions. Biochim. Biophys. Acta 773: 169.
Spain, J.D. 1982. Basic Microcomputer Models in Biology. Addison Wesley, Reading, MA.
Suurkuusk, J., Lentz, B.R., Barenholz, Y., Biltonen, R.L., and Thompson, T.E. 1976. A calorimetric and fluorescent probe study of the gel-liquid crystalline phase transition in small, single-lamellar dipalmitoylphosphatidylcholine vesicles. Biochemistry 15: 1393.
Uster, P.S., and Deamer, D.W. 1981. Fusion competence of phosphatidylserine-containing liposomes quantitatively measured by a fluorescence resonance energy transfer assay. Arch. Biochem. Biophys. 209: 385.
van Dijck, P.W.M., de Kruijff, B., Aarts, P.A.M.M., Verkleij, A.J., and de Gier, J. 1978. Phase transitions in phospholipid model membranes of different curvature. Biochim. Biophys. Acta 506: 183.
Wilschut, J., and Papahadjopoulos, D. 1979. Ca^{2+}-induced fusion of phospholipid vesicles monitored by mixing of aqueous contents. Nature 281: 690.

13 The Interaction of Trehalose with the Phospholipid Bilayer: A Molecular Modeling Study

Bruce P. Gaber
Indira Chandrasekhar
Nagarajan Pattabiraman

There is now ample evidence in the literature (Crowe et al., 1984) and from information presented in Chapters 11 and 12 that stabilization of organisms against dehydration is, at least in part, a membrane phenomenon. In particular the interaction of the membrane with water and with small molecules which can, in some way, substitute for water seems to be critical to our understanding of anhydrobiotic processes at a molecular level. Our approach to studying interactions involving membranes, water, and small molecules is to use computer modeling techniques to display accurate graphical representations of phospholipids and the bilayers they form, to model the interactions both graphically and by molecular mechanics, and ultimately to model the molecular dynamics of the processes involved. The results reported here indicate our progress with graphical modeling.

Traditionally our view of the cell membrane and the phospholipid bilayer of which it is composed has been iconographic at best. The membrane is usually depicted as a conceptual schematic of the fluid mosaic model or as a two dimensional sketch of a series of wiggly lines coupled to circles.

We have attempted to represent lipid molecules and the bilayer in a new light in which the structures are accurate representations of the physical chemistry of the molecules. The objective of this effort, at the most fundamental level, is to produce a model which is a more realistic guide to our thinking about the nature of the phospholipid bilayer. However, once such a model has been established, we soon observe aspects of bilayer structure that would otherwise have remained obscure. Furthermore, a chemically and physically accurate bilayer model is an obligatory starting point for subsequent attempts at higher level molecular modeling.

Database

The starting point for our modeling effort has been the construction of a data base consisting of the coordinates of lipids for which crystallographic data have been compiled. Lipids appear to be rather more recalcitrant than proteins in forming usable crystals. However, the work from several laboratories (see Table 13-1 for references) has resulted in a range of structures sufficient for our initial purposes. The phospholipids and their analogues crystallized to date and a summary of the properties of the crystals are compiled in Table 13-1.

By working with the coordinates for the lipid molecules and the space group in which the lipid crystallizes, it is possible to transform the initial fractional coordinates into orthogonalized coordinates representative of a bilayer.

For example, for dimyristoyl phosphatidylcholine (DMPC), the unit cell consists of four molecules organized as tail-to-tail pairs. Each pair consists of two molecules (type A and type B) which differ only in the conformation of the head group (this difference will be described more fully below). The space group P21 (Table 13-1) defines the operations which will replicate the unit cell molecules into a model of the bilayer. In the case of DMPC, we have constructed a model consisting of 36 lipid molecules (arranged 6 by 6) in each of the upper and lower monolayers. The size of the model is limited only by the number of atoms that can conveniently be manipulated computationally. For the DMPC model this is about 3,400 atoms, roughly the number in a protein of 44,200 daltons.

The data file is arranged in Protein Data Bank format such that each individual lipid is assigned a residue number. Thus any particular molecule or cluster of molecules may be selected for closer study. Atoms of each lipid are numbered following Sundaralingam (1972). In this manner, it is possible to identify a specific atom on a given molecule. For example, the asymmetric carbon of the lipid in the second row and the third column of the upper monolayer is designated C2 DML 23.

The solvent of crystallization, too, can be included in the model. Three waters of crystallization are closely associated with each of two conformers in the structure of the DMPC bilayer. In the crystal, a fourth water links the first three to the waters of the adjoining bilayer. The coordinates of these waters may be transformed exactly as are the lipid molecules, placing them in their crystallographically correct location in the bilayer.

Graphic Realizations

The actual graphic realization of the bilayer model has been achieved in two ways. Very high resolution space-filling raster images have been realized

Table 13-1
Structural parameters for various phospholipids and phospholipid analogues

Item	DLPE	DMPC	LPPC	LPA	PPE	DLG	DMPA	HPA1	HPA2
Crystal system	Monoclinic	Monoclinic	Monoclinic	Triclinic	Monoclinic	Monoclinic	Monoclinic	Triclinic	Monoclinic
Space group	P21/c	P21	P21/c	$P\bar{1}$	P21/a	P21	P21	$P\bar{1}$	P21
Unit cell (Å)	a = 47.70 b = 7.77 c = 9.95 γ = 92.00	a = 8.72 b = 8.92 c = 55.40 γ = 97.40	a = 24.82 b = 9.53 c = 10.94 γ = 99.70	a = 7.74 b = 5.54 c = 32.87 α = 92.50 β = 99.20 γ = 128.30	a = 7.66 b = 9.08 c = 37.08 γ = 90.20	a = 5.46 b = 7.59 c = 34.20 γ = 93.10	a = 5.44 b = 7.95 c = 43.98 γ = 114.20	a = 4.75 b = 5.72 c = 44.36 α = 91.00 β = 101.50 γ = 100.50	a = 4.75 b = 5.72 c = 88.72 γ = 100.80
Number of molecules per unit cell	4	4	4	2	4	2	2	2	4
Molecular area ($Å^2$)	38.6	38.9	52.1	33.6	34.8	41.4	24(43.30)[a]	26.69	26.71
Chain tilt	0	12.0	41.0	55.0	57.5	26.5	31.0	47.1	46.1
Chain packing mode	HS1	HS	M//	T//	T//	0/	T//	T//	T//
Chain cross section ($Å^2$)	19.3	19.0	19.7	19.3	18.7	18.5	18.6	18.2	18.5

Note: DLPE: 1,2 dilauroyl-DL-phosphotidylethanolamine; DMPC: 1,2-dimyristoyl-sn-glycero-3-phosphoroylcholine; LPPC: 3-dodecanoyl-propandiol-1-phosphoroylcholine; LPA: disodium 3-lauroyl-DL-glycero-1-phosphate dihydrate; PPE: 3-palmitoyl-DL-glycerol-1-phosphoroylethanolamine; DLG: 2,3-dilauroyl-D-glycerol; DMPA: monosodium 1,2-dimyristoyl-sn-glycerophosphate; HPA: hexadecyl-2-deoxyglycerophosphoric acid monohydrate.

[a] Area per head group is 24 $Å^2$. However, the molecules pack with two head groups from adjacent bilayers juxtaposed such that the effective head group area is 43.3 $Å^2$.

using the software package ''Amos'' written by Dr. R. Brown of the Naval Research Laboratory (Gaber and Brown, manuscript in preparation). ''Amos,'' written in Pascal, was specifically designed for lipid modeling and has an extended data structure which permits files of any size to be easily manipulated. This capability is particularly valuable for handling very large lipid arrays. The program permits the user to choose of atom color, size, and shading; stereo viewing directly on the color terminal and the production of stereo photographs are optional. Although the development of a very high-resolution image with ''Amos'' is computationally intensive, the program has a preview feature which permits rapid on-screen manipulation of skeletal and ball-and-stick models. Once a desired view has been chosen, it can easily be converted to an image that fully fills the space.

''Amos'' is currently implemented on a VAX 11/785 with a Ramtek 2020 as the graphic output device. Images of very high resolution (4000 × 4000 pixels) are produced by post processing on a Dicomed film recorder. Graphic drivers now under development will enable the ''Amos'' picture files to be displayed on the screens of personal computers. It is anticipated that substantial portions of the Pascal code will be usable on a personal computer, thus making high resolution molecular graphics directly available to the investigator.

For truly interactive molecular graphics of our bilayer model, we use an Evans and Sutherland PS 330. This vector device permits real time manipulation of the model such that it is possible to rotate, translate, and scale the model at will as well as to rotate, break and make bonds, and measure bond angles and lengths. As will be discussed below, the PS 330 is particularly useful for docking experiments in which a small molecule is juxtaposed with the bilayer. Solvent-accessible and van der Waals surfaces are also computed easily.

The solvent accessible, or Connolly surface, requires further explanation. The Connolly algorithm is based upon the concept (see Langridge et al., 1981; and Connolly, 1983) of mapping a molecular surface for which solvent access is possible. In its implementation, the algorithm computationally rolls a probing sphere the diameter of the solvent of interest across the van der Waals surface of the molecule. The locus of points of contact of the sphere with the van der Waals surface defines the ''contact surface.'' Locations where the probe partially or completely penetrates the molecular surface is called the reentrant surface. Reentrant surfaces appear as depressions, pockets, clefts, or cups where solvent could possibly be sterically accommodated.

Caveat

As we are dealing with idealized models derived from lipid crystals, caution should be exercised in extrapolating the model to the situation of fully hydrated bilayers. However, a comparison of NMR data with single crystal structures by Hauser et al. (1981) shows that the use of solid state data to represent and analyse a model membrane is not entirely unjustified. It should be noted that the models we develop are relatively simple in that they represent single lipid component bilayers and that, while the basic approach we have adopted permits us to incorporate other phospholipids, cholesterol, or proteins into our basic bilayer models, we have not yet done so. However, the bilayer model as presented here should prove instructive as to the nature of lipid structure and organization and should also provide a starting point for more sophisticated modeling.

Fundamentals of Phospholipid Structure

Much of what we know about the structure of DMPC and DLPE comes from the detailed analysis by Hauser et al. (1981) of crytallographic data (Elder et al., 1977; Pearson and Pascher, 1979). Many of the observations presented here are based on their work.

Dimyristoyl Phosphatidylcholine

The basic structure of a single DMPC is depicted as a skeletal- plus-van der Waals image in Plate 13-1a. The dominant structural element is the "monkey wrench" conformation adopted by the asymmetric arrangement of the acyl chains. Chain 1 forms an all-*trans* zigzag structure continuous with the three carbons of the glycerol. Chain 2, however, extends perpendicular from the glycerol moiety for two carbons and then turns and runs parallel to chain 1. (The jaws of the monkey wrench are formed from carbons 1 and 2 of chain 2 and from the choline phosphodiester bond.) The result of the chain asymmetry is an offset of approximately three methylene units between chains 1 and 2: an offset of about 3.7 Å. This conformational difference between chains 1 and 2 seems to be a fundamental element of phospholipid structure. It was first suggested by Seelig and Seelig (1975) on the basis of deuterium NMR studies and was later confirmed by Raman spectroscopy of selectively deuterated lipids (Gaber et al., 1978).

As noted earlier, the DMPC molecule crystallizes in two forms characterized by differences in the head group conformation. The type A and type B conformers are depicted in Plate 13-1b in which the skeletal conformation is emphasized by use of a ball-and-stick image (using ''Amos''). The two head group types display approximate mirror image symmetry. This implies, in effect, that the phospholipid head groups are oppositely oriented with respect to the hydrocarbon chains, in the two conformers. Another major difference between the conformers is that for type A the vector connecting the phosphorus and nitrogen lies at an inclination of about 17° relative to the bilayer surface, while in the type B molecule the P-N vector is at 27° relative to the bilayer. Despite this difference, the requirement for the zwitterionic head group to optimize both intra- and intermolecular interactions is observed as a constant distance between charge centers. For DMPC the distance between the center of gravity of the nonesterified oxygens on phosphorus and the choline nitrogen is 4.48 Å for DMPC type A and 4.68 Å for type B. A small difference exists in the orientation of the carbonyl oxygen in the ester linkage of the second chain. For the type A conformer, the C=O bond is almost parallel to the bilayer plane, whereas it points slightly upward for type B.

As already noted, the head group conformation of the type B head group tends to point up and away from the bilayer surface to a greater extent than does that of type A. Furthermore, in the bilayer, the conformers are offset with respect to one another by about 2.5 Å normal to the bilayer, with the type A molecule being displaced toward the midplane. The result is an alternating pattern in which the type B head group is highly exposed at the bilayer surface, while type A is almost buried. The tightly fitting arrangement keeps both head groups and chains in intimate contact. When the 12 molecules are viewed by looking almost directly down onto the head group region (Plate 13-1c), the orderly arrangement and tight packing of the bilayer is most striking. Rows of type A molecules are seen alternating with type B conformers, and considerable space appears to be available between successive tetrads of type head groups; we shall see that this space is partly occupied by bound water.

Dilauroyl Phosphatidylethanolamine

In the DLPE molecule we again encounter the monkey wrench conformation of the two acyl chains (Plate 13-1d). In contrast to DMPC, the orientation of the head group with respect to the bilayer normal is not sterically controlled by the bulky choline group. That is, even though the inclination of the P-N vector is 15° relative to the bilayer plane (compare 17° for type A DMPC) the head group is not forced upwards, as it is in both the DMPC

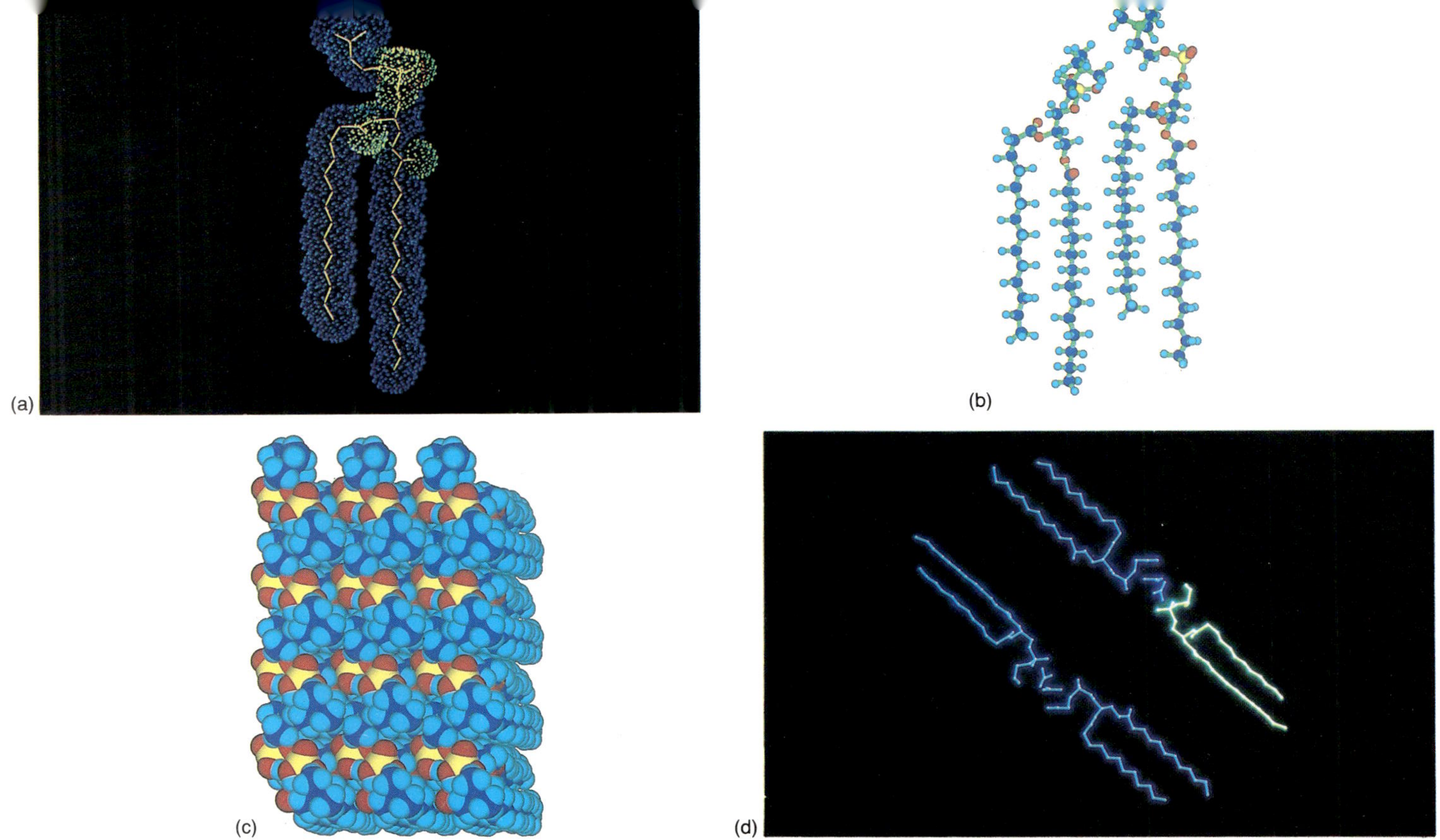

Plate 13-1. (a) A vector graphics depiction of a DMPC molecule with a van der Waals surface displayed on the atomic skeleton. See text for description of molecule. (b) A raster image of the two conformations which the DMPC molecule adopts in the crystalline state. See text for description. Type A conformer is at left, type B at right. Color codes: light blue = hydrogen; dark blue = carbon; red = oxygen; yellow = phosphorus. (c) The DMPC bilayer surface when viewed head-on; that is, down the Z-axis. This view emphasizes the tight packing of the head groups. Color codes as in Plate 13-1b. (d) A vector graphics image of the four DLPE molecules that constitute the unit cell. The L-DLPE (green) is related by the symmetry of the unit cell to its enatiomeric D-DLPE (blue) which is diagonally opposite.

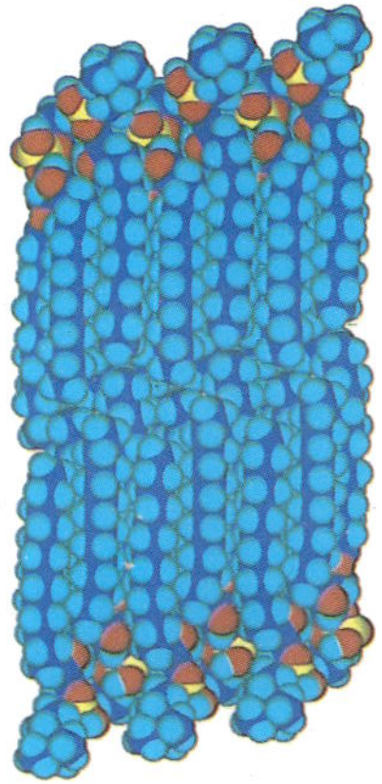

(a)

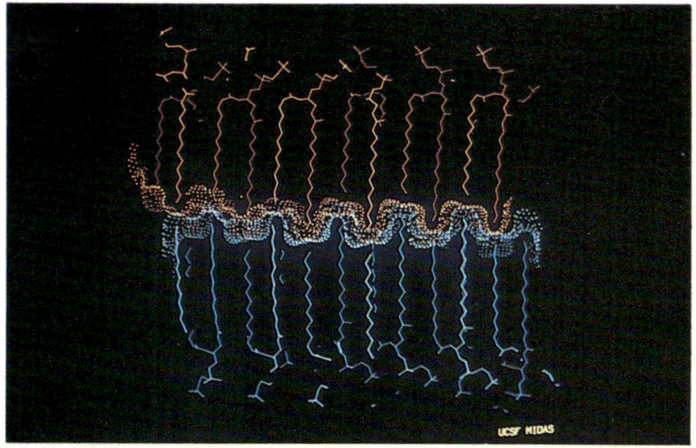

(b)

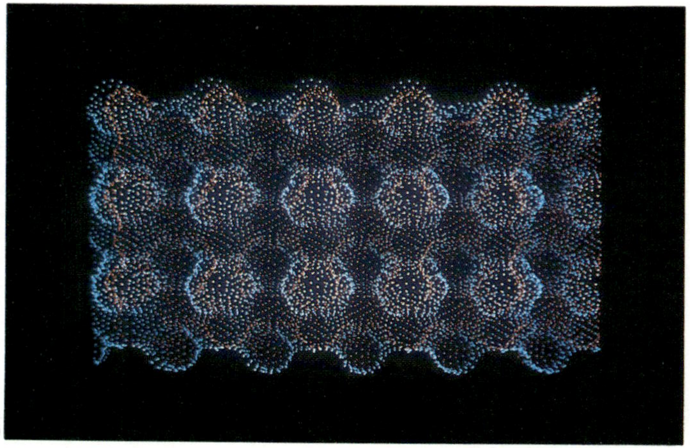

(c)

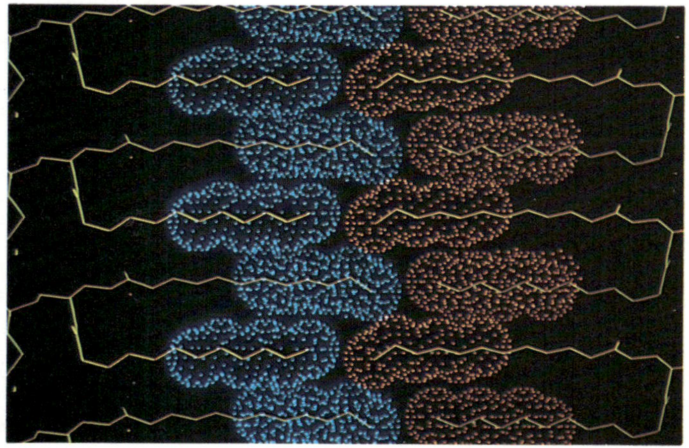

(d)

Plate 13-2. (a) A space-filling picture of the DMPC bilayer viewed down the x-axis of the system. Note the tilt of the hydrocarbon chains and the close packing of the hydrocarbon region. Color codes as in Plate 13-1b. (b) A vector graphics image showing the tight interdigitation of the two monolayers of DMPC at the bilayer midplane. Connolly surface indicated as dotted layer. (c) Cross sectional view of the Connolly surface in Plate 13-2b. (d) Vector graphics image of the DLPE bilayer with van der Waals surface; note the interdigitation at the bilayer midplane. View similar to (b).

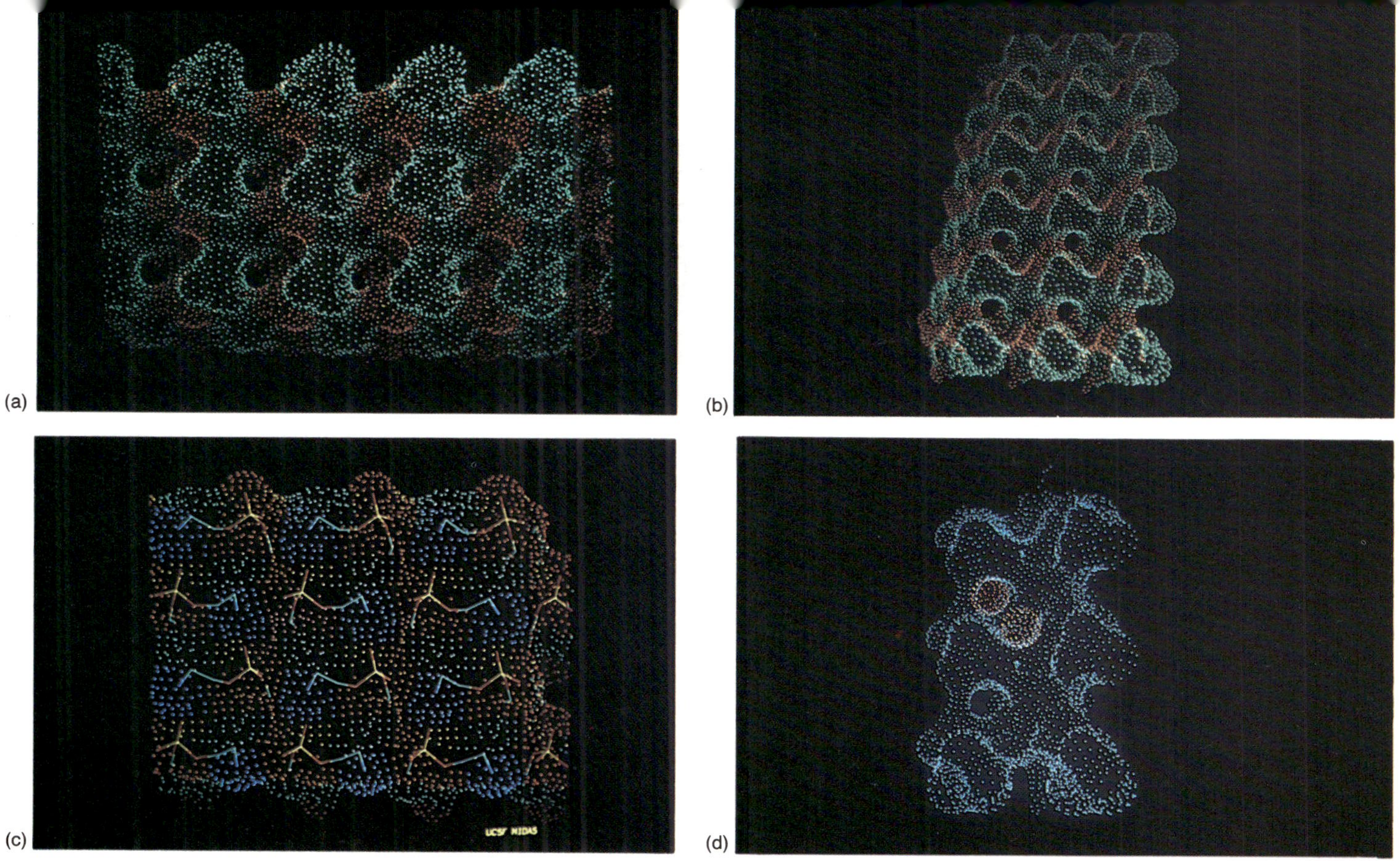

Plate 13-3. (a) Color-coded Connolly surface of the head group region in a DMPC monolayer. Code: red = hydrophilic domains, blue-green = hydrophobic domains. (b) Rotated and tilted view of Plate 13-3a, emphasizing the solvent accessible pockets on the surface of the DMPC bilayer. (c) Color-coded Connolly surface of DLPE bilayer. Compare with Plates 13-3a and 13-3b. Color code as in Plate 13-3a. (d) Solvent-accessible surface of the polar region of the DMPC bilayer with the two deeply buried waters of crystallization (red) included.

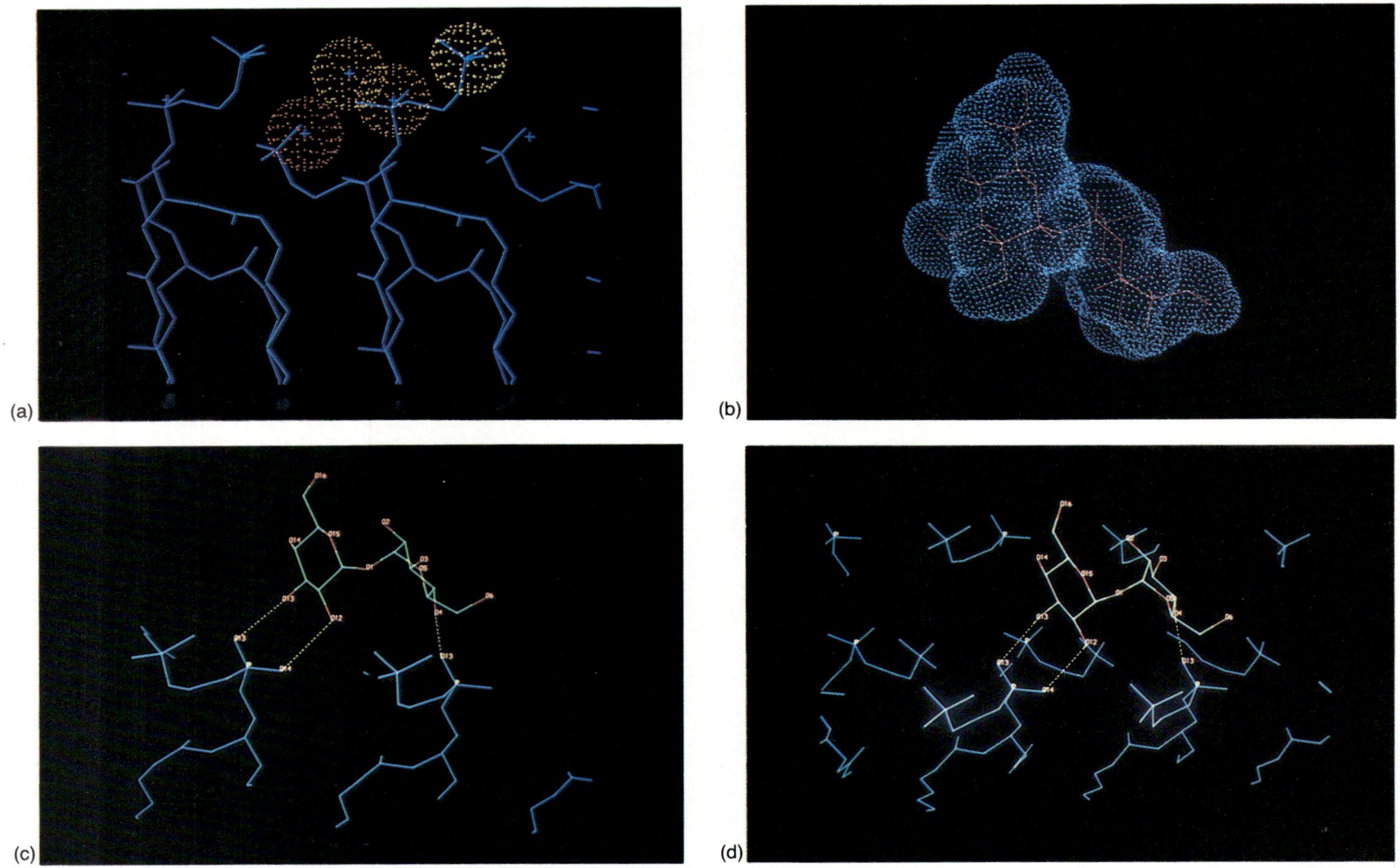

Plate 13-4. (a) View of the four waters of crystallization (dotted) associated with a DMPC in the bilayer. Increasing saturation of the yellow color of the water spheres indicates that the water is more deeply buried in the head group surface. (b) Skeletal image of trehalose with a van der Waals surface. (c, d) Views of the model for the interaction of trehalose (green) with the DMPC bilayer (blue). Proposed hydrogen bondings indicated by dotted lines.

conformers. Instead the ethanolamine group lies almost parallel to the bilayer plane. As in DMPC the nonester phosphate oxygens of DLPE lie at a distance of about 4.7 Å from the nitrogen of the amine group—a distance most favourable for electrostatic interaction between the two. Optimisation of the intramolecular contacts is achieved without the need for a chain tilt, as is made clear by the fact that the area of the head group is 38.6 Å^2, while the cross-sectional area of each acyl chain is 19.3 Å^2 (precisely half that of the head group).

The crystal structure of DLPE from which we derive our data is that of the racemic mixture of D- and L-DLPE. The two enantiometric forms of the lipid residue are present in the unit cell.

Organization of the Bilayer

The extremely compact structure of the DMPC bilayer can be appreciated from the spacing-filling side view in Plate 13-2a. When the bilayer is viewed along the X-axis, the 12-degree tilt of the chains is evident. This slight tilt is just sufficient to allow the cross-sectional area of the chains (38 Å^2) to be accommodated under the head group. Note that while the carbonyl oxygens of both chains 1 and 2 of type B and of chain 2 of type A may be regarded as being in the head group region, the carbonyl oxygen of chain 1 of the more deeply penetrating type A can be considered to fall in the hydrophobic region of the bilayer. Evident too, from this figure is the pronounced interdigitation of the chains.

The tight interdigitation at the bilayer midplane is shown again in Plate 13-2b, in which a Connolly surface (dotted) has been generated at the chain terminal methyl region for each monolayer. When the images are brought together, the Connolly surfaces appear to fit so well together that they almost merge as they snake across the bilayer midplane.

A cross-sectional view through the merging Connolly surfaces (Plate 13-2c) shows an orderly array of peaks and depressions representative of the alternating penetrating and retreating chains. Four distinct features can be recognized: (1) a high, bright knob denoting the deeply penetrating chain 1 of the type A conformer; (2) a secondary knob from type A chain 2; (3) a shallow depression from chain 1 of type B; (4) a deep depression from the least penetrating chain 2 of type B.

In contrast to DMPC, the DLPE molecule of given stereoisomeric form does not crystallize in two conformationally distinct forms. Hence the bilayer organisation in this case is different. The molecules do not optimize head group packing by an offset in the direction of the bilayer normal. This difference from DMPC is reflected not only in the packing of the acyl chains but also in the manner of the chain interdigitation in the bilayer midplane (Plate 13-2d). No longer are there four levels of chain interpenetration at the middle of the bilayer as in DMPC, but two, somewhat less tightly interdigitated levels.

Head Group Surface

Details of the head group packing and solvent interactions are studied most conveniently using an appropriate surface such as a Connolly surface. Our earlier discussion on the head group region (see Plate 13-1c), suggested that sufficient space was available for water interaction. The Connolly surface reveals two water-accessible regions of the DMPC bilayer (Plate 13-3a). The first consists of a series of deep holes in the surface, while the second is a set of pockets or ridges which appear to be nearer to the surface of the bilayer. The colour coding of the surface indicates hydrophobic areas (green) surrounding the choline methyls. The brighter green surface comprises the type B cholines; the lighter green, type A. Extended hydrophilic (red-orange) ribbons connect the phosphodiester regions. The pattern may be said to be a combination of undulating hydrophobic waves of cholines interspersed with almost sinusoidal hydrophilic ribbons. Rotating and tilting the image (Plate 13-3b) yield a view which emphasizes the water accessible holes and ridges and indicates their positions relative to the hydrophilic ribbons.

That surfacing techniques may be used to reveal patterns in bilayer structures is further illustrated by their application to the head group region of the DLPE bilayer (Plate 13-3c). Here the pattern is quite different from that of DMPC. Gone are the waves of undulating head groups and the ribbons of phosphates, now replaced by a patchwork quiltlike pattern. The pattern appears flat, as all the molecules of DLPE lie in the same plane. The hydrophobic areas (green) are isolated within the hydrophilic domains made up of the amino groups (blue) of the ethanolamine residues and just barely contiguous phosphodiester patches (red-orange). Pairs of D- and L-enantiomers are revealed in the alternating pattern of ethanolamine group orientation.

Water in the Bilayer

As the locations of the waters of crystallization for DMPC are known, they can be superimposed upon the Connolly surface describing the DMPC head group morphology (Plate 13-3d). Clearly the deep holes and the pockets are the locales for two of these water molecules. It is well known (Baer, 1953; see Hauser et al., 1980 for later references) that at least one water is bound to the bilayer under even the most strenuous conditions of dehydration. It is easy to speculate that the molecule buried deep within the hole of the solvent-accessible surface plays this role.

The manner in which the waters interact with neighboring lipids and with other water molecules is depicted in Plate 13-4a. We see that four waters of crystallisation are associated with each pair of lipid molecules. Of these, three are hydrogen bonded directly to the lipid head groups. Two of these are involved in linking the head groups of type A to type B molecules. The third water of crystallisation bonds to a fourth, which is in turn hydrogen bonded to the next bilayer. This elaborate hydrogen bonding network clearly plays a major role in the stabilisation of the bilayer. One might remark that the water molecules are primarily involved in linking the two conformationally distinct molecules. Stabilisation between two sets of identical conformers is electrostatic in nature, linking a nitrogen of a type A choline to two adjacent type B molecules. This three way interaction motif (i.e. B-A-B) will be repeated in our proposed model for the interaction of trehalose with the bilayer.

Bilayer-Trehalose Interactions

Crowe and Crowe (Chapters 11 and 12) have amply documented the physical and biochemical evidence for interaction of trehalose with phospholipids under circumstances of bilayer dehydration. Trehalose is a disaccharide in which the two glucose rings are linked together by an alpha-alpha linkage. The orientation of the glycosidic oxygen is axial with respect to both the sugar rings. The structure of trehalose derived from x-ray crystallography (Brown et al., 1972) reveals a compact structure in which the two glucose rings are not in the same plane (Plate 13-4b). The modeling is made more straightforward by the considerable steric hindrance to rotation about the glycosidic bond of trehalose. The hindrance restricts the range of conformations readily accessible to trehalose and therefore considerably reduces the number of conformers which must be tested to achieve a chemically realistic model.

We have assumed that a preliminary model of that interaction might be developed by positing, as suggested by Crowe et al. (1984), that some of the hydroxyl groups of trehalose are able to form hydrogen bonds in the locations which would be otherwise occupied by water. Thus we have used as our starting point the known locations of the waters of crystallization in the DMPC bilayer and have attempted to "dock" trehalose with the bilayer. We keep in mind, however, existing experimental evidence (Baer, 1953; Hauser et al., 1980) indicating that at least one water remains bound to the lipid even under conditions of dehydration.

The initial attempt at docking trehalose with the DMPC bilayer revealed an unfavorable steric interaction between one ring of the sugar and the choline head group of a type B DMPC. To remove this barrier, the bilayer was allowed to expand by increasing the spacing between rows of type A and type B molecules by 1.0 Å. This simple maneuver provided sufficient space in which to dock the trehalose to the phosphatidylcholine head group in the manner shown in Plates 13-4c, d.

The model consists of hydrogen bonds between the trehalose hydroxyls 2 and 4 to the nonesterified phosphate oxygens of a type B DMPC. A third hydrogen bond is formed from hydroxyl 6 of the second glucose moiety to the adjacent type B DMPC. Thus two lipid molecules (type B) are linked by a single trehalose in a scheme somewhat analogous to the hydrogen-bonded linkages formed by water. Furthermore, the approximate mirror image symmetry observed between the head group types places the choline group of the neighbouring type A molecule directly under the first sugar ring. The distances between the methyls of the choline and the sugar ring are appropriate for hydrophobic interactions between the two. Van der Waals surfaces (not shown) indicate that the ring and cholines just touch. The total interaction, therefore, consists of three lipid molecules linked by a single trehalose; two type B conformers bridged by hydrogen bonds and a type A linked through hydrophobic interactions. While the required spreading of the bilayer may disrupt the electrostatic stabilisation mentioned earlier, the trehalose compensates by mimicking the three-way hydrogen bond/electrostatic link described above for stabilisation by water.

The model implies a stoichiometry of one trehalose per three lipid molecules. Current experimental data from work on monolayers (private communication from Alan Rudolph) indicates a stoichiometry on the order of two to three phospholipid head groups per trehalose molecule. This is in gratifying agreement with our proposed model. We realize, however, that our model is at best tentative. It must be refined in terms of the energetics of the system. Work to this end is currently under way in our laboratory.

REFERENCES

Baer, E. 1953. On the crystallization, structure and infrared spectra of saturated L-α-lecithins. J. Amer. Chem. Soc. 75: 621-623.

Brown, G.M., Rohrer, D.C., Berking, B., Beevers, C.A., Gould, R.O. and Simpson, R. 1972. The crystal structure of α,α-trehalose dihydrate from three independent x-ray determinations. Acta Cryst. B28: 3145-3156.

Connally, M.L. 1983. Solvent-acessible surfaces of proteins and nucleic acids. Science 221: 709-713.

Crowe, J.H., Crowe, L.M. and Chapman, D. 1984. Preservation of membranes in anhydrobiotic organisms: the role of trehalose. Science 223: 701-703.

Elder, M., Hitchcock, P., Mason, R. and Shipley, G.G. 1977. A refinement analysis of the crystallography of the phospholipid, 1,2-dilauroyl-DL-phosphatidylethanolamine, and some remarks on lipid-lipid and lipid-protein interactions. Proc. Royal Soc. London A, 354: 157-170.

Gaber, B., Yager, P. and Peticolas, W. 1978. Conformational nonequivalence of chains 1 and 2 of di-palmitoylphosphotidylcholine. Biophys. J. 21: 161-176.

Harlos, K., Eibl, H., Pascher, I. and Sundell, S. 1984. Conformation and packing properties of phosphotidic acid: the crystal structure of monosodium dimyristolphosphotidate. Chem. and Phys. of Lipids 34: 115-126.

Hauser, H., Pascher, I., Pearson, R.H. and Sundell, S. 1981. Preferred conformations and molecular packing of phosphatidylethanolamine and phosphatidylcholine. Biochim. Biophys. Acta 650: 21-51.

Hauser, H., Pascher, I. and Sundell, S. 1980. Conformation of phospholipids crystal structure of a lysophosphotidylcholine analogue. J. Mol. Biol. 137: 249-264.

Hitchcock, P.B., Mason, R., Thomas, K.M., and Shipley, G.G. 1974. Structural chemistry of 1,2 dilauroyl-DL-phosphotidylethanolamine: molecular conformation and intermolecular packing of phospholipids. Proc. Nat. Acad. Sci. USA 71: 3036-3040.

Jarvis, L., Huang, C., Ferrin, T. and Langridge, R. 1985. UCSF MIDAS, User's Manual. Computer Graphics Labortaory, School of Pharmacy, University of California, San Francisco.

Jones, T.A. 1978. A graphics model building and refinement system for macromolecules. J. of Appl. Cryst. 11: 268-272.

Langridge, R., Ferrin, T.E., Kuntz, I.D. and Connally, M.L. 1981. Real time color graphics studies of molecular interactions. Science 211: 661-666.

Pascher, I. and Sundell, I. 1985. Interactions and space requirements of the phosphate head group in membrane lipids. The crystal structure of disodium lysophosphotidate dihydrate. Chem. and Phys. of Lipids 37: 241-250.

Pascher, I., Sundell, S., Eibl, H., and Harlos, K. 1984. Interactions and space requirements of the phosphate head group in membrane lipids: the single crystal structures of a triclinic and a monoclinic form of hexadecyl-2-deoxyglycerophosphoric acid monohydrate. Chem. and Phys. of Lipids 35: 103-115.

Pascher, I., Sundell, S. and Hauser, H. 1981a. Glycerol conformation and molecular packing of membrane lipids. The crystal structure of 2,3-dilauroyl-D-glycerol. J. Mol. Biol. 153: 791-806.

Pascher, I., Sundell, S. and Hauser, H. 1981b. Polar group interaction and molecular packing of membrane lipids. The crystal structure of lysophosphotidylethanolamine. J. Mol. Biol. 153: 807-824.

Pearson, R.H. and Pascher, I. 1979. The molecular structure of lecithin dihydrate. Nature 281: 499-501.

Pflugarth, J.W., Saper, M.A. and Quiocho, F.A. 1983. New generation graphics system for molecular modelling. Papers presented at the International Summer School on Cyrstallographic Computing, Kyoto, Japan.

Seelig, A. and Seelig, J. 1975. Bilayers of dipalmitoyl-3-sn-phosphotidylcholine: conformational differences between fatty acyl chains. Biochim. Biophys. Acta 406: 1-5.

Sundaralingam, M. 1972. Molecular structures and conformations of the phospholipids and sphingomyelins. Ann. NY Acad. Sci. 195: 324-355.

14 The Influence of Water on the Phase Properties of Membrane Lipids: Relevance to Anhydrobiosis

Martin Caffrey

As a lipid biochemist, I will describe one approach to studying the consequences for biological systems, cellular membranes in particular, of an imposed water stress. I will begin by explaining the perspective of someone studying lipid phase properties and how this perspective relates to anhydrobiosis. The theme of my paper is: "to know the complex we must understand the simple." In order for us to know how membranes in an intact organism respond to an imposed water stress, for example, we must understand how the individual lipid and other membrane components behave in isolation under such conditions. In this way we can build up a picture of the complex or natural biology of stress.

The focus of my attention is the lipid component of cellular membranes. To date a considerable amount of data on the phase properties of lipids has been collected. Unfortunately, this information is scattered throughout the literature. I have set out to bring together these data into a single location so that we can clearly establish our present state of knowledge regarding lipid phase behavior. I will report on this ongoing data compilation in two parts, considering, first, the tabulation of transition temperatures and enthalpies for individual lipid species, and second, miscibility properties in the form of temperature-composition phase diagrams. The possibility of establishing a

The author wishes to thank G. W. Feigenson (National Institutes of Health Grant HL-18255) for support and Ellen Patterson for carefully typing this manuscript. This work was supported in part by a Grant-in-Aid of Research from Sigma Xi, The Scientific Research Society, and the National Institutes of Health, through Grant No. 1R23AM36849-01, to the author.

permanent continuously updated lipid phase data compilation program will also be described.

Of central importance to the area of anhydrous biology is the effect of water stress on cellular membranes and in particular on their lipid component. A schematic view of a biological membrane is shown in Figure 14-1. It consists of a bimolecular lipid leaflet in and on which are situated a variety of proteins and other molecules. The lipid component of the biological membrane is itself made up of a vast array of different lipid species, each with its own unique miscibility characteristics and phase dependence on temperature, hydration, and ion concentration.

A variety of approaches can be employed to study cellular membrane lipid response to changes in water stress. At one extreme, experiments can be conducted with whole organisms, using a variety of physical techniques (Cameron et al., 1985). However, because of the complex nature of these systems, clear-cut answers are more the exception than the rule.

To simplify the system somewhat, we may choose to work with isolated cellular membranes or, alternatively, with total lipid extracted from the whole organism or from the isolated membranes. The total lipid extract can be further resolved into distinct lipid classes, such as the phosphatides, sulfolipids, glycolipids, and so forth, and each lipid class can be further separated into individual lipid types, such as phosphatidylcholines, phosphatidylethanolamines, and so on (Christie, 1982; Kates, 1982). If we simplify the system even more, each lipid type can be separated into chemically distinct lipid species. Here differences arise at the acyl or alkyl chain level as a result of variations in the number of chains, chain length and saturation,

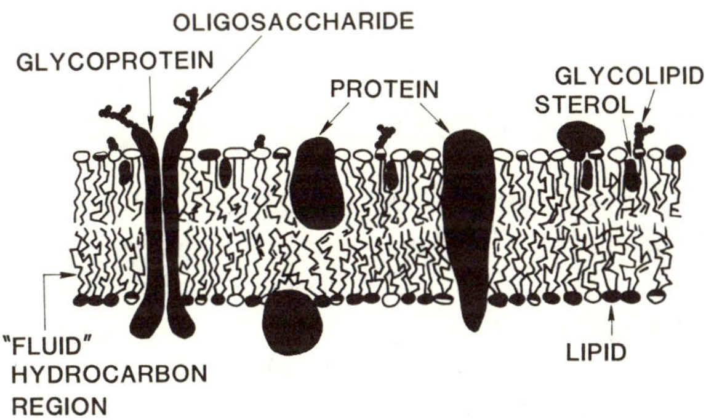

Figure 14-1. Schematic representation of a biological membrane. In this view the following features are illustrated: the bimolecular lipid leaflet nature of the membrane, the fluid character of the membrane interior; sterol-induced rigidification of acyl chains, lipid and protein asymmetry, lipid and protein glycosylation, specific protein-protein interactions, and different types of protein-membrane associations.

degree, position, and isomeric type of unsaturation, and the presence of branched chains, rings, and acetylenic, hydroxy, keto, and carboxylic acid groups. Thus, the total number of chemically distinct lipid species in a given biological membrane can be enormous, given the variability that exists at the hydrocarbon chain, lipid class, and lipid type level.

In order to know how a biomembrane of a given composition will respond to a reduction in available water to, say, 15% (w/w) at a temperature of 5°C, it would be most instructive to know how the components respond to such a treatment not only in isolation but also in mixtures. It is imperative, therefore, that we set about the task of determining how individual lipids behave in isolation first, then in mixtures, and subsequently how the response of these mixtures is modified by changes in such environmental factors that are likely to change in concert with the changing water status in the whole organism or model system.

Lipid Mesomorphism

In addition to existing in crystalline and liquid states, lipids exhibit a number of intermediate states or phases more ordered than a liquid but less so than the crystalline solid (Luzzati, 1968; Small, 1968; Shipley, 1973). Appropriately enough, such materials are called liquid crystals. The term *mesomorphism* describes the characteristic that multiple intermediate phase states are displayed. There are two principal types of mesomorphism, lyotropic and thermotropic. These terms refer to the formation of the various liquid crystal states by the manipulation of solvent content and temperature, respectively. Most membrane lipids exhibit both lyotropic and thermotropic mesomorphism.

Phase refers to that part of a system which is chemically and physically homogeneous throughout. The different lipid mesomorphic phases are distinguished by the manner in which the lipid and solvent molecules are arranged in space (Luzzati, 1968; Levine, 1973; Shipley, 1973). As an example, if we hydrate phosphatidylethanolamine under physiological conditions, it commonly adopts the lamellar phase shown schematically in Figure 14-2a. The molecules pack to form extensive planar bilayers which stack one on top of the other, each separated by a layer of water. This phase has long and successfully been used as a model for the biomembrane. The lamellar phase is analogous to the smectic phase of the thermotropic liquid crystals (Luzzati, 1968; Small, 1968; Shipley, 1973; Demus et al., 1983), having many modifications arising, at least in part, as a result of differences in hydrocarbon chain packing. Packing can be crystalline (L_c) or semicrystalline, as in the

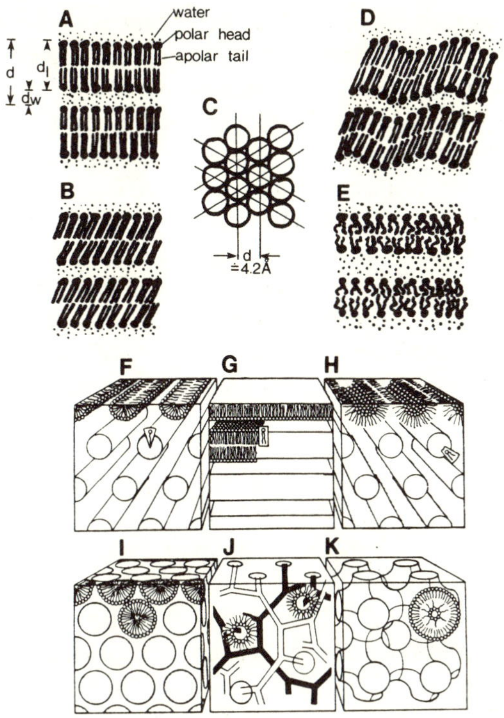

Figure 14-2. Schematic view of the various mesomorphic phases adopted by membrane lipids. (a) Lamellar gel phase (untilted chains, L_β); (b) lamellar gel phase (tilted chains, L'_β); (c) cross-sectional view of the hydrocarbon chains in a hexagonal close-packed arrangement (view is down the long axis of the chains); (d) ripple phase (P'_β); (e) lamellar liquid crystal phase (L_α); (f) normal hexagonal phase (H_I); (g) lamellar liquid crystal phase (L_α); (h) inverted or reversed hexagonal phase (H_{II}). The cubic phases are shown as consisting of (i) spherical aggregates, (j) rod-shaped aggregates, and (k) lamellar aggregates. Two-dimensional views are presented in (a-e), while in (f-k) the lipid aggregates appear from a three-dimensional perspective. (f-k from Rilfors et al., 1984, courtesy of *Biomembranes*.)

case of the recently discovered subgel (L'_c) phase (Chen et al., 1980; Ruocco and Shipley, 1982). Alternatively, acyl chains can be hexagonally or quasi-hexagonally packed, with chains parallel (L_β) or tilted (L'_β) with respect to the bilayer normal. Acyl chains may also be interdigitated (L^i_β), or if a periodic modulation exists in the lamellar plane, the phase is referred to as the ripple (P'_β) phase. In all cases described thus far, chains are rigid and fully extended. If sample temperature is now raised above the so-called main transition or chain melting temperature (T_t), an order/disorder transition at the level of the acyl chains can be brought about to form the lamellar liquid crystal (L_α) phase.

If we now change conditions such as water content, salt concentration, or temperature, in the case of certain lipids it is possible to effect a transition from the lamellar phase to one of the nonbilayer phases: the hexagonal and cubic phases (Figure 14-2). In the H_I, or normal hexagonal phase, water constitutes the continuous medium, whereas in the H_{II} or inverted hexagonal phase hydrocarbon forms the continuous medium. The cubic phase has many modifications that arise as a result of the packing together of micellar, rodlike, or lamellar aggregates.

A variety of factors determine the mesomorphic phase state, including temperature, hydration, pressure, proton and salt concentration, metal ions, state of dispersion (uni- versus multilamellar), and, in the case of lipid mixtures, the identity of the individual lipids in the mixture.

Lyotropic Effects

Many of the factors outlined above that ultimately determine lipid phase state are seen to change in a cell undergoing or recovering from anhydrobiosis. While we might regard the simple consequence of dehydration as the loss of water and the attendant lyotropic effects on membrane lipids, a variety of other factors must be considered. Thus in Figure 14-3 a change in hydration level leads to the "primary" effect on lipid, namely, the lyotropic phase change. This lyotropic effect can lead to lipid and protein phase separations in the plane of the membrane, to changes in membrane permeability, elasticity, and compressibility, and ultimately to a loss of membrane potential and cell death. Concomitantly water loss effects increases in the concentration of dissolved molecules such as salts, metabolites, and macromolecules. Changes occur in such cytoplasmic properties as ionic and osmotic strength, dielectric constant, pH, and viscosity. Finally, when solubility limits are reached, crystallization, glass formation, and/or denaturation ensues.

Volume changes occur with the loss of water, leading to closer membrane apposition facilitating fusion, vesiculation, and the possibility of altered metal ion binding to the membranes. There may be changed interaction between plasma membrane and the cell wall and between the plasma membrane and the endo- and exoskeletal structures. These surface effects could conceivably alter the lyotropic response of membrane lipids.

In summary, then, as water is lost, the primary lyotropic effect may be modified by other factors in the membrane environment that are changing at the same time. Consequently, we need to know not only the properties of the individual lipid components of the membrane in isolation and in mixtures, but also how these properties are influenced by pH, salt concentration, and

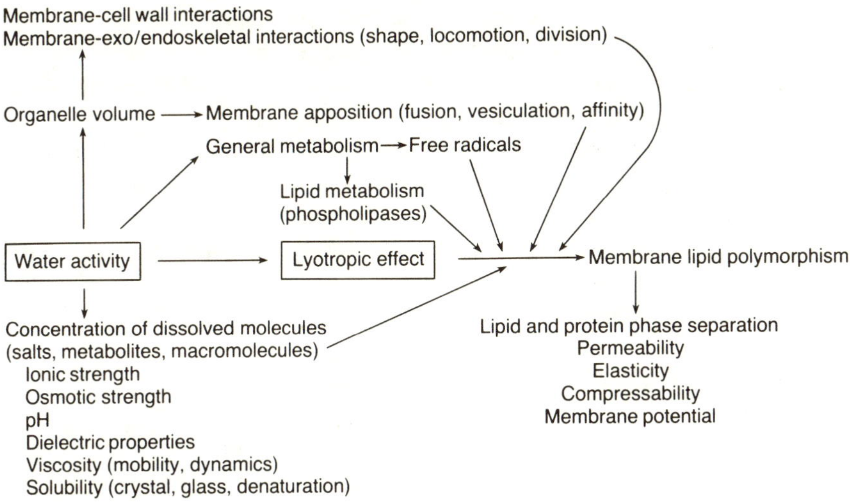

Figure 14-3. The various means by which a change in cellular water status might affect biomembranes with particular emphasis on the membrane lipid component. The view taken here is that a change in water activity has a primary lyotropic effect on membrane lipid phase behavior, leading to lipid and protein phase separation and ultimately to a loss of membrane potential and cell death. Other cellular properties which are sensitive to water stress are presented as modifying the primary lyotropic effect on membrane lipids.

viscosity if we are to understand how they behave in cellular membranes in their changing environment.

Lipid Systems

By applying basic knowledge of pure lipid systems, we should expect to reach an understanding of anhydrobiosis at a very fundamental level. More generally, a knowledge of the basic physicochemical properties of lipids might improve our understanding of other important and related phenomena such as cold hardiness and chilling injury, on the one hand, and, on the other, artifacts in electron microscopy. For example, preparatory steps for both staining and freeze-fracture electron microscopy involve sample dehydration, and the possibility of artifacts as a consequence of lyotropic effects is ever present. An electron-dense material is usually applied to stain the membrane preparation, and the sample is allowed to dry before examination under the microscope. Artifacts can arise both from the ionotropic effect of the staining material, usually a heavy metal, and from the lyotropic effect of drying

(Kilpatrick et al., 1985). In freeze-fracture electron microscopy, the freezing process involves a reduction in available water which, if freezing is not fast enough, can lead to changes in membrane morphology as a consequence of the lyotropic phenomenon.

The preservation and storage of biological materials through drying, freezing, or lyophilization involve in all cases a reduction in available water. Changes in the phase state of the lipid components can profoundly influence successful preservation. It is obvious, therefore, that our fundamental knowledge of lipid phase behavior can be put to a number of uses.

Reductionist Versus Whole Organism Approach

Simple systems by their very nature are more easily understood. Thus, cause-and-effect relations are more easily established with such systems. Simple systems also afford the advantage of promising an understanding at a most fundamental thermodynamic level, and the information so gained is expected to have high predictive potential. The principles of lipid phase behavior can thus be established and used to predict how a membrane of known composition responds to a given change in its environment. The information might also be used to direct the choice of routes through the relevant multicomponent phase diagram toward a final dry state, for example, that avoids undesirable intermediate phase fields or phase coexistence regions. Alternatively, the information might be used to indicate how membrane lipid composition might be manipulated to change the nature of the relevant phase diagram such that a direct treatment (heating, freezing, drying) is no longer damaging. In the preface to a metallurgical phase diagram compilation by Boyer and Gray (1977), the following statement is made: "A diagram showing how a particular steel responds to heating and cooling . . . enables the engineer to know before-hand which specific sequence of heating and cooling steps will develop the desired properties in a particular grade of steel." In like manner, a knowledge of the phase relations of lipid mixtures could be used to yield a parallel advantage.

On the negative side, simple systems are further removed from the reality that prevails in vivo, with regard to both composition and environmental factors. Operating at the simple system level can also mean failure to consider the minor and sometimes major membrane components that can significantly alter lipid phase properties. When we build up from the simple system, such components must obviously be considered.

Simple systems also neglect surface effects due to cell wall and exo- and endoskeletal structures, membrane potentials, and lipid (and protein) asymmetries known to exist in native membranes (Bretscher, 1985). In certain model systems, however, asymmetry is a necessary consequence of the state of dispersion, as in the case of the highly curved small unilamellar vesicles (Lentz et al., 1982).

Systematically building up from the simple to the native represents an enormous task because of the extremely large number of unique lipid species that can be generated. Consider for a moment the examination of binary lipid mixtures prepared from a single phospholipid type such as phosphatidylcholine, which has 2 acylation sites and with 2 different acyl chains available. Here 4 chemically distinct phosphatidylcholines can be procured, leading to 6 unique binary lipid mixtures. Extending this treatment to 3 different acyl chains, we get 9 chemically distinct phosphatidylcholines and 36, or more generally, $n^2(n^2 - 1)/2$ where n equals the number of different acyl chains available, unique binary mixtures. As an illustration, butterfat has 28 chemically distinct fatty acids which would imply in excess of a quarter million unique binary mixtures in the above analysis. Fortunately, nature spares us to some extent here, because fatty acyl chains are not usually randomly distributed on the lipid glycerol backbone. For example, in the case of egg yolk phosphatidylcholine, 70% of this lipid is of the palmityloleylphosphatidylcholine type. In the case of lung surfactant, the vast majority of the lipid exists as dipalmitylphosphatidylcholine (Bangham, 1980). Thus, using our knowledge of those natural systems, we can judiciously choose those systems with which to establish the principles of lipid phase behavior with a minimum of effort and with biologically relevant lipids.

As far as the more complex or native systems are concerned, the advantages of working with them are obvious in the sense that they are closer to the situation we seek to understand. Accordingly, there is no need for extrapolations; the systems embody all the environmental and compositional constraints that prevail in vivo. Indeed, the disadvantage of the whole organism approach is that cause-and-effect relations are difficult to establish and results are often times empirical and possess little if any predictive potential. Another problem is that the membrane usually represents a small fraction of total cell mass and measurements do not discriminate between different membrane types and nonmembranous lipid.

Some techniques allow determination of the phase properties of membrane lipids in living cells; for example, intact membranes of *Acholeplasma laidlawii* have recently been measured using infrared spectroscopy (Cameron et al., 1985).

A Compilation of Lipid Characteristics

I might make a useful contribution if I could demonstrate that indeed a relationship exists between the basic thermodynamic properties of lipids in model systems and their behavior in intact cellular membranes experiencing water stress. To this end I adopted the following strategy: (1) to compile the available lipid phase information in the literature in order to define our present state of knowledge in this area, (2) to evaluate these data, and (3) where possible to relate the data to membrane lipid phase behavior in native systems. The data-gathering process (step 1 above), however, has proven to be an enormous task and is not yet complete.

This lipid phase data compilation has two parts. (1) Transition temperatures and enthalpy changes associated with the mesomorphic phase changes undergone by each chemically distinct lipid species in the dry state and in the fully hydrated state and (2) miscibility properties of lipids alone, with water as a variable component, in excess water, and as influenced by other membrane components such as proteins. Data concerning the influence of pH, salt concentration, and so forth on lipid phase behavior is included where appropriate.

The Thermotropic Phase Properties of Single Lipid Species

The compilation includes a table of transition temperatures (T_t) and enthalpies (ΔH) of the major phase transitions for pure lipids both in the dry state and in the fully hydrated state. The table is modeled after that of Silvius (1982). It is arranged by lipid type and for each individual species within a particular lipid type the following information, when available, is reported: (1) lipid identity, (2) transition type, transition temperature, and enthalpy change, (3) method used to make the T_t and ΔH measurements, (4) scan rate and direction, (5) lipid purity and method of determination, (6) composition of the suspending medium, and (7) literature source. The table contains data for both dry lipid and fully hydrated lipid, and footnotes are extensively used for clarification or when additional relevant information is available. The effects of salt and pH are included as available.

Essentially all values available in the literature are included in the compilation. We are thus able to compare values as determined by different methods, degree of purity, method of sample preparation and equilibration, scan rate, and so on. The tabulation has been computerized and can very conveniently be updated. A sample page from this compilation is shown in Table 14-1.

Miscibility Properties of Lipids and Lipid/Water Systems

Miscibility data are most conveniently presented in the form of temperature-composition phase diagrams. These vary in complexity from binary to multicomponent. The number of associated diagrams reported in the literature falls off precipitously as the level of complexity increases.

There are three types of diagrams identified in the compilation, depending on the role of water in the system. These include diagrams (1) prepared in the absence of water, (2) where water is a variable component, and (3) where one of the lipid components is varied in the presence of excess water. Also included in the compilation are phase diagrams with protein as a variable component. For the sake of completeness a section has been set aside for miscellaneous phase diagrams which include calculated and pressure-temperature diagrams.

The majority of the phase diagrams included in this compilation must be viewed as strictly interpretive, since the actual data points used in constructing the diagrams are not always reported. Accordingly, reference to the original references(s) may be needed before the diagram is used to further an investigation.

The phase diagram compilation is organized under the following headings.

Lipid/water binary mixtures. These describe the phase behavior of individual lipid species as determined by hydration level. Phase diagrams of this type are germane to the effects of water stress on membrane lipid phase behavior.

The majority of the lipid/water phase diagrams available in the literature and included in the compilation belong to the saturated phosphatidylcholine/water type. There is also a considerable amount of information available on the phase properties of monoacylglycerides. The monoacylglycerides are of enormous commercial importance as food emulsifiers, which likely accounts for the degree to which their lyotropic phase properties have been studied. Still, even within this heavily examined group, new discoveries are being made. For example, just recently Longley and McIntosh (1983) and our laboratory group have found new cubic modifications for monoolein that occur at intermediate and full hydration levels and at room temperature.

Lipid/lipid binary mixtures. Among the abundant entries under this heading, the most extensively studied of all are mixtures with dimyristoylphosphatidylcholine, dipalmitoylphosphatidylcholine, or phosphatidylethanolamine as one of the lipid components.

A number of these systems satisfactorily obey the phase rule as developed by Gibbs over the entire composition range. This applies to lipids which differ in acyl chain length by some limiting value (two or four carbons). Once this limit is exceeded or a different lipid type is introduced in the mixture, markedly different types of phase diagrams are observed, indicating miscibility changes. Of relevance here is a recent paper by Sugar and Monticelli (1985) which analyzes binary mixtures of this type on the basis of the Landau

Table 14-1
Proposed format for the tabulation of lipid phase transition temperatures; phosphatidylethanolamines in excess aqueous phase

Lipid species	T_c	ΔH	T_t	ΔH	T_h	ΔH	Method	Rate	Purity	Aqueous phase	References
12:0/12:0			29.0	4.0			DSC	5.00	Pure (LC)	40 m*M* Tris Acetate/EG (1:1), 100 m*M* NaCl, pH 7.0	BBA 455: 576-587, 1976
	43.2	10.0	30.1	2.5			HSDSC	0.50	Pure (TLC)	0.1 *M* NH_4 Acetate, pH 5.5	BBA 728: 319-324, 1983
	41.0	12.0	28.0	3.0			HSDSC	0.50	Pure (TLC)	0.2 *M* Na-P, pH 9.5	BBA 728: 319-324, 1983
	44.0	13.3	30.5	3.5			DSC	1.00	Pure (TLC)	0.1 *M* Na-P, pH 7.0	BBA 728: 325-330, 1983
	41.1		28.5				DSC	1.00	Pure (TLC)	0.03 *M* NaBorax, pH 9.4	BBA 728: 325-330, 1983
			30.5	3.7			DSC	2.50	Pure (LC)	Water or 10μM EDTA	Biochem. 22: 1280-1289, 1983
			30.5	4.3			DSC	1.00	Pure (TLC)	Water	Biochem. 22: 5436-5442, 1983
14:0/14:0			49.5	5.8			HSDSC	0.50	NR	10 m*M* Na-P, pH 7.0	PNAS 73: 3862-3866, 1976
			47.5	6.4			DSC	5.00	Pure (LC)	40 m*M* Tris Acetate/EG (1:1), 100 m*M* NaCl, pH 7.0	BBA 455: 576-587, 1976
	53.2	14.8	49.0	5.7			DSC	1.00	Pure (TLC)	0.1 *M* Na-P, pH 7.0	BBA 728: 325-330, 1983
			49.6	5.8			DSC	2.50	Pure (LC)	Water or 10 μM EDTA	Biochem. 22: 1280-1289, 1983
			49.9	6.6			DSC	1.00	Pure (TLC)	Water	Biochem. 22: 5436-5442, 1983
			20.0				FI (NPN)	1.00	Pure (TLC, FA)		Biophys. J. 10: 261-271, 1979
16:0/16:0			60.0	8.5			DSC	5.00	Pure (LC)	40 m*M* Tris Acetate/EG (1:1), 100 m*M* NaCl, pH 7.0	BBA 455: 576-587, 1976
			63.5	8.8			DSC	10.00	Pure (TLC)	Water	FEBS Lett. 47: 158-161, 1974
			66.0				Raman		Pure (TLC)	Water	Biochem. 17: 3944-3949, 1978
	64.9	18.5	63.2	8.8			DSC	1.00	Pure (TLC)	0.1 *M* Na-P, pH 7.0	BBA 728: 325-330, 1983
			64.0	7.9	123.0	0.30	DSC	2.50	Pure (LC)	Water or 10 μM EDTA	Biochem. 22: 1280-1289, 1983

	63.5	8.6			DSC	1.00	Pure (TLC)	Water	Biochem. 22: 5436-5442, 1983
16:1cΔ^9/16:1cΔ^9	−33.5	4.3			DSC	5.00	Pure (LC)	40 m*M* Tris Acetate/EG (1:1), 100 m*M* NaCl, pH 7.0	BBA 455: 576-587, 1976
	26.0				P-NMR			0.1 *M* NaCl, 1 m*M* EDTA, pH 7.0, 10 m*M* PIPES	BBA 814: 195-198, 1985
	24.8	5.2	71.4	0.40	HSDSC	0.20	NR	20 m*M* PIPES, 1 m*M* EDTA, 150 m*M* NaCl, 0.02 mg/ml NaN_3, pH 7.4	CPL 36: 387-393, 1985
L-16:0/16:0	62.2	9.6	65.3	0.20	HSDSC	0.50	>99%	50 m*M* Borate buffer, pH 8.0	Biochem. 23: 3553-3558, 1984
DL-16:0/16:0	60.9	8.7	81.7	0.20	HSDSC	0.50	(TLC, GLC)	50 m*M* Borate buffer, pH 8.0	Biochem. 23: 3553-3558, 1984
18:0/18:0	71.0				DSC				Blume Thesis, Univ. Freiburg, 1976
	74.0	10.5	101.0	0.90	DSC		Pure (LC)	Water or 10 μM EDTA	Biochem. 22: 1280-1289, 1983
18:0/18:0	70.4	10.5			DSC	1.00	Pure (TLC)	Water	Biochem. 22: 5436-5442, 1983
18:1cΔ^9/18:1cΔ^9	−16.0	4.7			DSC	5.00	Pure (LC)	2 m*M* EDTA, 25 m*M* Tris Acetate, pH 7.0	BBA 513: 21-30, 1978
	−16.0	4.5			DSC	5.00	Pure (LC)	40 m*M* Tris Acetate/EG (1:1), 100 m*M* NaCl, pH 7.0	BBA 455: 576-587, 1976
			14.0	0.29	HSDSC	0.80	NR	20 m*M* PIPES, 1 m*M* EDTA	CPL 36: 387-393, 1985
			8.0	0.29	HSDSC	0.20	NR	150 m*M* NaCl, 0.02 mg/ml NaN_3, pH 7.4	CPL 36: 387-393, 1985
18:1cΔ^9/18:1tΔ^9	35.0	7.0			DSC	5.00	Pure (LC)	40 m*M* Tris Acetate/EG (1:1), 100 m*M* NaCl, pH 7.0	BBA 455: 576-587, 1976
	38.5	4.3			DSC	1.25	Pure (TLC)	100 m*M* NaCl, 10 m*M* Tris, 1 m*M* NaN_3, 1 m*M* EDTA, pH 7.4	JBC 254: 8256-8262, 1979
	37.5				ESR (DTBN)		Pure (TLC)	100 m*M* NaCl, 10 m*M* Tris	JBC 254: 8256-8262, 1979
	36.5				ESR (TEMPO)		PURE (TLC)	1 m*M* NaN_3, 1 m*M* EDTA, pH 7.4	JBC 254: 8256-8262, 1979
	37.5				HSDSC	1.50	NR	100 m*M* Na-P, pH 6.6	JBC 252: 4749-4751, 1977
	37.2	8.8	65.6	0.50	HSDSC	0.80	NR	20 m*M* PIPES, 1 m*M* EDTA, 150 m*M* NaCl, 0.02 mg/ml NaN_3, pH 7.4	CPL 36: 387-393, 1985
	38.3	8.4	63.5		HSDSC	0.40	Pure (TLC, GLC)	200 m*M* NaCl, 5 m*M* Tris 5 m*M* Tes, 1 m*M* EDTA, pH 7.4	Biochem. 24: 4400-4408, 1985
18:0/18:1cΔ^9	30.4				LS		Pure (LC)		H-S ZPC 357: 21-33, 1976
rac-1,2-di-c-hexadecyl glycerol-3-phosphoryl ethanolamine	67.5				DSC				Lipid-Protein Interactions, 276-281, Wiley & Sons, 1981

theory for a homologous series of phosphatidylcholines and phosphatidylethanolamines, and the manner in which they can go from one phase diagram type to another.

Tertiary and quaternary phase diagrams. At the next higher level of complexity, we have the tertiary phase diagrams of the following types: (1) three lipid components in excess water and (2) two lipid components with water as a third component. Only a limited number of such phase diagrams are reported in the literature, although the mono- and disaccharidediacylglyceride/water systems have been studied in considerable detail because of their relevance to chloroplast membranes. Tertiary phase diagrams are usually shown as triangular diagrams at a series of fixed temperatures while the relative amounts of the three components are altered.

Quaternary phase diagrams are either of four lipid components in excess water or three lipids with water as the fourth component. Again, very few examples of this type are to be found in the literature. Much of the work in this area is due to D. M. Small, Boston University, who has extensively studied representative lipids from the fatty deposits of diseased biological systems. In addition to containing water, most of these phase diagrams include one or more of the following lipid components; bile acid, bile salt, sterol, and phospholipid. These studies have provided useful insights into the conditions which favor aberrant fatty accumulations such as gallstones.

A View to the Future

It is my opinion that a compilation of lipid phase data is timely and worthwhile. It provides us with ready access to phase data and allows us to review what has been accomplished to date.

This compilation program would benefit from the establishment of an online computerized central depository where lipid phase data are collected from around the world and are evaluated and edited. Such a format is shown in Figure 14-4 as an example. Periodic updating, careful indexing, and dissemination would be other important functions.

I believe that a knowledge of the fundamental physicochemical properties of lipids will greatly augment our understanding of the phase behavior of lipids in a biological setting. A compilation of lipid phase data should allow us to peruse the existing and newly acquired data and, it is to be hoped, to establish trends, general forms of behavior, and principles of lipid phase miscibility, and ultimately to relate this fundamental information to the situation that prevails in cellular membranes and other lipidic accumulations in vivo.

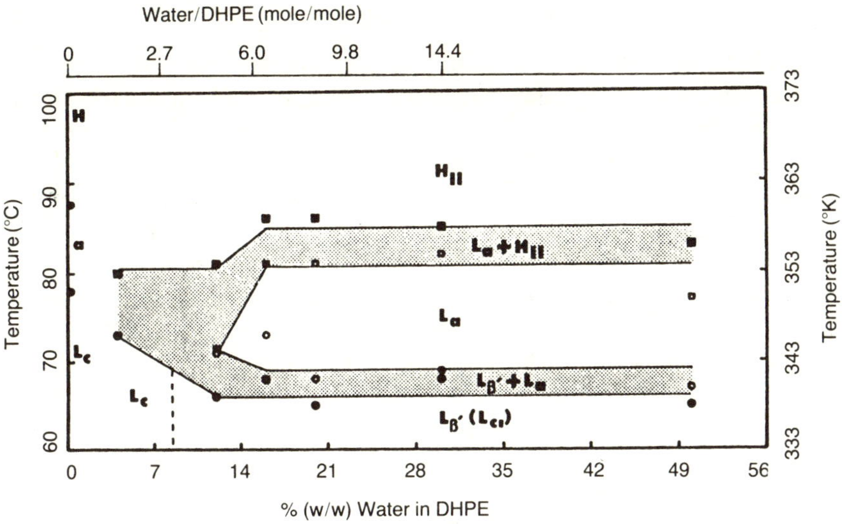

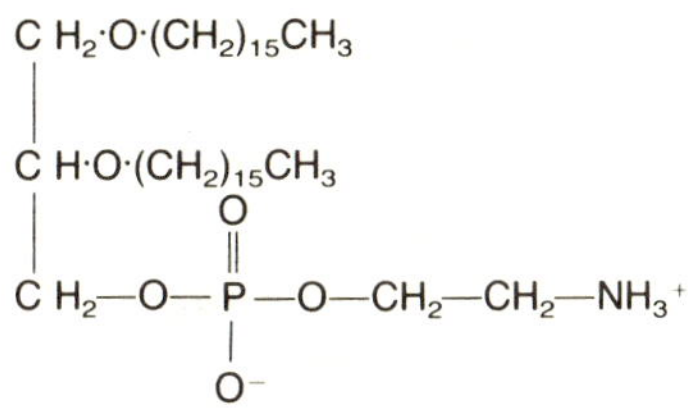

MW = 664

Components: 1. Dihexadecylphosphatidylethanolamine (DHPE).
2. Water (Milli-Q).

Variables: Water content.

Constants: Pressure.

Notation: As described by Luzzati [Biological Membranes: Physical Fact and Function (Chapman, D., ed.) Vol. 1, pp. 71-123, Academic Press, New York, 1968]. In addition, α denotes a phase of unknown lattice type with disordered acyl chains; *H* denotes a liquid crystalline phase with two low-angle spacings in the ratio $1{:}1/\sqrt{3}$; L_{CI} denotes a lamellar phase with two sharp, wide-angle reflections at $(0.47\ nm)^{-1}$ and $(0.43\ nm)^{-1}$; L_c denotes a lamellar phase with three sharp wide-angle reflections at $(0.46\ nm)^{-1}$, $(0.41\ nm)^{-1}$, and $(0.38\ nm)^{-1}$.

Method: Time-resolved x-ray diffraction, 5-10°C/min, heating.

Sample Preparations: DHPE (Fluka Chem. Corp.), Purity ≥ 98% (TLC), dispersed dry powder in water above Tt, multiple T-cycling to ensure equilibration.

Comments: Mole water/mole lipid = %(w/w)water × 36.9/(100−%(w/w)water).
Phase behaviour between 0 and 4% water was not explored.
The L_{β}' phase is metastable.

Source: M. Caffrey, Biochemistry 24: 4826-4844, 1985.

Figure 14-4. Proposed standard format for presenting binary lipid-water phase diagrams in the lipid phase data compilation.

The type of trends I refer to include the tendency, with phosphatidylcholines and phosphatidylethanolamines at least, for the chain order/disorder transition temperature (T_t) in the lamellar phase to fall and with phosphatidylethanolamines for the lamellar/hexagonal transition temperature (T_h) to rise with increasing hydration. Another trend relates to salt effects. For example, high salt concentration tends to decrease T_h and increase T_t for the phosphatidylethanolamines. This effect has, at least in part, been attributed to the dehydrating effect of salt, which ties in well with observed lyotropic effects (Seddon et al., 1983). However, in the case of the monoacylglycerides, high salt concentration stabilizes the hexagonal phase (Caffrey, 1986) but apparently by a different mechanism because dehydrating the monoacylglycerides has the opposite effect (i.e., it tends to increase T_h; Lutton, 1965). As another example, monogalactosyldiacyldiglyceride (MGDG) exists in a hexagonal phase, while digalactosyldiacyldiglyceride (DGDG) is lamellar when hydrated. Mixtures of the two lipids display the expected intermediate characteristics (Larsson and Puang-Ngern, 1979). However, equimolar mixtures of dioleylphosphatidylcholine and monoolein at 5% water form the hexagonal phase most unexpectedly, since both in isolation under the same conditions exist in the lamellar phase (Gutman et al., 1983). This, then, points to an additional feature of lipid phase behavior, namely, that specific lipid associations are important. These are but a few examples of the types of trends that we may hope to see emerge from an examination of compiled lipid phase data.

The compilation will soon be up to date. It remains for us, then, to work through the data, to evaluate it, and to seek insights into what takes place in native cellular membranes with the onset of and departure from the anhydrobiotic state.

REFERENCES

Bangham, A. 1980. Breathing Made Easy. New Scientist. Pp. 408-410.

Boyer, H.E. and Gray, A.G. 1977. Atlas of Isothermal Transformation and Cooling Transformation Diagrams.

Bretscher, M.S. 1985. The molecules of the cell membrane. Sci. Amer. Pp. 100-108.

Caffrey, M. 1986. Kinetics of transitions involving the lamellar, cubic, hexagonal, and fluid isotropic phases of hydrated monoacylglycerides monitored by time-resolved x-ray diffraction. Biophys. J. 49: 502a.

Caffrey, M. and Lew, R.R. Submitted for publication. The Ca^{2+}-induced phase transformation in soybean root microsomes is a consequence of phospholipase activation. Plant Cell Physiol.

Caffrey, M., Werner, D.A. and Priestley, B.G. In preparation. A crystalline lipid phase in a dry biological system: evidence from x-ray diffraction of *Typha latifolia* pollen.

Cameron, D.G., Martin, A., Moffatt, D.J. and Mantsch, H.H. 1985. Infrared spectroscopic study of the gel to liquid-crystal phase transition in live *Acholeplasma laidlawii* cells. Biochemistry 24: 2355-2359.

Chen, S.C., Sturtevant, J.M. and Gaffney, B.J. 1980. Scanning calorimetric evidence for a third phase transition in phosphatidylcholine bilayers. Proc. Natl. Acad. Sci., USA 77: 5060-5063.

Christie, W.W. 1982. Lipid Analysis. Pergamon Press, New York.

Demus, D., Diele, S., Grande, S. and Sackmann, H. 1983. Polymorphism in thermotropic liquid crystals. Adv. Liq. Cryst. 6: 1-107.

Gutman, H., Arvidson, G., Fontell, K. and Lindblom, G. 1983. ^{31}P and ^{2}H NMR studies of phase equilibria in the three component system: monoolein-dioleoylphosphatidylcholine-water. Proc. Int. Symp. on Surfactants in Solution. Lund.

Kates, M. 1982. Techniques in Lipidology. North Holland/Elsevier, Amsterdam.

Kilpatrick, P.K., Miller, W.G. and Talmon, Y. 1985. Staining and drying-induced artifacts in electron microscopy of surfactant dispersions. J. Coll. Inter. Soc. 107: 146-158.

Larsson, K. and Puang-Ngern, S. 1979. The aqueous system of monogalactosyl diglycerides and digalactosyl diglycerides: significance to the structure of the thylakoid membrane. *In* Adv. Biochem. Physiol., Plant Lipids (L.A. Appelqvist and C. Liljenberg, eds.) Elsevier/North Holland, Biomedical Press. Pp. 27-33.

Lentz, B.R., Madden, S. and Alford, D.R. 1982. Transbilayer redistribution of phosphatidylglycerol in small, unilamellar vesicles induced by specific divalent cations. Biochemistry 21: 6799-6807.

Levine, Y.K. 1973. X-ray diffraction studies of membranes. Prog. Surf. Sci. 3: 279-352.

Longley, W. and McIntosh, T.J. 1983. A bicontinuous tetrahedral structure in a liquid crystalline lipid. Nature 303: 612-614.

Lutton, E.S. 1965. Phase behavior of aqueous systems of monoglycerides. J. Am. Oil Chem. Soc. 42: 1068-1070.

Luzzati, V. 1968. X-ray diffraction studies of lipid-water systems. *In* Biological Membranes: Physical Fact and Function (D. Chapman, ed.) Vol. 1. Academic Press, New York. Pp. 71-123.

Rilfors, L., Lindblom, G., Weislander, A. and Christiansson, A. 1984. Lipid bilayer stability in biological membranes. Biomembranes 12: 205-245.

Ruocco, M.J. and Shipley, C.G. 1982. Characterization of the sub-transition of hydrated dipalmitoylphosphatidylcholine bilayers. Biochim. Biophys. Acta 691: 309-320.

Seddon, J.M., Cevc, G. and Marsh, D. 1983. Calorimetric studies of the gel-fluid (L_β-L_α) and lamellar-inverted hexagonal (L_α-H_{II}) phase transitions in dialkyl- and diacylphosphaditylethanolamines. Biochemistry 22: 1280-1289.

Shipley, G.G. 1973. X-ray diffraction studies of biological membranes and membrane components. *In* Biological Membranes (D. Chapman and D.F.H. Wallach, eds.), vol. 2. Academic Press, New York. Pp. 1-89.

Silvius, J.R. 1982. Thermotropic phase transitions of pure lipids in model membranes and their modification by membrane proteins. *In* Lipid-Protein Interactions (P. Jost and O.H. Griffith, eds.). Wiley and Sons, New York.

Small, D.M. 1968. A classification of biological lipids based upon their interaction in aqueous systems. J. Am. Oil Chem. Soc. 45: 108-119.

Small, D.M. 1986. Handbook of Lipid Research, vol. 4. The Physical Chemistry of Lipids: From Alkanes to Phospholipids (D.J. Hanahan, ed.). Plenum Press, New York.

Sugar, I.P. and Monticelli, G. 1985. Interrelationships between the phase diagrams of the two-component phospholipid bilayers. Biophys. J. 48: 283-288.

15 The Binding of Water to Dry Biological Systems

John A. Rupley
Linda Siemankowski

Structural and chemical studies on solutions and ordered preparations of proteins, nucleic acids, and membranes have given us a relatively full molecular-level understanding of many of their functions. Few would disagree with the statement that we understand in detail the catalytic activities of lysozyme, chymotrypsin, and at least a few other enzymes. We know the structure of the macromolecule and of its complexes with substrate at atomic resolution, the atoms of macromolecule and substrate participating in bond rearrangement and catalysis, the nature of the intermediate enzyme-substrate complexes (generally more than one) along the reaction path, and the nature of the transition states, particularly that for the rate-determining step.

Nevertheless there are weaknesses in our understanding of function, and arguably the most important is in the understanding of how water participates. It is evident that change in solvent affects the rates of reactions and the thermodynamics of interactions (hydrophobic, hydrogen bonding, and electrostatic), which are important in biology. However, water is generally not taken into consideration as part of an otherwise detailed mechanistic presentation. For example, descriptions of the structures of enzyme-substrate intermediates or of enzyme transition states include water not at all or as a homogeneous background that modulates electrostatic and other properties. A treatment in which the solvent is considered homogeneous is incorrect; the character of the protein surface orders and affects the thermodynamic properties of the adjacent water in ways that depend on the local structure of the surface; the protein surface, particularly that at an active site, is not homoge-

This work was supported by a grant from the National Institutes of Health.

neous. The failure to include water in a detailed mechanism is not surprising, when there has not been general agreement on simple questions such as how much water is bound to a macromolecule or even what comprises ''bound'' water.

One way of thinking about macromolecule hydration is in terms of two phases, macromolecule and water, and phase diagrams that describe the dependence of system properties on composition. As water is added incrementally to a sample of dry macromolecule, it dissolves in the macromolecule phase. The properties of this phase change with the composition of the system. At some point sufficient water will have been added to saturate the macromolecule, and there will be phase separation into fully hydrated macromolecule and bulk water. The problem is to describe as completely as possible the character of the interface between the macromolecule and bulk water. This interface extends into the water and into the macromolecule. It is necessary to describe the structure of the interface, the properties of the interface, and the extent of the interface, i.e., how much water and macromolecule have properties different from the pure substances.

Protein Hydration: The Water

Continuing with the above picture of two phases, we denote the point of phase separation as the point of full hydration. The amount of water in the interface region is the amount added to the system up to the point of full hydration (i.e., the amount dissolved in the protein phase). The point of phase separation is thermodynamically defined and can be experimentally determined. Figure 15-1, from Yang and Rupley (1979), shows the specific heat of the lysozyme-water system over the full range of composition, from dry protein to dilute solution. As one moves from right to left, increasing the amount of water in the system, the heat capacity changes nonlinearly with system composition until 0.725 weight fraction protein. Beyond this point, at higher water contents, the heat capacity changes linearly with system composition, through the region of heavily hydrated powder and through the region of dissolution of the powder and into the region of dilute solution. Where there is linear change in heat capacity with system composition, one can be adding water only with the heat capacity of bulk water to the system. The partial specific heat capacity of the water in the region below 0.73 weight fraction protein is constant and is the value expected for bulk water, $4.18\,J{\cdot}K^{-1}{\cdot}g^{-1}$.

The phase diagram of Figure 15-1 defines full hydration for lysozyme at 25°C as 0.275/0.725 = 0.38 h (g of water/g of protein). This estimate of full

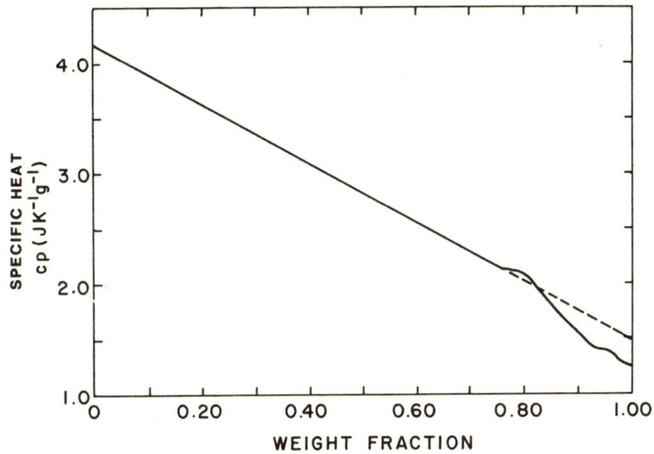

Figure 15-1. Specific heat of the lysozyme-water system from 0 to 1.0 weight fraction protein, at 25°C. From left to right, water fraction decreases until it reaches zero at a protein weight fraction of 1.0. (Redrawn with permission from Yang and Rupley, 1979. Copyright 1979 American Chemical Society.)

hydration is arguably the best available. The interactions of protein surface with solvent and of solvent with solvent are hydrogen bonding or hydrophobic. Because the heat capacity changes strongly with change in either type of interaction, one expects it to see all of the interface and to give an upper limit on the extent of the interface. Other thermodynamic measurements that, like the heat capacity, can be carried out over the full range of system composition, give a phase diagram showing bulk water properties for added solvent at lower hydration (0.1 to 0.2 g/g for the volume and enthalpy; Rupley et al., 1983). If water within the interface has the same thermodynamic property as bulk water, it will not show as part of the protein phase and thus not as part of the interface. An estimate of the point of full hydration from thermodynamic data requires imposition of a molecular model. Estimates of this kind surely give a lower limit on the extent of the interface. As noted, the heat capacity most probably is as accurate a measure as is possible of its extent.

Figure 15-2 illustrates changes which occur during the process of hydrating dry lysozyme. Curve (d) gives values of the apparent specific heat capacity derived from the data of Figure 15-1. Comparison of these results with IR spectroscopic and heat capacity results obtained in Careri's laboratory shows that, first, there is excellent correlation between different measurements of the hydration process, and second, there are three stages to the hydration process. In the lowest hydration region, between 0 and about 0.07 g/g, about

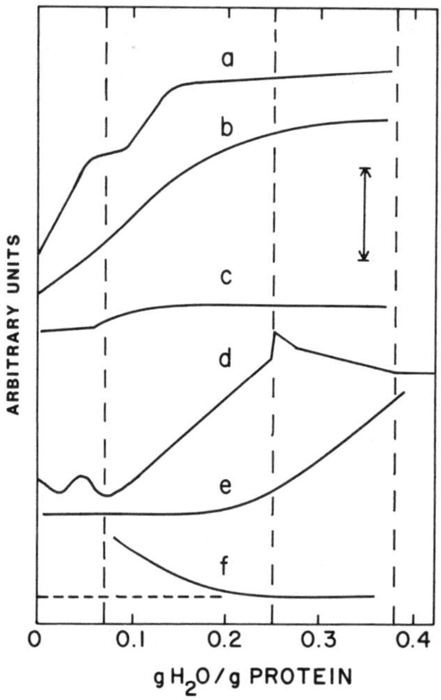

Figure 15-2. Properties of the lysozyme-water system as a function of hydration level (g of water/g of protein). (a) Infrared absorbance at the carboxylate band maximum. (b) Change in the intensity of the amide I' band. (c) Change in the frequency of the OD stretching band of adsorbed D_2O. (d) Excess heat capacity. (e) Enzymatic activity. (f) Diamagnetic susceptibility. (Redrawn by permission from *Nature*, vol. 284, pp. 572-573. Copyright © 1980 Macmillan Journals Limited.)

20% of the total hydration water is bound at charged sites and exhibits properties substantially different from bulk those of water; between 0.07 and about 0.25 g/g, half of the hydration water binds principally to polar groups in clusters that fluctuate in size or arrangement; the final hydration stage is a condensation of water over the weakly interacting, unfilled portions of the surface, to give monolayer coverage.

There is a discontinuity in several properties between the low- and the middle-hydration regions. The heat capacity behavior suggests (1) that in the low-hydration region the bound water is dispersed about the surface, such dispersion being consistent with the distribution of the bound water at one or two molecules per ionizable group and (2) that in the middle region the water is bound in mobile clusters, with more freedom than in bulk water. Thus the discontinuity appears to be a two-dimensional condensation of water on the surface of the molecule. Both this condensation and the high-hydration condensation that leads to complete coverage of the surface are predicted by statistical mechanical modeling of adsorption of a self-interacting adsorbate like water on a heterogeneous surface like that of a protein.

The pattern of change in properties with hydration shown in Figure 15-2 is evident in other thermodynamic properties. The free energy (sorption) isotherm is characterized by a "knee," which is the low-hydration region, a plateau, which corresponds to the middle region, and a sharp upswing in water bound at g/g values >0.85, which corresponds to the start of the condensation at above 0.25 g/g. At high water activity the composition of the system changes so strongly with water activity that the sorption isotherm cannot be measured through the full range of system composition. The volume and enthalpy of a protein-water system follow curve (e) of Figure 15-2, for which bulk water values of the property are found when about half of the hydration shell has been added.

At the point of full hydration of a protein, the average properties of the bound solvent differ only slightly from those of bulk solvent. For transfer of a mole of water from bulk water to the interface region, the changes in thermodynamic parameters, averaged over all of the interface water, are $\Delta G = -0.5$ kcal/mol, $\Delta H = -3.0$ kcal/mol, $\Delta Cp = 2.5$ cal/K-mol, $\Delta V = -1.7$ ml/mol. These changes are small. The free energy difference is less than the ambient thermal energy, and the heat and volume changes are 10-20% of the bulk water values. The motional properties of the interface water at full hydration have been examined by NMR and dielectric measurements on partially hydrated protein powders, yielding the conclusion that at full hydration rotational freedom in the interface is only slightly (perhaps two times) less than in bulk water.

The estimate of the total amount of water from the heat capacity measurements described above, 0.4 g of water/g of protein, corresponds to 300 mol of water/mol of lysozyme. As noted, this value is larger than estimates from other time-average measurements, and there is reason to believe that it is a true estimate of the amount of water required for full hydration. Certainly it must be sufficient for monolayer coverage of the surface. The surface area of lysozyme, from the crystal structure, is 6000 Å^2. Thus each molecule of water covers 20 Å^2 of protein surface. The average "area" of a molecule of bulk water is 10 Å^2. Thus the surface of the protein must order the surrounding solvent. A 20 Å^2 area per water is consistent with the local geometry found in the layers of water perpendicular to the c-axis of Ice I crystals, and it is the largest effective area found for a water molecule. This observation, that on the average each water of the interface covers a 20 Å^2 area, taken together with the observation of the preceding paragraph, that the properties of the interface water are not much different from those of bulk water, suggests that the protein surface selects from among the locally ordered network arrangements existing in bulk water those that can give high surface coverage per water molecule. The selection would be directed by the hydrogen-bonding sites of the protein surface, which have on average about 25 Å^2 of surface area each. The motional freedom of the interface water indicates that any one region of the protein surface would be covered by a fluctuating set of such locally ordered networks, similar to what obtains for a particular molecule of bulk water. Thus the interface water retains the geometry and number of hydrogen bonds found in bulk water, accounting for the smooth meshing of the interface with the bulk solvent (there is no perturbed multilayer water for proteins) and for the closely similar properties of the interface and bulk water.

Protein Hydration: The Protein

Experiments that define the properties of the protein region of the interface suggest the following points. First, removal of water from about a protein need not affect conformation. ESR spectra measured for a high concentration of spin probe bound to lysozyme show that the average distance between spin centers is invariant to hydration from 0.02 to 0.41 g/g. Other experiments that lead to the conclusion that conformation is invariant to hydration have been summarized elsewhere (Rupley et al., 1980).

Second, surface motions are affected strongly by hydration. For the spin probe tempone bound to lysozyme, motion is frozen in the low-and middle-hydration regions, and it increases sharply at 0.25 g/g (i.e. it is correlated with the condensation event leading to completion of the hydration monolayer; Figure 15-3). The rate of surface motion shows 15th order dependence on water activity, as should be expected if it is correlated with the condensation event. Although the change in characteristic rotational time of the spin probe is 90% complete at 0.38 g/g, the point of full hydration seen through the heat capacity, there is a further small change at higher hydration.

The development of the enzymatic activity of lysozyme, also a surface property, follows closely the development of surface motion seen through the spin probe measurements (Figure 15-3). In the enzymatic mechanism of lysozyme deduced from solution studies in various laboratories, the rate-determining step involves movement of the reducing-terminal half of the hexasaccharide substrate deeper into the cleft in a motion synchronous with bond rearrangement. Thus the correlation of surface motion with enzymatic activity is consistent with the requirements of the mechanism. The water acts at least as a lubricant for the motions important for catalysis. One imagines water may play additional and possibly even more important roles in enzyme catalysis. Chapter 16 addresses such possibilities.

Third, internal motions of the protein, like the conformation of the protein, appear to be little affected by removal of the solvent. Figure 15-4 shows the dependence on hydration level of the rate of amide proton exchange with solvent, for lysozyme at pH 7 and 25°C. The effect of hydration on the rate is

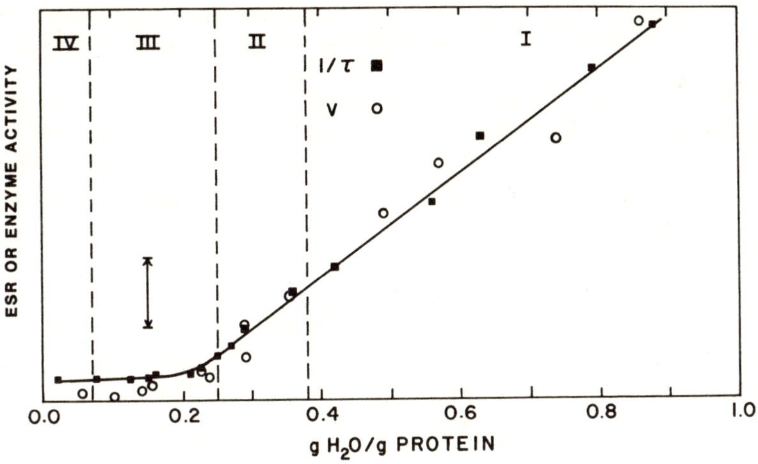

Figure 15-3. Dependence of the rotational relaxation of an ESR probe (log l/tau; closed symbols) and of the enzymatic activity of lysozyme (log *V*; open symbols) on hydration level. (Redrawn from Rupley et al., 1983, courtesy of *Trends in Biochemical Science*.)

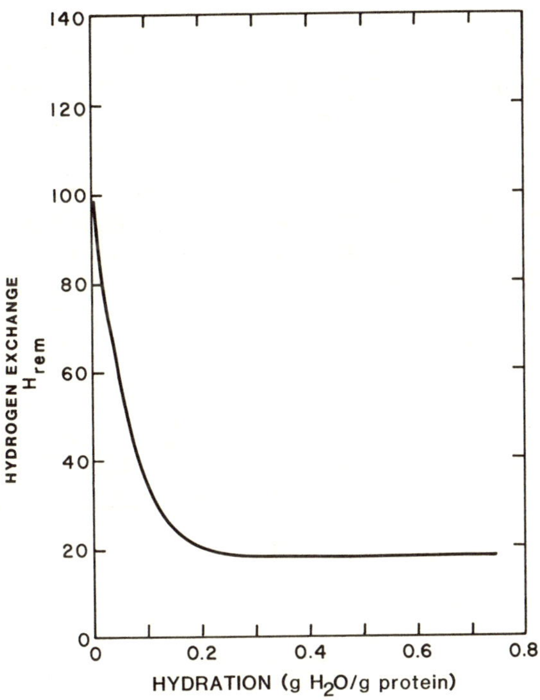

Figure 15-4. Dependence of the amide hydrogen exchange on hydration level of lysozyme at pH 7. (Redrawn with permission from Schinkel et al., 1985. Copyright 1985 American Chemical Society.)

substantially complete at 0.15 to 0.2 g/g (i.e. at less than half-coverage of the protein with water. Exchange of the labile amide hydrogens of a protein with solvent is a global measure of the internal mobility of a protein. Folding of the polypeptide chain converts the freely exchanging peptide hydrogens of the disorganized molecule into a highly dispersed set of labile protons, which exchange at rates differing from those of the denatured molecule by as much as 10^8. Analysis of the exchange of lysozyme in partially hydrated powders as a function of pH shows that more than 100 of the 128 amide protons behaved as shown in Figure 15-4. The hydrogen exchange rate shows about 4th order dependence on water activity. The low order in water may reflect entirely the chemical events of the exchange process, and the internal motions of the protein even at low hydration may be independent of the surface.

Nucleic Acid Hydration

The hydration of nucleic acids can be examined by the same methods used to describe protein-water interactions (P.-H. Yang and J.A. Rupley, unpublished data). Figure 15-5 gives the apparent specific heat of Mg^{++}-tRNA as a function of hydration level for the water-poor region of the phase diagram. Measurements for the Mg^{++}-tRNA water system over the full range of composition define, as do the measurements of Figure 15-1 for lysozyme, a point of full hydration, which is 0.85 g of water/g of nucleic acid. This point is given in Figure 15-5 by the vertical line defining the start of region I, within which there are separate phases of bulk water and fully hydrated protein. The heat capacity results for Mg^{++}-tRNA do not show the rise and fall in the heat capacity observed for lysozyme (Figure 15-2) in the low hydration region. The latter effect has been attributed to a reaction heat

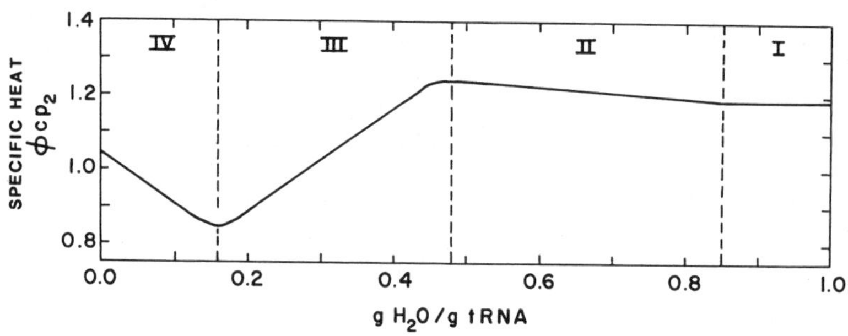

Figure 15-5. Apparent specific heat of tRNA^{phe} (excess specific heat per gram of tRNA) as a function of hydration level. Compare with curve (d) of Figure 15-2.

contributed by a shift in protons from carboxylic acid to basic groups, following normalization of the pK order with the first water added to the dry protein. There can be no comparable proton transfer process for nucleic acids, which are unprotonated salts. The heat capacity results, except for the reaction heat contribution described above, are qualitatively similar for protein and nucleic acid. The picture of the hydration process developed for proteins carries over to nucleic acids. Specifically, in the low hydration region (IV), the partial specific heat capacity of the water is icelike for both nucleic acid and protein (Table 15-1); the water is bound largely at charged sites and dispersed about the surface. In the middle hydration region (III) the partial specific heat capacity of the water is higher than for bulk water, as is to be expected if there are mobile clusters of water on the surface. With increasing hydration level the clusters grow in size and extent, up to the point (0.27 g/g for the protein and 0.48 g/g for the nucleic acid) at which there is condensation of water over the weak-binding sites to complete the hydration shell.

Important quantitative differences between the hydration processes of nucleic acid and protein are listed in Table 15-1. First, the amount of water added in each of the three steps of the hydration process is substantially greater for the nucleic acid. This is particularly notable for region II, the region of high water activity, in which solvent condenses on the surface to complete the hydration shell at 0.38 g/g for the protein and 0.85 g/g for the

Table 15-1
The hydration of $tRNA^{Phe}$ and lysozyme

Hydration range	Partial specific heat ($J\ K^{-1}\ g^{-1}$)	Numbers of waters	Exposed atoms
tRNA			
0.00-0.16 g/g	2.900	220	190 charged
0.16-0.48	5.500	445	500 polar
0.48-0.85	4.000	515	150 nonpolar
0.85 dilute solution	4.179	—	
Total	—	1180	10.7 Å^2/water
Lysozyme			
0.000-0.075 g/g	3.300	60	75 charged
0.075-0.270	5.800	160	143 polar
0.270-0.380	3.350	85	233 nonpolar
0.380 dilute solution	4.183	—	
Total	—	305	21.6 Å^2/water

nucleic acid. In this region the nucleic acid binds 3.5 times more water than the protein (0.37 versus 0.11 g of water/g of macromolecule).

The quantitative differences between nucleic acid and protein hydration can be understood by comparing the character of the surfaces, as described by analyses of the x-ray diffraction results for lysozyme (Shrake and Rupley, 1973) and $tRNA^{Phe}$ (Thiyagarajan and Ponnuswamy, 1979). Table 15-1 gives for tRNA and lysozyme the number of exposed atoms of charged, polar, and nonpolar type. The picture of the hydration process given above suggests that charged atoms should dominate hydration in region IV and polar atoms in region III. The table shows that the hydration of tRNA and lysozyme is very similar in the low- and middle-hydration regions, as judged by the approximately one water bound/exposed charged atom in region IV and one water bound/exposed polar atom in region III. Hydration in region II, however, is greatly different for lysozyme and nucleic acid. The large amount of region II water bound by the nucleic acid accounts for the greater overall hydration of the nucleic acid.

The point of full hydration for the tRNA sample (0.85 g/g) is twice the level for full hydration of a protein (0.38 g/g for lysozyme and comparable values for other globular proteins). The nucleic acid hydration corresponds to a coverage of 10 $Å^2$ of surface per water molecule, compared with 20 $Å^2$ coverage on a protein. The special arrangements of the water selected by the protein surface apparently are not those selected by the nucleic acid surface. The density of charged and polar groups on the tRNA surface is about 50% greater than on the protein. The greater density of sites of strong interaction would require different arrangements of the surface water, which would be characterized by, in particular, closer spacing of the surface water and a lower area covered per water molecule.

It is possible that the high hydration of region II reflects the buildup of multilayer water about the tRNA (Falk, 1970; Texter, 1978). For protein, the special arrangements of the surface water must mesh well with the arrangements of bulk water, or more than a monolayer of water would be perturbed by the protein surface, which is not observed. The high density of strong surface sites on the nucleic acid may produce a monolayer of bound water that cannot mesh smoothly with the bulk solvent. Also, the grooves of double helical regions may serve as foci for multilayer water. Although other measurements have suggested this possibility, it is not established that there in fact is a significant amount of multilayer water on the fully hydrated nucleic acid. An average surface coverage of 10 $Å^2$ per water molecule in a putative monolayer is not unreasonable.

The description of nucleic acid hydration through heat capacity measurements is in accord with other measurements. The sorption isotherm determined for Na^+-tRNA (Falk, 1970) gives a hydration level of 15.6 mol of water/mol of nucleotide at the highest hydration studied (relative humidity 0.93), compared with full hydration at 15.5 mol/mol determined from the heat capacity endpoint. The hydration of tRNA appears to be closely similar to that of DNA, for which there is a larger body of experimental results (Texter, 1978).

Membrane Hydration

Hydration of a membrane is of primary importance for its structure and function. The balance between lipid-water and lipid-lipid interactions is the basis of bilayer formation. Biological membranes consist of approximately equal weights of protein and lipid. Some part, and commonly a major part, of a membrane protein is exposed to the solvent. Membrane function is often presumed to involve water channels, generally within protein elements.

Continuing with the picture of hydration in terms of phases, for a membrane there are three phases, water, protein, and lipid, and three interfaces, water-protein, water-lipid, and lipid-protein. In the hypothetical process of transfer of a membrane protein from dilute solution into the membrane phase, a substantial portion of the surface water would be shorn away and replaced by lipid, to establish the lipid-protein interface. The protein surface within the membrane is hydrophobic, and it appears to be commonly assumed that there are few if any water molecules bound to the protein within the lipid-protein interface. Although this is possible, it is not clear that the exclusion of water from within the membrane is probable. It is not reasonable to expect greater perfection of backbone structures for membrane proteins than for other proteins, and imperfections would favor the introduction of water into the lipid-protein interface.

The purple membrane of *Halobacterium* is attractive for the study of membrane hydration (Stoeckenius, 1985) for several reasons: (1) because the purple membrane is 75% protein, estimation of the water bound to the protein component of the membrane can be done with relatively high accuracy; (2) a low resolution structure of the membrane protein has been derived from electron diffraction and imaging; (3) because the structural studies show that bacteriorhodopsin is an intrinsic membrane protein with about one-quarter of its surface exposed to solvent, hydration measurements can examine the possibility of water bound to the protein within the lipid-protein interface; (4) because lightly hydrated purple membrane shows light-dependent spectroscopic changes like those of functional importance in solution, dehydration does not destroy the integrity of the membrane.

Figure 15-6 compares the sorption isotherm of the purple membrane of *Halobacterium halobium* (lower curve) with sorption isotherms for the proteins lysozyme (middle curve) and bovine serum albumin (top curve). The ratio of hydration of purple membrane to the hydration of serum albumin is 0.4 in the plateau region of the sorption isotherm and 0.5 and 0.6 at low and high hydration, respectively.

Because interest centers on the region with high water content when sorption isotherms are used to define the point of full hydration, the following measurements were made to compare accurately purple membrane and typical globular proteins at high humidity. Concentrated solutions of serum albumin and lysozyme and a concentrated suspension of purple membrane (three samples of each substance) were slowly dehydrated in the same controlled atmosphere vessel at 25°C until equilibrated for the first measurement at $p/p_o = 0.95$. Subsequent measurements were at 0.05 intervals to $p/p_o = 0.75$, to obtain desorption regime data. The samples were then equilibrated at the same water activities in ascending order, to obtain the adsorption regime data. The results appear in Table 15-2, together with standard deviations of the hydration levels and the hydration ratios.

The data of Table 15-2 show that both adsorption and desorption regimes give the same ratio. Apparently hysteresis in the sorption process, including

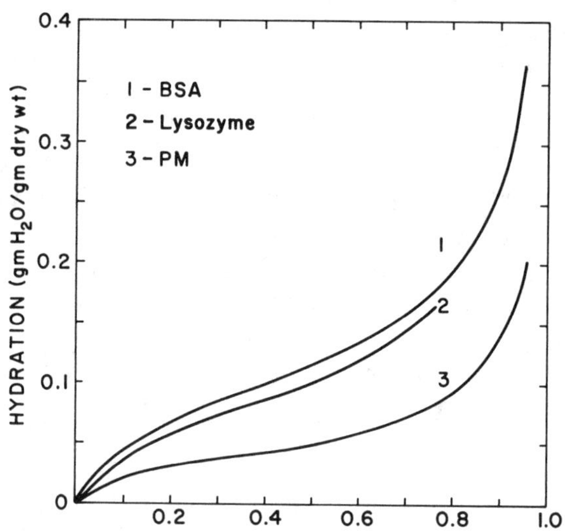

Figure 15-6. Comparison of the sorption isotherm of the purple membrane of *Halobacterium halobium* (curve 3) with isotherms for the typical globular proteins bovine serum albumin and lysozyme (curves 1 and 2).

possible changes in state of the membrane preparation with dehydration, is not significant in determining the relative hydration of membrane and globular protein. Serum albumin adsorbs slightly more water than most proteins. Thus the ratio of hydration of purple membrane to globular protein from the data of Table 15-2 is at least 0.6 (the value for purple membrane/serum albumin at 0.95 relative humidity) and possibly greater than 0.7 (the value for purple membrane/lysozyme, extrapolated to unit water activity).

To obtain from these results an estimate of the amount of water bound to the membrane protein, one must correct for the hydration of the lipid (Hauser, 1975). Sorption isotherms on egg lecithin samples give 0.4 g of water/g of lipid at full hydration. Calorimetric and NMR measurements of nonfreezing water give 0.25 g of water/g of lipid. For proteins and nucleic acids, estimates of the hydration based on nonfreezing water are close to those found by other methods such as calorimetry. From the composition of the purple membrane (one-quarter lipid, three-quarters bacteriorhodopsin) and the hydration of the purple membrane determined from the sorption isotherm, one calculates for the high and low estimates of lipid hydration that the membrane protein binds, respectively, 48% or 68% of the amount of water bound by a globular protein. If we assume that the one-quarter of the bacteriorhodopsin exposed at the membrane surface is fully hydrated, like a globular protein, then the interior three-quarters of the surface would be one-third to one-half covered with water.

Neutron diffraction studies on purple membrane equilibrated with water or deuterium oxide have shown that no water channels span the membrane

Table 15-2
The hydration of purple membrane, serum albumin, and lysozyme

		Hydration[a]			Hydration ratio	
p/p_o	Sorption regime	PM	BSA	LYS	PM/BSA	PM/LYS
0.95	desorption	0.1944 (.0066)	0.3234 (.0004)	0.2820 (.0060)	0.600 (.020)	0.689 (.029)
	adsorption	0.1779 (.0045)	0.2944 (.0002)	0.2634 (.0011)	0.604 (.015)	0.675 (.017)
0.90	desorption	0.1404 (.0015)	0.2640 (.0001)	0.2459 (.0018)	0.532 (.006)	0.571 (.007)
	adsorption	0.1287 (.0021)	0.2386 (.0001)	0.2186 (.0013)	0.540 (.009)	0.589 (.010)

[a]gm H_2O/gm dry weight and standard deviation. The standard deviations appear in parentheses.

(Zaccai and Gilmore, 1979). The sorption isotherm suggests that an appreciable portion of the protein-lipid interface contains water. With proper geometry of the water sites and mobility of the bound water, the water internal to the membrane could participate in function.

REFERENCES

Careri, G., Gratton, E., Yang, P.-H. and Rupley, J.A. 1980. Correlation of IR spectroscopic, heat capacity, diamagnetic susceptibility and enzymic measurements on lysozyme powder. Nature 284: 572-573.

Falk, M. 1970. A gravimetric study of hydration of polynucleotides. Can. J. Chem. 44: 1107-1111.

Hauser, H. 1975. Lipids. Water Compr. Treatise, Volume 4, 209-303. (Franks, F., ed.) Plenum, New York.

Rupley, J.A., Gratton, E. and Careri, G. 1983. Water and globular proteins. Trends Biochem. Sci. 8: 18-22.

Rupley, J.A., Yang, P.-H. and Tollin, G. 1980. Thermodynamic and related studies of water interacting with proteins. ACS Symp. Ser., 127 (Water Polym.), 111-132.

Schinkel, J.E., Downer, N.W. and Rupley, J.A. 1985. Hydrogen exchange of lysozyme powders. Hydration dependence of internal motions. Biochemistry 24: 352-366.

Shrake, A. and Rupley, J.A. 1973. Environment and exposure to solvent of protein atoms. Lysozyme and insulin. J. Mol. Biol. 79: 351-371.

Stoeckenius, W. 1985. The rhodopsin-like pigments of halobacteria: light-energy and signal transducers in an archaebacterium. Trends Biochem. Sci. (Pers. Ed.) 10: 483-486.

Texter, J. 1978. Nucleic acid-water interactions. Prog. Biophys. Mol. Biol. 33: 83-97.

Thiyagarajan, P. and Ponnuswamy, P.K. 1979. Solvent accessibility study on $tRNA^{phe}$. Biopolymers 18: 2233-2247.

Yang, P.-H. and Rupley, J.A. 1979. Protein-water interactions. Heat capacity of the lysozyme-water system. Biochem. 18: 2654-2661.

Zaccai, G. and Gilmore, D.J. 1979. Areas of hydration in the purple membrane of *Halobacterium halobium*: a neutron diffraction study. J. Mol. Biol. 132: 181-191.

16 Dielectric Properties of Nearly Dry Biological Systems at Megahertz Frequencies

Giorgio Careri
Andrea Giansanti

In the past few years we have measured the dielectric properties of several intact biological systems in their final stages of dehydration (Careri and Giansanti, 1984; Baffioni et al., 1985; Careri et al., 1985). The main difference between our recent work and previous similar studies relates to our ability to assess the status of water in the sample by simultaneously measuring the evaporation rate and to our ability to store our data in a great amount. Still, our technique suffers from the intrinsic limitations of the multilayer capacitor. Using this set of physical data, we have detected some phenomena of biological interest that we believe should be reported in the present proceedings.

In this chapter we give, first, a fairly detailed account of our experimental technique, with a few remarks about its possible use in other potentially similar applications. The next section deals with the experimental findings, divided into two subsections, one concerned with the detection of protonic currents induced by hydration in dry biological systems and the other with new data relating to the onset of dielectric relaxation in intact biological objects as compared with simpler systems. The chapter concludes with a few remarks on this physical approach to the study of biological systems.

Experimental Apparatus

The apparatus was designed to monitor simultaneously changing evaporation rates and dielectric properties in hydrated specimens under a constant dry-air flux.

In working with electrodes directly in contact with hydrated samples (Pethig, 1979), one often encounters artifacts induced by resistive effects. To avoid this undesirable complication, we put our samples on glass dishes inside a parallel-plate capacitor and thus eliminated contact. Moreover, since the samples in our apparatus are open to the capacitor-balance chamber, hydration equilibrium can be established rapidly. On the other hand, it is not always a simple task to derive the intrinsic dielectric properties (ϵ' and ϵ'') of the samples because of the Maxwell-Wagner effect that occurs in this composite capacitor (Pethig, 1979). However, for our geometry, according to the two-layer model (Daniel, 1967), if we assume that the two layers are of equal thickness, with one being dry air, and if we discount the contribution of the thin layer of glass, the measurable dissipation time τ_δ is:

$$\tau_\delta = \epsilon_o (1 + \epsilon) \sigma^{-1} (\epsilon/1 + \epsilon)^{\frac{1}{2}},$$

where ϵ_o is the vacuum permittivity, ϵ the dielectric constant, and σ the DC conductivity of the sample. Of course, the above expression is useful when the effect of hydration registers a different magnitude either in ϵ or in σ.

Figure 16-1 shows the layout of the apparatus. The case of the sample holder is Plexiglas, and a removable Palmer gauge makes it possible to set the distance between the circular brass electrodes to reach the desired capacitance of the empty holder. The electrode diameter is 2.5 cm.

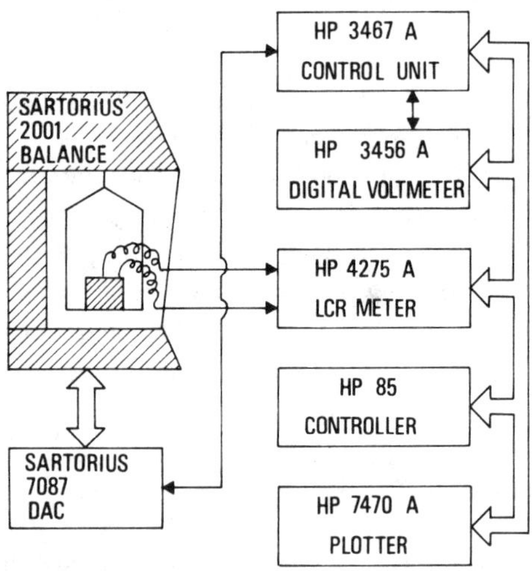

Figure 16-1. Schematic layout of the apparatus for measuring dielectric properties of biological materials at varying moisture levels.

The condenser is placed on the pan of an analytical balance and connected by two thin wires and a shielded fixture to the LCR meter. The balance, a Sartorius 2001 MP, with analog output (sensitivity 0.1 mg, maximum load 170 g), is enclosed in a metal box thermostated by water circulation. To achieve a convenient rate of evaporation from the samples, the chamber inside the balance is continuously swept by a flow of dry air ranging between 100 and 200 l/hr. Temperature of the sample area is measured by YSI 44033 thermistors and can be kept constant to 0.1°C.

Data acquisition and treatment are processed using a chain of Hewlett Packard instruments linked by the standard HP IB interface controlled by an HP 85 personal computer. Control unit HP 3467A and Digital Voltmeter HP 3456A are used to record temperature continuously from thermistors and weight from the analog output of the balance. The HP 4275A LCR multifrequency bridge is used to measure capacitance and loss factor in the range of frequencies from 10 KHz to 10 MHz. All raw data are automatically recorded every few minutes on magnetic tape for further elaboration. A typical run lasts about a day to give more than 300 points per sample. This is an essentially continuous record of the dielectric properties as a function of the sample's hydration level h, which is expressed in g H_2O/g dry weight and is evaluated after drying the sample at 105°C. The evaporation rate E is immediately evaluated from the weight loss in a time unit and is here expressed in g H_2O/min. The evaporation rate is related in a simple and direct way to the water status in the sample, because of its experimental dependence upon the surface activation energy of H_2O in the sample.

Detection of Protonic Conductivities

As the following will show, the detection of an isotopic effect in Maxwell-Wagner dispersion can be used to offer direct evidence for the presence of protonic conductivity in an intact biological specimen. Of course, using Equation 16-1, hydration-induced protonic conductivity $\sigma(h)$ can be detected only when the low-frequency dielectric constant $\epsilon(h)$ is comparatively less sensitive to $\sigma(h)$. This is actually a general occurrence in several biomacromolecules, where $\sigma(h) = A \cdot 10^{Bh}$, with B near 20 (Pethig, 1979). Therefore, although our results will be displayed in this subsection as τ_δ versus h curves, these data will be discussed as if they represented σ versus h curves. One limitation of this procedure is the need to compare data for samples of the same density because of the well-known dependence of the dielectric constant on the density of the medium.

Wheat Seeds

Dielectric techniques are currently used to detect the water content in stored seeds (Barrs, 1968), but specific mechanisms responsible for the observed dispersion do not seem to have been reported. In order to investigate the decline in dielectric with lower water content, we have studied two species of wheat, *Triticum turgidum* and *Triticum aestivum* (Careri and Giansanti, 1984). A sample of 150 seeds was hydrated by the isopiestic method for a week in moist air at 50°C, then placed in a glass dish. Using a Maxwell-Wagner analysis of the resulting data, the changing dielectric can be attributed to protonic currents, which are greatly reduced as water is withdrawn.

In the narrow h range here investigated, we can write Equation 16-1 as

$$\log \tau_\delta = -Bh + C, \tag{16-1}$$

where C is essentially constant. In order to help identify the nature of the charge carriers responsible for conductivity $\sigma(h)$ in wheat seeds, we performed several runs with D_2O-hydrated samples. As shown in Figure 16-2,

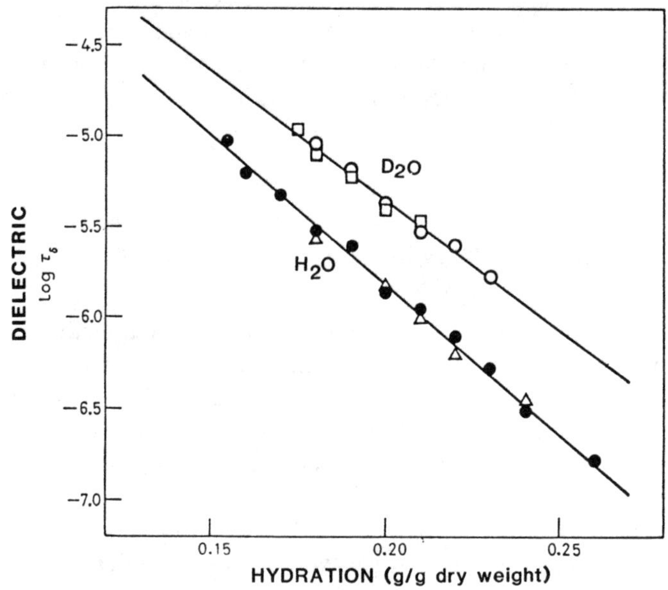

Figure 16-2. Characteristic dissipation time τ_δ plotted versus water content h. Upper data refer to D_2O-, lower data to H_2O-hydrated *Triticum* (*Lin*) seeds. The data have been fitted with regression lines: $\log \tau_\delta(H_2O) = -16.4\,h - 2.5$, and $\log \tau_\delta(D_2O) = -14.2\,h - 2.5$. (From Baffioni et al., 1985, courtesy of *Lett. Nuovo Cim.*)

the B value displayed in these samples differed from the one found in the H_2O-hydrated samples. From these data we find $B(H_2O) = 16.4$ and $B(D_2O) = 14.2$ (reciprocal hydration), and the two regression lines meet at $h = 0$. Therefore, we can conclude that, in the hydration range investigated here, protons are a major component of the moving charges in wheat seeds and that protonic conductivity displays an exponential increase with increasing hydration. The enhancement of protonic currents in imbibed seeds may play a physiological role in the early stage of the germination process when a series of biochemical events is controlled by hydration in a way that is still obscure.

Lysozyme Powders

It is well known that water-macromolecule interactions can influence enzymatic activity and other properties of globular proteins. To extend our previous studies on the hydration of lysozyme powders (Rupley et al., 1983), we have considered the hydration-dependent dielectric losses in the hydration range critical for the onset of the enzymatic function of this protein. Recently it has been shown that hydrated protein powders exhibit dielectric dispersion at three different frequencies (Bone et al., 1981). The first dispersion, $\Delta\epsilon_1$, occurs at frequencies near 1 Hz. The corresponding change in the DC conductivity has been attributed to protons because of an isotope effect (Behi et al., 1982). The second and third dispersions, denoted $\Delta\epsilon_2$ and $\Delta\epsilon_3$, occur at higher frequencies, 10^5 and 10^{10} Hz, respectively. $\Delta\epsilon_3$ can be safely assigned to Debye relaxation of water bound on the macromolecule surface (Harvey and Hoekstra, 1972).

In dielectric measurements on lysozyme powders of varied hydration levels over a range of frequencies centered on $\Delta\epsilon_2$ (Careri et al., 1985), we found an isotope effect, a strong pH dependence, and an effect produced by complexing of the lysozyme with a substrate. Together these indicate the existence of a proton conduction process that involves ionizable side-chain groups of the protein surface and that may be of particular significance for the enzymatic and other functional properties of proteins. The lysozyme powder samples used in this work were prepared from crystallized and salt-free protein. Preparations were carried out with controlled levels of other components (salt or substrate) or with full acetylation of the protein amino groups (to block their protonation). Samples were prepared in groups to minimize differences except in the parameter varied, so as to allow precise comparisons.

For all samples, when hydration levels were less than $h_0 = 0.07 \pm 0.05$, the evaporation rate E (g/min) was close to zero. For $h > h_0$, the evaporation rate for all samples depended linearly upon h, with $E = K\,(h - h_0)$, $K =$

$3.88 \pm 0.43 \times 10^{-3}$ (g/min)$^{-1}$. Physically, h_0 is the hydration value below which the ionizable side chains respond to water binding and above which water begins to be adsorbed on other sites of the macromolecule surface (Rupley et al., 1983). The vanishing E at h_0 was used to fix the h scale of samples without determining dry weights.

In the frequency range studied here, the hydration dependence of ϵ(h) for a protein sample is expected to be much lower than the hydration dependence of σ(h): for lysozyme powders an increase in h from 0.1 to 0.2 decreases, by a factor less than 3, the difference in permittivity at 10 KHz and 9.95 GHz.

Figure 16-3 shows the good reproducibility obtained for different runs on the same sample and for samples from different preparations. The addition of salt at low concentration, about equal to the concentration of the counterions in the preparation, has no effect on τ_8; as a result we know that counterion contributions are not important for the conductivity measured in this frequency range. On the other hand, we found that the addition of equimolar substrate markedly reduces τ_8. This is a specific effect associated with formation of the enzyme-substrate complex, since the addition of more substrate, to 3 $\times$ equimolar, increases τ_8, as expected.

Figure 16-3 shows the effect of isotopic substitution at pH 7, for native lysozyme. The isotope effect has been measured at other pH, with similar results. The values are plotted according to the weight of water adsorbed. When correction is made for the 10% different molecular weights of D_2O and H_2O, by conversion to a mol-based hydration level, so that comparison is made at equal coverage of the protein surface of adsorbed water, the effect of deuterium substitution is a 1.47 times the change in τ_8. This is exactly the expected change for a rate process involving protons, which should be slower by the square root of 2 in D_2O than in H_2O.

The presence of a clear isotope effect in our data is a strong indication that protonic conduction is the dominating contribution to the dielectric relaxation $\Delta\epsilon_2$ in the frequency range studied here. This process must not be confused with the DC conductivity, which also involves protons and is responsible for the relaxation $\Delta\epsilon_1$ mentioned above (Bone et al., 1981). In our opinion, the presence of two conduction processes operating in two different frequency ranges can be understood if the $\Delta\epsilon_1$ relaxation is assigned to proton transport through the bulk powder sample, while the $\Delta\epsilon_2$ relaxation is attributed to proton displacements on a single macromolecule only. As we understand it, the response of the protein molecule to a field with frequency in the $\Delta\epsilon_2$ range is due to intramolecular movement of protons between protein sites, which produces a shifting charge distribution. This rate process could comprise dissociation of protons from a protein site and transfer to an

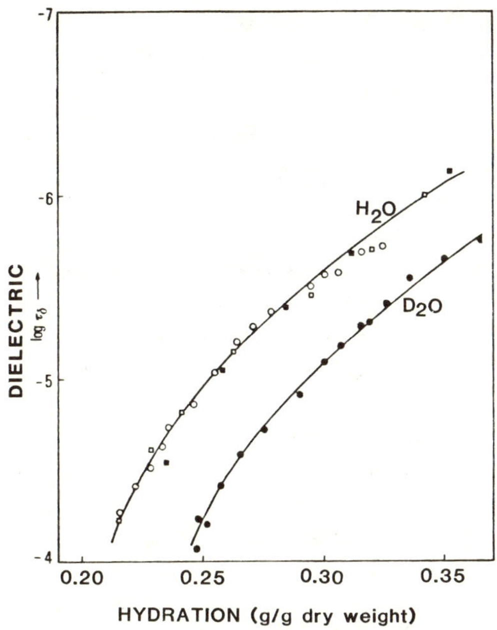

Figure 16-3. The dielectric relaxation time τ_δ (see text) versus hydration h for different lysozyme powder samples of pH near 7. Filled and open squares are for H_2O-hydrated samples from protein preparation 1, pH 7.0, with 0 and 10 mol of NaCl/mol of protein, respectively. Open circles are data for two different runs on H_2O-hydrated samples from protein preparation 3, pH 7.35. Filled circles are for D_2O-hydrated samples, run under the conditions for the data shown by open circles. (From Careri et al., 1985.)

adjacent site, the transfer necessarily being mediated by solvent, taking into consideration the circa 15 Å average separation between proton binding sites.

Regardless of the detailed mechanism, τ_δ on bound H^+—about 7—indicates an event consisting of cooperative proton movements, with relaxation time in the microsecond range. At the present, we are studying the percolative character of these cooperative proton movements. The possible biological significance of our picture is indicated by the effect of substrate binding at the active site. The substantial effect of saccharide binding on τ_δ suggests that a large part, perhaps half, of the proton flow over the protein surface at pH 7 is channeled through the active site. This observation gives a basis for coupling proton-dependent reactions at the active site with protonic processes of other regions of the enzyme surface. The involvement of protons in biological processes is general, and the cooperative channeling of proton flows through active sites may be of fundamental significance.

Hydration Limitations of the Dielectric Relaxation

An abrupt onset of dielectric dispersion induced by an increasing water content h is known to occur in several systems of biological interest, particularly in most hydrated biomolecular powders (Pethig, 1979). So far, this phenomenon has not been studied as a function of the water content status. With our apparatus, which allows both dielectric data and the evaporation rate to be simultaneously measured in the course of time during a continuous dehydration process, we can accurately study the abrupt decrease of the dielectric properties in intact samples. This kind of study is in progress at present in our laboratory, and some preliminary findings will be reported here.

In the following, the threshold hydration h_T for the onset of dielectric relaxation is defined as the difference between the water content h_c (where the capacity c[h] abruptly increases with increasing h) and the water content h_E (where the evaporation rate becomes suddenly different). Thus $h_T = h_c - h_E$. All the quantities can be inferred from the time course of raw data E(t), C(t), and h(t), as shown for a sample of starch in Figure 16-4, or else from the derived plot of E(h) and C(h) as shown in Figure 16-5. We have chosen the capacitance of the composite capacitor as the best experimental quantity to this end, because an abrupt increase in C(h) or C(t) can be attributed only to the intrinsic dielectric behavior of the sample induced by hydration and because the absolute value of C is not relevant in identifying the threshold h_c. Moreover, since low-frequency C data display a clear threshold at h_c, we prefer to consider C(h) as representative dielectric data, rather than $\tau_\delta(h)$, because at low hydration the peak of the dissipation **tgδ** vs. frequency often occurs at a frequency lower than 10 KHz, which makes it impossible to determine with our apparatus.

In all samples of starch, cellulose, and lysozyme, we have detected well-defined values of h; as expected, Figures 16-4 and 16-5 show abrupt onset of capacitance at 0.15 and 0.12, respectively. These are representative examples of this behavior. Since in these samples the conductivity $\sigma(h)$ is much more sensitive to the hydration than is the dielectric constant $\epsilon(h)$, we can identify the sudden increase displayed by C(h) as the onset of a conduction process which produces the detected charge displacement in the sample.

In physics, the abrupt increase of a sample's conductivity as a function of its composition is a known phenomenon called percolation, and it usually occurs when the sample is a network of conductive and nonconductive bonds (Zallen, 1983).

Given this situation, it can be shown that, above a critical composition of

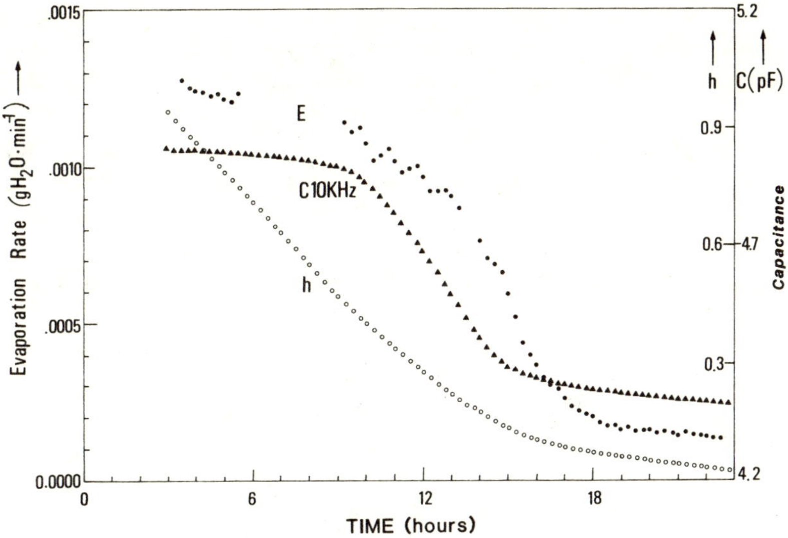

Figure 16-4. Time course of capacitance at 10 KHz (▲), evaporation rate (●), and hydration (○) in a typical experiment on starch.

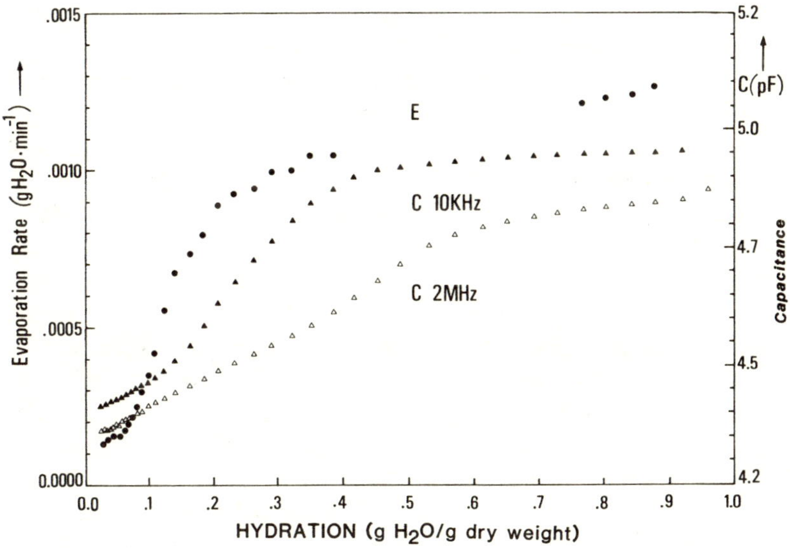

Figure 16-5. From the experiment reported in Figure 16-4: capacitance at 10 KHz (▲), 2MHz (△), and evaporation rate (●) in starch are plotted as a function of hydration level h.

the sample, there is a cooperative transition toward a state where many microscopic conductive paths are created, thus allowing macroscopic conductivity to appear in the sample. In the cases considered here, since for $h > h_E$ the water molecules adsorbed in the samples become less firmly bound, we are likely to face a dynamic random distribution of water molecules which allows water-assisted conduction of a percolative nature. Notice that only the "free" water molecules at $h > h_E$ are relevant to this end, because only these molecules can take part in the statistical distribution on which the percolative transition is grounded.

The physical situation outlined above can be adapted to the more thoroughly studied case of lysozyme. Here for $h < h_E$ the water molecules are firmly bound to ionizable side chains (Rupley et al., 1983), whereas for $h > h_E$, clusters of water molecules arrange themselves statistically on the protein surface. These mobile water molecules are the ones that contribute to the percolative transition in the observed proton conductivity (Careri et al., 1985), because the single proton displacements are assisted by the water-water hydrogen bonds of these water molecules, which are also hydrogen bonded to the amide backbone of the protein surface (Rupley et al., 1983). At present, in collaboration with John Rupley, we are trying to formulate a quantitative description of this interesting cooperative proton transfer on the enzyme surface, which takes place only when hydration values are remarkably close to the critical ones for the onset of enzymatic activity.

In collaboration with L. Azzone and F. Zoratti of the University of Padua, and with F. Bruni of our laboratory, we are investigating the time course of E(t) and C(t) during the dehydration of the aqueous solutions of several membrane preparations. Because of the well-known large contribution of $\epsilon(h)$ to C(h) in membranes (Jendrasiak and Mendible, 1976), these data cannot be interpreted as those considered above. However, as one can see from Figure 16-6, phospholipid membrane preparations present a new feature, namely the absence of a measurable value of h_0. Furthermore, Figure 16-6 makes it clear that phospholipids display a C(t) curve that differs sharply from the one for macromolecular systems considered above. Evidently, we are facing in membranes a sudden jump of C(t) and not the abrupt but then gradual onset of C(t) that is displayed by macromolecules. Because this sudden jump of C(t) in membranes occurs simultaneously with the sudden change in the evaporation rate, we can conclude that only when enough water is present so that some becomes weakly bound does the membrane assume its own structure, thus raising the capacity of the sample almost to its final value.

At present, we are carrying out similar studies on a variety of biological systems, including wheat seeds, leaf tissues from *Geranium* and *Zebrina*,

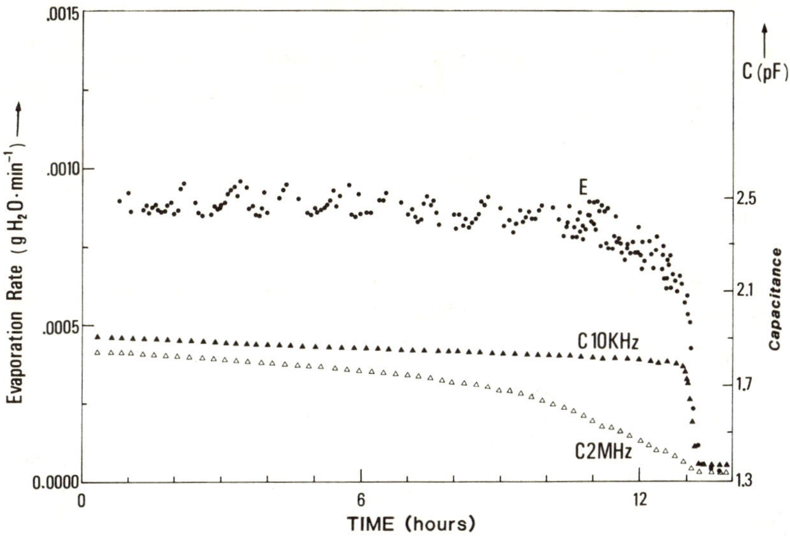

Figure 16-6. Time course of capacitance at 10 KHz (▲) and 2 MHz (△), and of evaporation rate (●) in an experiment on phospholipids from rat liver.

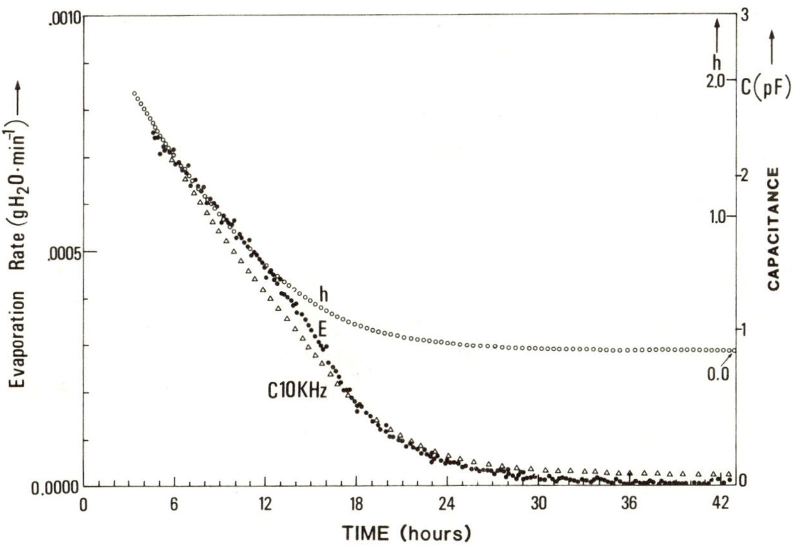

Figure 16-7. Time course of capacitance at 10 KHz (△) and of evaporation rate (●) and hydration (○) in a typical experiment on *Geranium* leaf tissue.

and the seaweeds *Ascophillum nodosum* and *Fucus* (Careri et al., in press). A representative set of these data is shown in Figure 16-7. So far, we have not detected an appreciable value of h_c in any of these samples.

Conclusions

The preceding discussion has dealt with the possibilities and limitations of this experimental approach. The attractive feature is the ability to detect the conductivity of the dielectric constant of an intact sample when one of these two dielectric properties is more sensitive to hydration. Moreover, these dielectric properties can easily be related to the binding state of the water in the sample, as displayed by the evaporation rate. This technique is particularly useful at lower frequencies, where electrode effects can induce serious artifacts. It is better suited to samples in powdered or granular form, in order to avoid shrinking during dehydration. This procedure is also well suited for a quick evaluation of the dehydration-induced dielectric behavior of intact biological samples and can easily be extended to include temperature changes, light absorption, composition of the drying flux, and so forth. Moreover, if one is aware of the problems intrinsic to a multilayer capacitor, it is even possible to devise the appropriate treatments for quantitatively evaluating the time constant of the sample's observed dielectric relaxation.

For the time being, the results reported in the subsections above show our ability to detect the presence of protonic currents. Furthermore, the sharp decrease observed in dielectric properties when hydration in intact biological samples is monotonously decreased suggests some useful applications in biological experiments.

REFERENCES

Baffioni, C., Careri, G., and Giansanti, A. 1985. Leaf tissue dessication process: a dielectric-gravimetric study. Lett. Nuovo Cim. 42: 295-298.

Barrs, H.D. 1968. Water Deficit and Plant Growth, Vol. I. Kozlowski, T.T., ed. Academic Press, London, New York. p. 235.

Behi, J., Bone, S., Morgan, H. and Pethig, R. 1982. Effect of deuterium-hydrogen exchange on the electrical conduction in lysozyme. Int. J. Quant. Chem.: Quant. Biol. Symp. 9: 367-374.

Bone, S., Eden, J., and Pethig, R. 1981. Electrical properties of proteins as a function of hydration and NaCl content. Int. J. Quant. Chem.: Quant. Biol. Symp. 8: 307-316.

Careri, G., Geraci, M., Giansanti, A. and Rupley, J.A. 1985. Protonic conductivity of hydrated lysozyme powders at megahertz frequencies. Proc. Nat. Acad. Sci. USA 85: 5342-5346.

Careri, G. and Giansanti, A. 1984. Deuterium effect in the dielectric losses of wheat seeds. Lett. Nuovo Cim. 40: 193-196.

Careri, G., Giansanti, A. and Fensom, D. In press. Dielectric-gravimetric measurements on *Fucus vesiculosum* and *Ascophillum nodusum* during dehydration. J. Exp. Bot.

Daniel, V.V. 1967. Dielectric Relaxation. Academic Press, London, New York.

Harvey, S.C. and Hoekstra, P. 1972. Dielectric relaxation spectra of water adsorbed on lysozyme. J. Phys. Chem. 76: 2987-2994.

Jendrasiak, G.L. and Mendible, J.C. 1976. The effect of the phase transition on the hydration and electrical conductivity of phospholipids. Biochim. Biophys. Acta 424: 133-148.

Mayer, A.M. and Shain, Y. 1974. Control of seed germination. Ann. Rev. Plant Physiol. 25: 167-193.

Pethig, R. 1979. Dielectric and Electronic Properties of Biological Materials. Wiley and Sons, New York.

Rupley, J.A., Gratton, E. and Careri, G. 1983. Water and globular proteins. Trends in Biochemical Sciences 8: 18-22.

Zallen, R. 1983. The Physics of Amorphous Solids. Wiley and Sons, New York.

17 Hydration, Dehydration, and the Stresses and Strains in Membranes

Joe Wolfe
Michael F. Dowgert
Bernard Maier
Peter L. Steponkus

Desiccation and rehydration inevitably impose large changes in the volume of cells and organelles and thus produce large strains (changes in the geometry or size) in the membranes bounding those entities. How large are the stresses associated with varying strains in the membrane? Can or do they disrupt the semipermeability of the membrane? To what extent are membrane stresses affected by cytoplasmic solutes which are concentrated during dehydration? To answer these questions one needs to know: (1) the mechanical properties of membranes, both dynamic and static, and (2) the stresses and strains imposed upon them during desiccation and hydration. These two areas are the subject of this chapter. Direct mechanical experiments on membranes have thus far been confined to plasma membranes in aqueous suspension and to artificial membranes. Electron micrographic studies of organelles, however, indicate similar behaviour to that of the unsupported plasma membrane. When considering dehydration, it is important to distinguish between the behaviour of a membrane separating two fluid phases, such as that of an organelle or that of a wall-less cell in suspension, and a membrane which is pressed against a locally rigid phase such as a cell wall. To make this distinction clear, we shall briefly describe the hydraulics of the two cases.

The Mechanical Stability of Membranes

Researchers working with macroscopic reconstituted membranes are well aware that they are as fragile as the soap bubbles to which they are such close

This material is, in part, based on work supported by the U.S. Department of Energy under Contract DE-ACO2-81ER10917 and Grant DE-FG02-84ER1324. The authors valued the comments of J. R. Smith.

structural analogues. All membranes whose mechanical properties have been studied to date are lysed by an isotropic surface tension in the range 3 to 10 mN/m. The plasma membrane of the red blood cell is the strongest (Evans and Skalak, 1979)—but this strength is probably a property of the cytoskeleton rather than of the fluid mosaic layer which provides the semipermeability. The plasma membrane of plant protoplasts is ruptured by tensions of about 4 to 6 mN/m. This is similar to the range of tensions at which pure lipid bilayers are ruptured, suggesting that the bilayer regions of living membranes are the "weak links" or provide the sites for rupture. We shall first discuss the mechanical properties of lipid bilayers.

Membranes are held together by the surface tension of water. To stretch a membrane—or to make a hole in it—more water-membrane interface must be created. To make more water surface, hydrogen bonds between water molecules must be broken and the water molecules reoriented: these requirements contribute respectively positive enthalpy and negative entropy to the large surface free energy of water. The membrane molecules are attracted to the water surface (even the unfortunately named hydrophobic substances are attracted to water by van der Waals forces), but on the whole this attraction is less than that between water and water.

Figure 17-1 shows an idealised representation of the surface forces acting

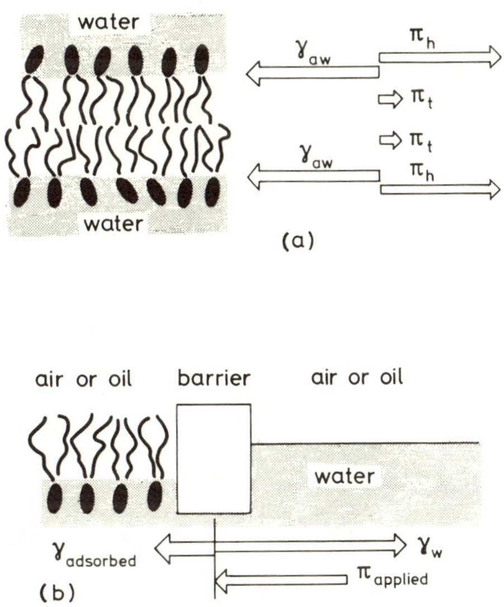

Figure 17-1. The surface forces acting on a lipid bilayer are represented in (a). Increasing the area of contact between the CH_2 groups of the tails and water is opposed by a surface tension about equal to that between alkane and water, γ_{aw}. The restriction on the number of chain configurations increases at low area per molecule a, giving a repulsion Π_t. The head groups contribute a repulsion Π_h which includes dynamic pressure and specific interactions. In an unstressed membrane, these surface forces add to zero. In monolayer experiments, a barrier separates a pure interface between water and oil or air from an interface at which a lipid monolayer is adsorbed. The latter has a lower surface tension, and the difference is provided by the barrier. This difference is measured and called Π. If one assumes that the surface between the lipid tails and the phase above resembles that between an alkane and the phase above, then $\Pi = \Pi_h + \Pi_t$. With this assumption, a monolayer resembles one side of a bilayer when $\Pi = \gamma_{aw}$.

on a lipid bilayer. Stretching of the membrane increases the area of contact between water and the CH_2 groups of the lipid chains; the interfacial tension between bulk alkane and water is about 50 mN/m, so a tension of about this size is assumed to act at each side of the membrane (see Israelachvili et al., 1976). The interaction energies giving rise to the surface tensions are sketched as a function of area in Figure 17-2. The surface tension between two pure phases is numerically equal to the free energy per unit area of that interface, so this effect is represented as a straight line in the figure. In a membrane whose overall tension is near zero, this ''hydrophobic'' contribution is balanced by negative surface tensions or lateral pressures provided by

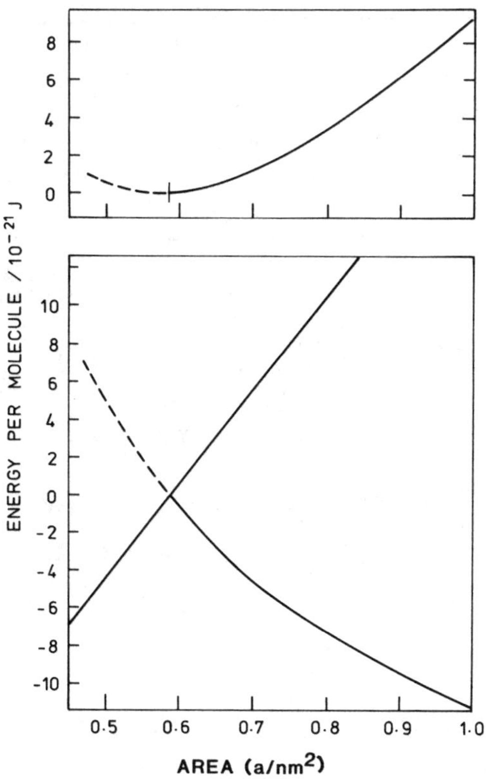

Figure 17-2. The energy per molecule is plotted against the area per molecule for dioleoyl-lecithin at an oil-water interface at 3.1°C. The straight line is the surface tension of the pure oil-water interface, and the area-dependent repulsive interaction has been obtained by integration of Π(a) from a monolayer studied by Yue et al. (1975). The two terms are added to give the total energy, displayed in the top curve. The dashed lines represent extrapolation beyond the collapse pressure.

the repulsive interactions between the lipid molecules at sufficiently close approach. The van der Waals attraction between lipid tails has been implicitly included in the interfacial energy term.

Repulsive interactions between tails can occur for two reasons: first, steric repulsion is possible when large, imposed forces push them together until the area per tail approaches the cross section of an all-*trans* chain. Compression to smaller areas would require enormous pressures. Second, a largely entropic interaction occurs at larger areas per molecule in the fluid state. In the frozen state, only one configuration of the lipid tail (all *trans* for a saturated chain) is possible, and this gives a minimum internal entropy. Larger areas per molecule permit more possible configurations (subject to the condition that the density of the interior is about that of liquid alkane because any holes require large van der Waals energies); larger areas thus imply higher internal entropy and lower free energy. These repulsive forces cannot be measured directly; however, they can be calculated from the partition function, and this has been done explicitly by Gruen (1981). This calculation indicates a contribution to the surface tension of -7 mN/m in a wet, unstressed membrane.

The remaining negative tension or lateral pressure, π, needed to oppose the interfacial tension must be supplied by the interaction between head groups. Part of this may be accounted for by the two-dimensional pressure exerted by particles confined to an interface. A two-dimensional van der Waals equation of state relating π to area a, for molecules with a hard disc area α and no long-range interactions, is

$$\pi = kT/(a - 4\alpha), \tag{17-1}$$

where k is Boltzmann's constant and T is the thermodynamic temperature. This kinematic pressure (probably overestimated by Equation 17-1) plus the specific head interactions must add to about 43 mN/m in an unstressed bilayer. Using the numbers given above, a hard disc area less than 0.13 nm^2 requires a repulsive specific interaction among the heads, and a larger area requires an attraction. To our knowledge, it is not yet resolved whether the specific interaction between neutral head groups is attractive or repulsive.

A quantity which is roughly equivalent to the sum of all the area-dependent interactions is measured as a function of area in Langmuir-Blodgett monolayers. The reduction by an adsorbed lipid monolayer of the surface tension between water and alkane or water and air is given the name *lateral pressure* π. Such monolayers are used to model half a bilayer at the appropriate area per molecule. Monolayers at $\pi < 50$ mN/m are poorer analogues.

The Membrane Stress-Strain Relation

The slope of the energy curve (Figure 17-2) gives the tension of half a bilayer: a membrane at zero bifacial tension γ has an area given by the minimum in energy. The second derivative is proportional to the area modulus of elasticity, k_A defined by:

$$\Delta\gamma = k_A \cdot \Delta A / A, \qquad (17\text{-}2)$$

where A is the membrane area. It follows that k_A equals $-2a$ times the slope of a monolayer isotherm at an equal area per molecule—and this area is achieved at about $\pi = 50$ mN/m. Monolayers at the air- or alkane-water interface collapse at about $\pi = 50$ mN/m, and so the energy curve to the left of the minimum is an extrapolation using the first and second terms of a Taylor expansion. Under negative γ, Equation 17-2 will describe area compression accurately only for small reductions in a.

The tension in a membrane separating two liquid phases is related to the pressure difference Δp across it and its curvature, C, by the Laplace-Young relation:

$$\Delta p = \gamma C. \qquad (17\text{-}3)$$

(The curvature of a sphere with radius r, for example, is $2/r$.) Using a micropipette abutted to the membrane, one can determine the tension in the membrane and the membrane area from the suction in the pipette and from geometrical measurements.

This technique has been used to measure the elastic properties of several membranes. For a lecithin bilayer at 25°C, k_A has the value 140 mN/m (Kwok and Evans, 1981), and for the plasma membrane of isolated plant protoplasts it is 230 mN/m (Wolfe and Steponkus, 1980). It is plausible that the difference is due to intrinsic proteins which are less easily deformed by tension than are lipids. If the proteins are incompressible, k_A for a fluid mosaic membrane is k_A of the bilayer divided by the fraction of the membrane area occupied by the lipid bilayer. The two moduli quoted above are reconciled if incompressible protein comprises about 40% by area of the plasma membrane.

The area elastic modulus and the area per molecule in bilayers should, from the argument above, be changed by any agent which affects monolayer isotherms at the collapse point. The addition to the water subphase of certain carbohydrates (known to render tissues less vulnerable to desiccation) shifts monolayer isotherms to the right and makes the slope less negative (Crowe et al., 1984; Johnstone et al., 1984). We expect such products to lower the k_A of membranes under analogous conditions, although the effect is yet unmeasured.

The effect of temperature has been studied in both mono- and bilayers. At sufficiently low temperatures and high π, the lipid chains undergo a transition from a "lipid" to a "solid" phase and, for a pure, one-component monolayer, $d\pi/da = 0$ (Yue et al., 1975). Under analogous conditions, k_A for a one-component bilayer can approach zero: the area of a membrane can be changed by about 14% without change in tension (Evans and Kwok, 1982). For membranes composed of several miscible membrane species, the area modulus would not be zero, but calculations made for ideal mixtures indicate a k_A in the region of phase coexistence which may be several times smaller than that for single phases (Marčelja and Wolfe, 1979). The effect of intrinsic membrane proteins is to extend the temperature range of the transition but reduce the effect on k_A (deLacey and Wolfe, 1982). It should be remembered that the monolayer analogy is not complete, and caution should be used in relating thermal behaviour of bilayers and monolayers, particularly those at the air-water interface (see Albrecht et al., 1978; Gruen and Wolfe, 1982).

Beyond the phase coexistence region, k_A is much larger than the tensions which lyse the membrane. Thus, for all membranes studied to date, a purely intrinsic or elastic stretching of only a few percent lyses the membrane. Yet it is observed that isolated protoplasts can suffer much larger changes in area than this during osmotic expansion (Wiest and Steponkus, 1978) or during deformation by micropipette suction (Wolfe and Steponkus, 1980). The membrane appears smooth and spherical in both light and electron microscopy before expansion, so the membrane itself suffers a large change in area. This change must involve shifts in the number of molecules composing the membrane rather than large changes in the area per molecule, and so we proposed the existence of a reservoir of membrane material.

Using micropipette measurements of tension and area as functions of time, and taking into account the small, elastic stretching, one can determine the rate of incorporation of new material into the plane of the membrane as a function of tension (Wolfe et al., 1985). We define A_0 as the area which would be occupied by the material in the membrane at any particular time if the tension in the membrane were zero. The proportional rate of area incorporation, Z, is defined as $(1/A)\cdot(dA_0/dt)$. For the plasma membrane of the isolated protoplast, Z is a strong function of γ (Figure 17-3). For γ greater than about 3 mN/m, $Z(\gamma)$ is well approximated by an exponential: Z doubles for each increase in γ of 0.41 mN/m. At very low γ this approximation does not apply. For this membrane, there is a finite positive tension γ_r at which Z

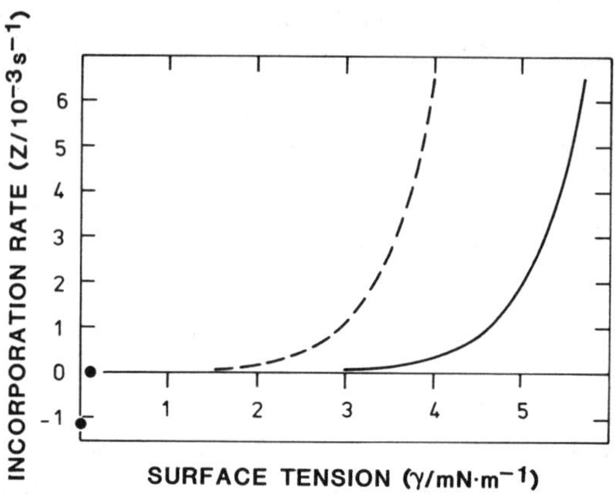

Figure 17-3. The proportional rate of incorporation of material into the plasma membrane of protoplasts from Rye leaves is plotted against the surface tension in the membrane. (The dashed line is for protoplasts isolated from cold-acclimated plants; the solid line for nonacclimated.) Values of Z below 0.01%/s and tensions below about 0.05 mN/m cannot be measured reliably. The point (0.1, 0) represents equilibrium. The negative Z at $\gamma = 0$ is estimated from the time taken for a flaccid membrane to regain spherical shape. The preexponential factor in $Z(\gamma)$ varies widely in a population, but the form and characteristic tension are similar.

$= 0$. This "resting tension"—the tension measured in a membrane which has equilibrated at a particular degree of expansion—is typically 0.1 mN/m. When $\gamma = 0$, Z is negative (i.e. material is deleted from a flaccid membrane). When a protoplast is volumetrically contracted by increasing the osmotic pressure of its supporting medium, it has at first an irregular shape and thus (from Equation 17-3), $\gamma = 0$. Over several minutes it gradually becomes spherical, with a new, smaller area; finally the membrane regains its resting tension. A reexpansion to the original size from this contracted, spherical state may lyse the plasma membrane if the membrane cannot incorporate sufficient new material sufficiently rapidly (Wolfe and Steponkus, 1983a).

The form of $Z(\gamma)$ suggests that $\gamma = \gamma_r$ represents a dynamic equilibrium between deletion and incorporation, which are strong decreasing and increasing functions of γ, respectively. Such a dynamic exchange determines the resting tension of artificial bilayer lipid membranes in equilibrium with a reservoir of lipid dissolved in a contiguous alkane phase containing inverted lipid micelles (called the Plateau-Gibbs border; Laver, 1983). In such a case, where the reservoir is a well-defined phase, the resting tension can be related to thermodynamic properties of the membrane and the lipid aggregates forming the reservoir (Gruen and Wolfe, 1982). What is the reservoir of membrane material for a living membrane? Protoplasts from leaves of rye are observed to internalize vesicles when viewed with digitally enhanced optical microscopy during osmotic contraction (Dowgert and Steponkus, 1984a), and electron microscopy shows many more cytoplasmic vesicles in contracted protoplasts than are seen in isotonic media (Gordon-Kamm and Steponkus, 1984a). Whether such vesicles are the source of material incorporated during expansion is not yet known.[1] It is interesting that, when large (~30 μm) lecithin vesicles are contracted osmotically, they remain spherical but extrude tiny secondary vesicles (Boroske et al., 1981). The implication is that exchange with a reservoir may be rather common in membranes containing lipid bilayer portions.

What tensions cause membranes to lyse? The repulsive interactions between membrane components fall to zero as the area per molecule (and thus the tension) increases. The interfacial tension between hydrocarbon interior and water is constant at about 50 mN/m per side, and so one might expect the membrane to resist tensions up to about 100 mN/m; however, tensions much smaller than this cause lysis. Rupture of the plasma membranes of isolated protoplasts is probabilistic, and the chance of lysis per unit time increases roughly exponentially with tension. The expected lifetime is 43 s at $\gamma = 5.0$ mN/m, and it is halved by each increase of 0.5 mN/m. Stochastic lysis of soap films and membranes has been explained in terms of transient or fluctuating pores. The energy of formation of a circular pore of radius r is lowered under an applied tension by $\gamma \cdot \pi r^2$. The effect which limits pore expansion is the energy of deformation of its perimeter—roughly proportional to r. The energy of a pore is thus parabolic in r (for sufficiently large r). At a critical radius the energy is a maximum; for larger r, the negative quadratic term dominates, and pores grow indefinitely to cause lysis. The critical radius for lysis and the maximum energy decrease with tension; at some value the energy to form a critical pore becomes comparable with the thermal energy, and lysis follows (Petrov et al., 1980). In this model, molecular packing at the pore perimeter strongly affects membrane stability. The presence of lipids which in aggregates have large interfacial area and small hydrocarbon volume (such as lysolipids), or of aqueous components which have a similar effect on molecular geometry, lowers the perimeter energy and thus the tension for lysis.

1. The picture is complicated by cold acclimation. When contracted osmotically, protoplasts isolated from acclimated plants form dense, osmiophilic extrusions surrounded by a membrane continuous with the plasma membrane. The greater capacity of acclimated protoplasts to survive rehydration from the contracted state is apparently due to the ability to reincorporate this material without producing large membrane tensions.

The Mechanics of Osmotic Contraction and Expansion

Changes in the hydration and therefore in volume of vesicles, organelles, or cells must impose changes in the area or shape of the membranes bounding them. We shall describe first the relatively wet state, in which water molecules are much more numerous than solutes and in which "bulk" volumes of solution exist: that is, the structures bounding the solution are in some places separated by distances greater than the range of the repulsive forces between those structures—typically a few nanometres.

Membranes in Excess Solution

Biological membranes are moderately permeable to water, and so membrane-bound vesicles, organelles, and cells dehydrate if the chemical potential of water outside is lowered. At equilibrium the chemical potential of water has the same value on either side of a membrane. The chemical potential of water outside may be lowered by increasing the osmotic pressure or decreasing the hydrostatic pressure of a solution or (at sufficiently low temperature) by freezing some of the external solution or by reducing the humidity of the atmosphere. If pressure differences are much smaller than the bulk modulus of water (2.04 GPa), then the chemical potential per unit volume or water potential, Ψ, is everywhere constant at equilibrium. For a solution with gauge pressure P and osmotic pressure Π, $\Psi = P - \Pi$.

Unsupported biomembranes separating two fluid phases are incapable of supporting negative surface tensions unless they are curved, with radii not much larger than their thickness (Wolfe and Steponkus, 1983b). The small range of γ (0 to about 5 mN/m) and Equation 17-3 give the possible range of pressure excesses inside organelles or cells bounded by such a membrane. Provided that the radius is, say, 1 μm or more, $\Delta P \leq 10$ kPa, which is much smaller than biological osmotic pressures. Thus organelles or cells bounded by unsupported membranes must, at equilibrium, have nearly the same osmotic pressure inside and out. For chloroplasts (Nobel, 1969), isolated plant protoplasts (Wiest and Steponkus, 1978), and a range of other membrane-bound cells and organelles in the wet state, the volume and the external osmotic pressure are related by the Boyle-van't Hoff relation:

$$\Pi(V - b) = RTn, \qquad (17\text{-}4)$$

where n is the number of moles of osmotically active solute and R the gas constant. At high osmotic pressure, the volume approaches a constant, b, the osmotically inactive volume. This volume includes the volume of internal solutes and structures and water of hydration; it is a larger fraction of the fully hydrated volume in chloroplasts than in vacuoles. At relatively high hydration, $b \ll V$, so changes in osmotic pressure cause approximately reciprocal changes in volume.

Dehydration in the wet phase produces large strains in unsupported membranes. Protoplasts suspended in a 0.53 Osmolal sorbitol solution and then transferred to a 1.20 Osmolal solution suffer an area reduction of 40%, and electron microscopy shows smooth plasma membranes before and after contraction (Gordon-Kamm and Steponkus, 1984a). The stress in the plasma membrane, however, is little changed: in both cases there is a small isotropic tension. The chloroplasts in these protoplasts suffer a similar contraction, and their envelope membranes are also smooth in both suspensions. These organelles are not spherical, and so the area reduction is not readily calculated, but linear dimensions are reduced by more than 20%. The observation that these membranes have varying curvature and Equation 17-3 need not imply zero tension in the membrane, because the interior of the chloroplast is not fluid. The observed shape is readily explained if the membrane has a small γ and is stretched over the inner structure. Direct mechanical studies have not been made on organelles, so one cannot be sure that the chloroplast membrane, like the plasma membrane, exchanges material with a reservoir. If the contraction was intensive, however, it would require the membrane to thicken in proportion to its contraction and to have an area elastic modulus much lower than that of any membrane studied to date. This eventuality seems improbable.

Rehydration of an osmotically contracted protoplast causes increases in tension that may lyse the plasma membrane. Because $\gamma \ll k_A$, $A/A_0 \approx 1$, and using this approximation, Equation 17-2 may be differentiated to give:

$$\frac{d\gamma}{dt} = k_A \left[\frac{1}{A}\frac{dA}{dt} - Z(\gamma) \right], \tag{17-5}$$

that is, the rate of increase in tension is proportional to the rate of area increase less the rate of incorporation. For any diminution in osmotic pressure of the external medium, $A(t)$ can be determined from the hydraulic conductivity L_P of the membrane. Equation 17-5 can then be solved to yield a lengthy expression for $\gamma(t)$ (Wolfe et al., 1985).

The initial expansion rate of a spherical osmometer (radius r) exposed to a step reduction in external osmotic pressure from Π_I to Π_F is

$$\frac{1}{A}\frac{dA}{dt} = \frac{2}{r} L_P (\Pi_I - \Pi_F), \tag{17-6}$$

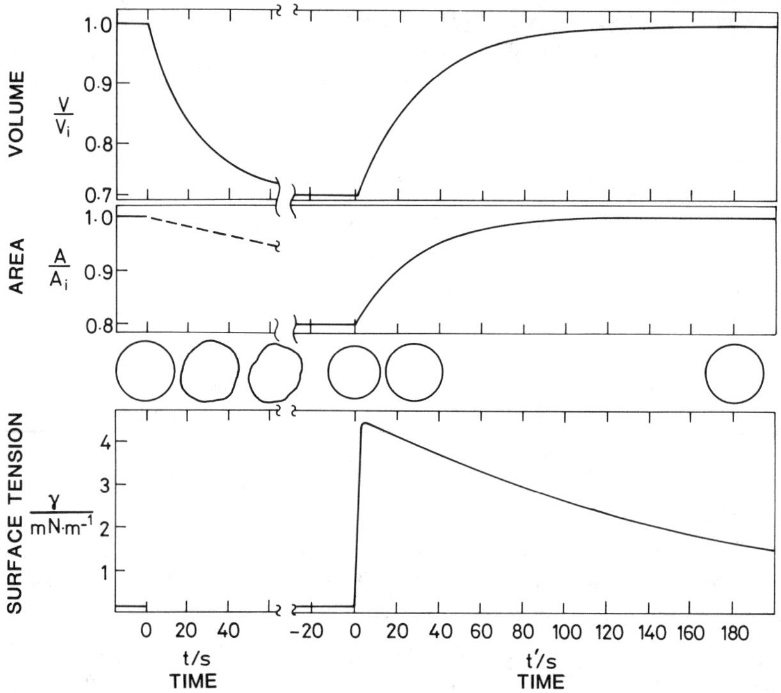

Figure 17-4. Calculated values of volume, membrane area, and surface tension are plotted against time for a protoplast subjected to a cycle of de- and rehydration. A step increase in external osmotic pressure is applied at $t = 0$, and a step decrease several minutes later at $t' = 0$. Sketches represent the shape of the protoplast. The contraction phase is difficult to study quantitatively, and the dashed line is an estimate.

and it decreases slowly thereafter. If γ is initially low, Z is very small, the membrane behaves almost elastically, and γ increases rapidly. As γ increases, material begins to be incorporated until the rates of expansion and incorporation are equal and a maximum γ is reached. γ subsequently decreases over many minutes (Figure 17-4). The maximum γ is determined by Equation 17-6 and $Z(\gamma)$; see Figure 17-3. From $\gamma(t)$ and the dependence on γ of the probability of lysis, the chance of survival of any given expansion can be calculated (Wolfe et al., 1985).

The Cell Wall

What is the osmotic response of a cell whose plasma membrane is supported by a cell wall? First, if the pressure in the cytoplasm is positive, this turgour pressure distends the wall, and the volume of the cell is limited by the global mechanical properties of the surrounding wall. It appears that the cell wall starts to collapse inwards under fairly small negative pressures, so the volume of a nonturgid cell is determined by the osmotic response of the protoplast. The symplast volume of a plant leaf gives excellent agreement with Equation 17-4 down to at least $\Psi = -3.5$ MPa. (See e.g. Melkonian et al., 1982.) The water in the cell wall has a very low concentration of free solutes, so its pressure is negative below full hydration. This suction is resisted by the mechanical forces between the microfibrils, and the wall material itself shrinks and swells with water potential. The cytoplasm has an osmotic pressure of typically 1 MPa at full hydration, and larger osmotic pressures when dehydration concentrates the internal solutes. From (17-4), we know that the hydrostatic pressure across the plasma membrane is

$$P = \Pi_o(V_o - b)/(V - b), \tag{17-7}$$

where the subscript refers to a suitable reference state such as full hydration. Thus the pressure difference across the plasma membrane[2] is typically 1 MPa at full hydration and is larger at lower hydration.

In addition to being in contact with the solid components of the cell wall (and exerting this pressure upon it), the membrane must contact the solution in the wall. That solutes pass readily through the wall demonstrates that it is perforated by interstices—gaps between the microfibrils. By measuring the permeation of the wall by polymers with a range of sizes, Carpita et al. (1979) estimated that the smallest gaps have radii of about 4 nm. From (17-3) it is clear that, wherever this membrane separates two liquid phases, it must be highly curved: taking $\gamma < 5$ mN/m, a spherical section would have $r < 10$ nm at full hydration and less at lower hydration. This is comparable with the membrane thickness (about 8 nm), and so it is not helpful to treat the membrane as a two-dimensional surface and use (17-3). Here the pressure difference will be resisted by the curvature stress in the membrane: it will press the membrane against the wall and bulge it into holes and crevices until it has a radius at which the energy to bend it further equals the work done by the pressure in further extrusion.

The bending energies of membranes with radii of nanometres have not to our knowledge been measured; at this curvature, deformation at a molecular

2. The walls separating adjoining cells are subjected to this pressure from both sides. For a cell adjacent to the air, the very large pressure that the membrane exerts on the wall of a nonturgid cell is equal to the pressure difference across the menisci of the air-water interface on the atmosphere side of the wall. Such a miniscus can support a pressure of 350 MPa in a wetted pore of radius 4 nm; however, it is likely that the water would cavitate at pressures less negative than this.

scale is involved. Surfactant molecules which can form highly curved convex interfaces with water do not form stable membranes (Israelachvili et al., 1980), and so curvature energies must be large. As the interfacial area on the convex side is increased, the surface tension opposing further deformation approaches that of an alkane-water interface, which can support a pressure of 350 MPa in a cell wall pore (see note 2). The question is further complicated by the likelihood that many membrane proteins would be found at the holes in the wall. This is a matter not of teleology but rather of electrostatics: the proteins have several charges, and in general their Born energies would be lower near water (which has a high dielectric constant) than near cellulose.

What happens to the area of a plasma membrane appressed to a cell wall as the cell is dehydrated beyond the point of zero turgour? We know of no quantitative study, but some deductions can be made. The internal surface area of the wall is reduced as the wall shrinks, its fibres drawn more tightly together by the increasing suction in the wall. This area reduction would be offset to some extent by the increasingly irregular surface of the membrane as it is pressed further into the small pores of the wall (see e.g. the electron micrographs of Opik, 1980, and Chabot and Leopold, 1985). The wall folds or crumples to achieve the low enclosed volume of the dry state, and so the change in the plasma membrane area is not large, although it probably exceeds the 2 to 3% possible with intrinsic stretching.

Membranes in Concentrated Phases

At low hydration, the interior of cells and organelles cannot be treated as a bulk solution suspending the internal structure. Water which is in equilibrium with an atmosphere with relative humidity (RH) must have pressure P and osmotic pressure Π such that

$$P - \Pi = (RT/\bar{V}) \ln(\text{RH}), \qquad (17\text{-}8)$$

where $\bar{V}$ is the molar volume of water. At RH = 90% and T = 293° K, ($P - \Pi$) is thus −14 MPa. An osmotic pressure of this magnitude requires that numbers of solute molecules are comparable with those of water, and the approximations used to derive (17-4) are invalid. A negative pressure in the solution between two membranes requires that the membranes repel each other with an equal force per unit area; such large repulsions occur at distances of at most a few nanometres (Lis et al., 1982; Horn, 1984). Thus at a very low hydration, removal of water is initially opposed by osmotic pressure as the remaining solutes are concentrated, but once the internal structures are separated by a few nanometres, removal of water does work against the repulsive forces between membranes and macromolecules. The shape of membranes in this state depends on the geometry of the objects around them, which are often other membranes; so it is unsurprising that lamellar regions are often found with spacings of a few nanometres. This is not, however, the only amphiphile configuration found at low hydration.

Reservoir Geometries

Gordon-Kamm and Steponkus (1984a, 1984b, 1984c) have reported the ultrastructural response of protoplasts to dehydration induced either by freezing of the suspending medium or by increasing its concentration at constant temperature (the effects of the two procedures are similar). Contraction in the wet state produces cytoplasmic vesicles adjacent to the plasma membrane, as described above. Further dehydration (in 5.37 Osmolal sorbitol solution) produces aparticulate lamellae beneath the plasma membrane. (These are traditionally interpreted as pure lipid lamellar phases.) It is inferred that these are formed by the appression of the cytoplasmic vesicles as water is withdrawn. (The reverse of this process—production of lipid vesicles from lamellae upon wetting—has been described by Harbich and Helfrich, 1984.) Coexisting with the lamellae (hereinafter L) at severe dehydration are the tightly packed cylinders identified as hexagonal II phase (hereinafter H_{II}).

Models of varying complexity for the L-H_{II} transition in water-pure lipid mixtures have been proposed (Jonsson, 1981; Kirk et al., 1984) and the transition is described in detail in Chapter 14. We shall give a simple analysis of the transition using the lipid packing concept of Israelachvili, Mitchell, and Ninham (1976) (hereinafter the *Canberra model*). Such an approach makes sweeping simplifications, but it retains the essential physics, and it has been prosecuted with considerable success in explaining the various lipid aggregations in excess water, including living membranes (Israelachvili et al., 1980). The Canberra model assumes that the resultant of the surface forces in Figure 17-1 acts at or near the tail-hydrophilic interface and that there are only small torques about this surface. The tails are assumed to behave as an incompressible fluid with average volume v per lipid, subject to the further constraint that no point in the tail region may be further from the interface than a distance l_c. In wet phases, it is assumed that the area (a) per molecule at the interface is constant at a_0: this is true if the resultant surface forces are $\ll k_A$. In the dry state, however, membranes may support substantial negative γ because large normal forces prevent them from relieving the

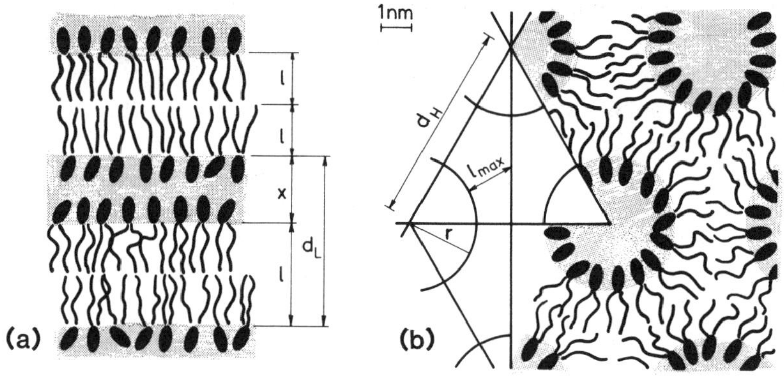

Figure 17-5. A portion of lamellar phase of lipids is represented in (a). Part (b) shows the idealised geometry of a hexagonal II phase with the same composition and same area a of tail-water interface per molecule. The longest distance from this interface is l_{max}. The geometry is to scale and the distance between head groups is depicted as $\sqrt{a}$.

stress by bowing. A pressure of −5 MPa between two surfaces at 1 nm separation,[3] for example, can produce $\gamma = -5$ mN/m.

A portion of a lamellar phase is sketched in Figure 17-5a; the degree of hydration is cited as f, the number of water molecules per lipid, x is the distance between the tail-hydrophilic interfaces of opposing membranes, and $2l$ is the thickness of the tail region. A head group and f water molecules together occupy, on average, a volume $v_a = a \cdot x/2$. A dehydration which reduces v_a by dv_a reduces a, x, or both. A reduction of x by dx at constant a requires work per molecule:

$$dW = -Pa \cdot dx,$$

where P is the (negative) pressure in the water and $-P$ is the force per unit area between the bilayers. A reduction of a by da at constant x requires work per molecule:

$$dW = k_A(a/a_0 - 1) \cdot da$$

3. It is interesting to ask at what pressure, and thus at what separation, the lateral stress in the membrane ceases to be negligible in comparison with the area modulus. The measurements of Lis et al. (1982) show a repulsion of typically 15 MPa at a mean water thickness of about 0.8 nm, which gives $(-\gamma)/k_A = 10\%$ and corresponds to a water content of about 20%. The very strong dependence of repulsion on separation implies that, at water contents much above this, the repulsion is much smaller, and so the lateral stress in the membrane is negligible. At less than about 20% water content, the lateral stress in the membrane is no longer negligible in comparison with the area modulus.

plus a term representing the increase in potential energy of the intermembrane forces due to the increase in molecular area density. While x is large and the intermembrane forces small, dehydration will reduce x at virtually constant a. P increases approximately exponentially with decreasing x (Lis et al., 1982; Horn, 1984), so at sufficiently small x, a must also decrease upon dehydration in the lamellar phase. As $v = al =$ constant, however, so a reduction in a implies an increase in l. This is opposed by the energy curve (Figure 17-1), which must become very steep as l approaches its value for a straightened tail, l_c. Thus a cannot easily be reduced below v/l_c, and we expect it to approach this value at low f.

Figure 17-6 shows the dependence on hydration f of the geometric parameters defined by Figure 17-5 for a lipid-water system which undergoes L-H_{II} transitions: brain lipids (Luzzati and Husson, 1962). (We know of no detailed study of the phase diagram of lipids from seeds or other dry systems.) In both cases it is seen that there is a large, almost linear reduction in x and only a small reduction in a with dehydration in the L phase: a decreases and l increases by 14% for $f = 34$ to $f = 13$. Further dehydration would cost a lot of energy: l cannot increase beyond its critical length l_c, so the separation x must be reduced against the increasingly strong repulsion.

How does this compare with the energy of the H_{II} phase? First note that the area of the tail-hydrophilic interface in the two phases is similar except at very low f. The hydrophilic geometry, however, is very different. In the L phase, the separation in the normal direction is $x = 2v_a/a$; in the H_{II} phase it is $2r = 4\pi r^2/2\pi r = 4v_a/a$. Thus for any particular value of the water-to-lipid ratio f and area a, the normal separation of the interfaces is twice as great in the H_{II} phase as in the L phase, as Figure 17-5 shows. Of course the different geometries make such a comparison complicated in detail, but we shall simply postulate that this geometry has a lower interaction energy. For a more detailed analysis, see Jonsson (1981), Kirk et al. (1984), or Maier (1985). If the H_{II} phase can lower the repulsion energy at the same hydration and the same a, why is it not found at higher f? Van der Waals forces prohibit holes in the tail region, so no point in the tail phase may be more than l_c from the interface. The largest distance l_{max} from the interface in a H_{II} is given as a dashed line in Figure 17-6 as a function of f with constant a: we propose that the range of hydration over which H_{II} can exist in a lipid-water system is that for which $l_{max} \leq l_c$. In the Canberra model, l_c depends on the length and degree of unsaturation of the tails, whose composition in this preparation is unknown. l_c could be determined from measurements of aggregates in the wet state. (Data from beef heart mitochondrial lipids [Gulik-Krzywicki et al., 1967] may be similarly analysed [Maier, 1985].) The largest hydration f_c at

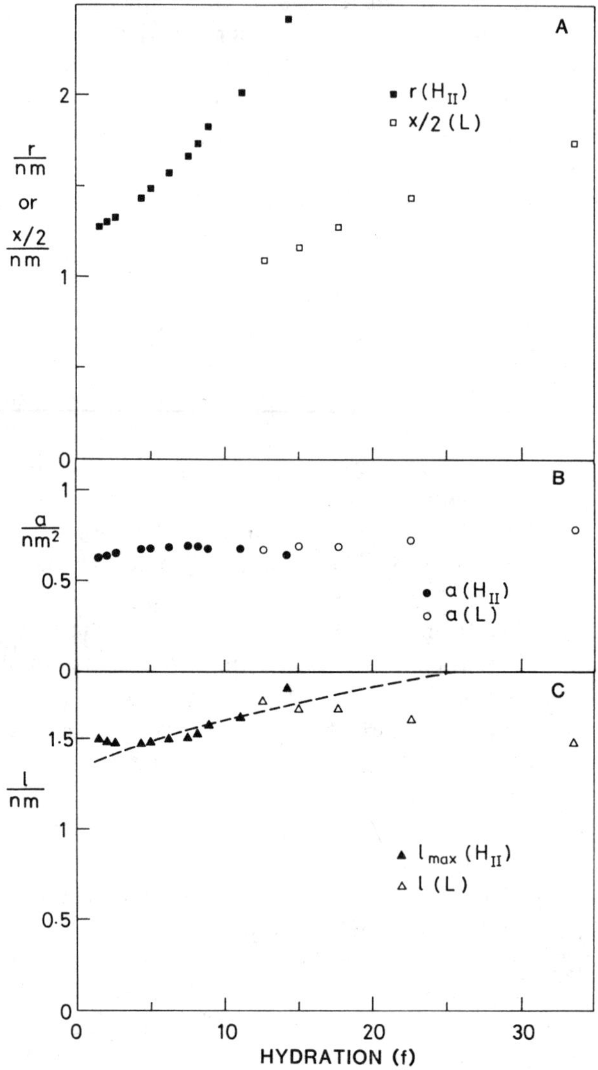

Figure 17-6. The geometric parameters (defined in Figure 17-5) of lamellar (open symbols) and hexagonal II phases (closed symbols) of brain lipids are plotted as a function of f, the ratio of the numbers of water and lipid molecules. Part (a) shows the "wet thickness" r or $x/2$, (b) shows a, and (c) shows the "tail thickness" l_{max} or l. The dashed line is the geometrical calculation for l_{max}, assuming constant a.

which a H_{II} phase can occur can be related to the lipid parameters l_c, a, and the head and tail volumes v_h and v by simple geometry. The equation is

$$l_c = \frac{2}{a}\left[\sqrt{\frac{2\pi}{3\sqrt{3}}} \cdot \sqrt{(v_w f_c + v_h)(v + v_w f_c + v_h)} - v_w f_c - v_h\right], \quad (17\text{-}9)$$

where v_w is the volume of a water molecule.

The transition from L to H_{II} is biologically significant: lamellae clearly have a topology more suited to retaining intracellular solutes than the long cylinders of H_{II}, and Gordon-Kamm and Steponkus (1984b) have correlated the appearance of H_{II} with dehydration-induced loss of semipermeability. The model described above is still under development, but Equation 17-9 suggests that a molecular explanation is possible for the changes induced by acclimation, which preclude the formation of H_{II} at comparable hydration. Increasing a or decreasing l_c in (17-9) yields a lower f_c; perhaps the effect of carbohydrates which increase a (Crowe et al., 1984; Johnston et al., 1984) is to preclude H_{II} in the dry state.

REFERENCES

Albrecht, O., Gruler, H. and Sackmann, E. 1978. Polymorphism of phospholipid monolayers. J. Physique 39: 301-313.

Boroske, E., Elwenspoek, M. and Helfrich, W. 1981. Osmotic shrinkage of giant egg-lecithin vesicles. Biophys. J. 34: 95-109.

Carpita, N., Sabularse, D., Montezinos, D. and Delmer, D.P. 1979. Determination of the pore size of cell walls of living plant cells. Science 205: 1144-1147.

Chabot, J.F. and Leopold, A.C. 1985. Ultrastructural aspects of chilling injury in soybean axes. Am. Jour. Bot. 72: 1120-1126.

Crowe, J.H., Whittam, M.A., Chapman, D. and Crowe, L.M. 1984. Interactions of phospholipid monolayers with carbohydrates. Biochim. Biophys. Acta 769: 151-159.

de Lacey, E.H.B. and Wolfe, J. 1982. Elastic response of bilayers with intrinsic proteins. Biochim. Biophys. Acta 692: 425-430.

Dowgert, M.F., and Steponkus, P.L. 1984. Behavior of the plasma membrane of isolated protoplasts during a freeze-thaw cycle. Plant Physiol. 75: 1139-1151.

Evans, E. and Kwok, R. 1982. Mechanical calorimetry of large dimyristoylphosphatidylcholine vesicles in the phase-transition region. Biochemistry 21: 4874-4879.

Evans, E.A. and Skalak, R. 1979. Mechanics and thermodynamics of biomembranes. C.R.C. Press, Boca Raton.

Gordon-Kamm, W.J. and Steponkus, P.L. 1984a. The behavior of the plasma membrane following osmotic contraction of isolated protoplasts: implications in freezing injury. Protoplasma 123: 83-94.

Gordon-Kamm, W.J. and Steponkus, P.L. 1984b. The influence of cold acclimation on the behavior of the plasma membrane following osmotic contraction of isolated protoplasts. Protoplasma 123: 161-173.

Gordon-Kamm, W.J. and Steponkus, P.L. 1984c. Lamellar-to-hexagonal II phase transitions in the plasma membrane of isolated protoplasts after freeze-induced dehydration. Proc. Natl. Acad. Sci. USA 81: 6373-6377.

Gruen, D.W.R. 1981. A mean field model of the alkane-saturated lipid bilayer above its phase transition. I. Development of the model. Biophys. J. 33: 149-166.

Gruen, D.W.R. and Wolfe, J. 1982. Lateral pressures and tensions in lipid monolayers and bilayers. Biochim. Biophys. Acta 688: 572-580.

Gulik-Krzywicki, T., Rivas, E. and Luzzati, V. 1967. Structure et polymorphisme des lipides: étude par diffraction des rayons X du systeme formé de lipides de mitochondries de coeur de boeuf et d'eau. J. Mol. Biol. 27: 303-322.

Harbich, W. and Helfrich, W. 1984. The swelling of egg lecithin in water. Chem. Phys. Lipids 36: 39-63.

Horn, R.G. 1984. Direct measurement of the force between two lipid bilayers and observation of their fusion. Biochim. Biophys. Acta 778: 224-228.

Israelachvili, J.N., Marčelja, S. and Horn, R.G. 1980. Physical principles of membrane organization. Quart. Rev. Biophys. 13: 121-200.

Israelachvili, J.N., Mitchell, D.J. and Ninham, B.W. 1976. Theory of self-assembly of hydrocarbon amphiphiles into micelles and bilayers. J. Chem. Soc., Faraday Trans. II 72: 1525-1568.

Johnston, S.J., Coppard, E., Parera, G.V. and Chapman, D. 1984. Langmuir film balance study of the interactions between carbohydrates and phospholipid monolayers. Biochem. 23: 6912-6919.

Jonsson, B. 1981. The thermodynamics of amphile-water systems: a theoretical analysis. Ph.D. thesis, Chemical Center, Lund.

Kirk, G.L., Gruner, S.M. and Stein, D.L. 1984. A thermodynamic model of the lamellar to inverse hexagonal phase transition of lipid membrane-water systems. Biochem. 23: 1093-1102.

Kwok, R. and Evans, E. 1981. Thermoelasticity of large lecithin bilayer vesicles. Biophys. J. 35: 637-652.

Laver, D.R. 1983. The molecular organization of artificial lipid bilayers: a study of their dielectric impedance and surface tension. Ph.D. thesis, Univ. New South Wales, Sydney.

Lis, L.J., McAlister, M., Fuller, N., Rand, R.P. and Parsegian, V.A. 1982. Interactions between neutral phospholipid bilayer membranes. Biophys. J. 37: 657-666.

Luzatti, V. and Husson, F. 1962. The structure of the liquid- crystalline phases of lipid-water systems. J. Cell Biol. 12: 207-219.

Maier, B. 1985. Lamellar-hexagonal II phase transitions in surfactant-water systems. B.Sc. Thesis, Univ. New South Wales, Sydney.

Marčelja, S. and Wolfe, J. 1979. Properties of bilayer membranes in the phase transition or phase separation region. Biochim. Biophys. Acta 557: 24-31.

Melkonian, J.J., Wolfe, J. and Steponkus, P.L. 1982. Determination of the volumetric modulus of elasticity of wheat leaves by pressure-volume relations and the effect of drought conditioning. Crop Science 22: 116-123.

Nobel, P.S. 1969. Light-induced changes in ionic content of chloroplasts in *Pisum sativum*. Biochim. Biophys. Acta 172: 134-143.

Opik, H. 1980. The ultrastructure of coleoptile cells in dry rice grains. New Phytol. 85: 521-529.

Petrov, A.G., Mitov, M.D. and Derzhanski, A.I. 1980. Edge energy and pore stability in bilayer lipid membranes. *In* Advances in Liquid Crystal Research and Applications, L. Bata, ed. Akadémiai Kiadó, Budapest.

Wiest, S.C. and Steponkus, P.L. 1978. Freeze-thaw injury to isolated spinach protoplasts and its simulation at above-freezing temperatures. Plant Physiol. 62: 599-605.

Wolfe, J., Dowgert, M.F. and Steponkus, P.L. 1985. Dynamics of membrane exchange of the plasma membrane and the lysis of isolated plant protoplasts during rapid expansions in area. J. Membrane Biol. 86: 127-138.

Wolfe, J. and Steponkus, P.L. 1980. The stress-strain relation of the plasma membrane of isolated plant protoplasts. Biochim. Biophys. Acta 643: 663-668.

Wolfe, J. and Steponkus, P.L. 1983a. Mechanical properties of the plasma membrane of isolated plant protoplasts. Plant Physiol. 71: 276-285.

Wolfe, J. and Steponkus, P.L. 1983b. Tension in the plasma membrane during osmotic contraction. Cryo Lett. 4: 315-322.

Yue, B.Y., Jackson, C.M., Taylor, J.A.G., Mingins, J. and Pethica, B.A. 1975. Phospholipid monolayers at non-polar oil/water interfaces. J. Chem. Soc. Faraday Trans. I 72: 2685-2693.

18 Structural Transitions Induced by Osmotic Stress

Adrian Parsegian
Donald Rau
Joshua Zimmerberg

Biological materials suffering even a slight lowering of solvent water activity can undergo dramatic changes in organization and assembly. These transitions, viewed in the context of "anhydrous biological systems," suggest to us that organisms adapted to transient extreme dryness might not only use protectants to prevent damage but might also preserve their vitality by undergoing intentional structural changes.

Since the very early observations on the alignment of tobacco mosaic virus particles at only about 1% volume fraction (R. Best, 1940; Bernal and Fankuchen, 1941) and the theoretical explanation of this alignment of rods in terms of steric interactions (Onsager, 1949), one has been aware that large particles will assemble under the very slightest energetic provocation when that energy is measured as a change in the activity of water. The recognition of solvation forces in colloidal systems (Churaev and Derjaguin, 1985) and the preoccupation with solvophobic or hydrophobic interactions governing the assembly of macromolecules (Kauzmann, 1959; Lauffer, 1975) imply that the properties of the water solvent are at issue in understanding molecular arrangement. Our own work on the measurement of forces between molecules unambiguously demonstrates that the dominant interactions as polar molecules approach contact are forces of solvation. It is almost a truism that changes in the activity of solvent water will determine the state of organization of biological molecules.

The data on the systems to be mentioned below—sickle cell hemoglobin, DNA double helixes, lipid bilayers—have been well reported elsewhere and will be only briefly reviewed here. We will try to extract from those studies some idea of the kinds of stress that can be expected to produce structural reorganization inside a cell and to compare the magnitude of stresses seen in these measurements with those characteristic of strong dehydration.

The strategy in our studies has been to create thermodynamically well defined changes in a system—usually by decreasing the activity of water but also by varying temperature, pressure, and the activity of liganding species that affect molecular assembly—and then to observe the structural consequences of these imposed thermodynamic changes. Proteins and nucleic acids can be driven to assemble, lipids can freeze their hydrocarbon chains, and transmembrane ionic channels can be forced to close. Both the prevention of these kinds of responses and their practical exploitation could be intentionally incorporated into the design of a cell forced to endure dehydration in its natural life.

Structural Transitions under Stress of Dehydration

By a rather simple device, we have been able to create a phase diagram for a protein assembling under osmotic stress. Schematically (Figure 18-1), the method is to enclose the material in dialysis tubing and then to immerse this package into a large solution of the desired ionic composition and temperature which also contains dextran or some other large polymer that cannot permeate the tubing. This large polymer exerts osmotic pressure on the contents, which then shrink or swell until the outward osmotic pressure of the inside protein suspension equals that from outside. Thus the osmotic pressure of the dextran solution allows us to determine that of the protein of interest even if it is in a state of aggregation that would have made a direct measurement impossible. After equilibration, we measure the resulting protein concentration to obtain the data needed to plot pressure versus volume relations much as one does for a condensing gas.

Hemoglobin

The mutant sickle cell hemoglobin, when deprived of oxygen, can polymerize into rods of indefinite length (see e.g. Noguchi and Schechter, 1985). We have put this protein under stress at several temperatures to create pressure-concentration or pressure-volume diagrams (Prouty et al., 1985). At 20, 30, and 37°C there are sharp discontinuities in concentration coinciding with sudden changes in viscosity that indicate a change in structure. These data can be replotted in the units of traditional phase diagrams; see Figure 18-2.

By presenting the data this way we are able to learn about protein assembly in much the same way that physical chemists have traditionally studied

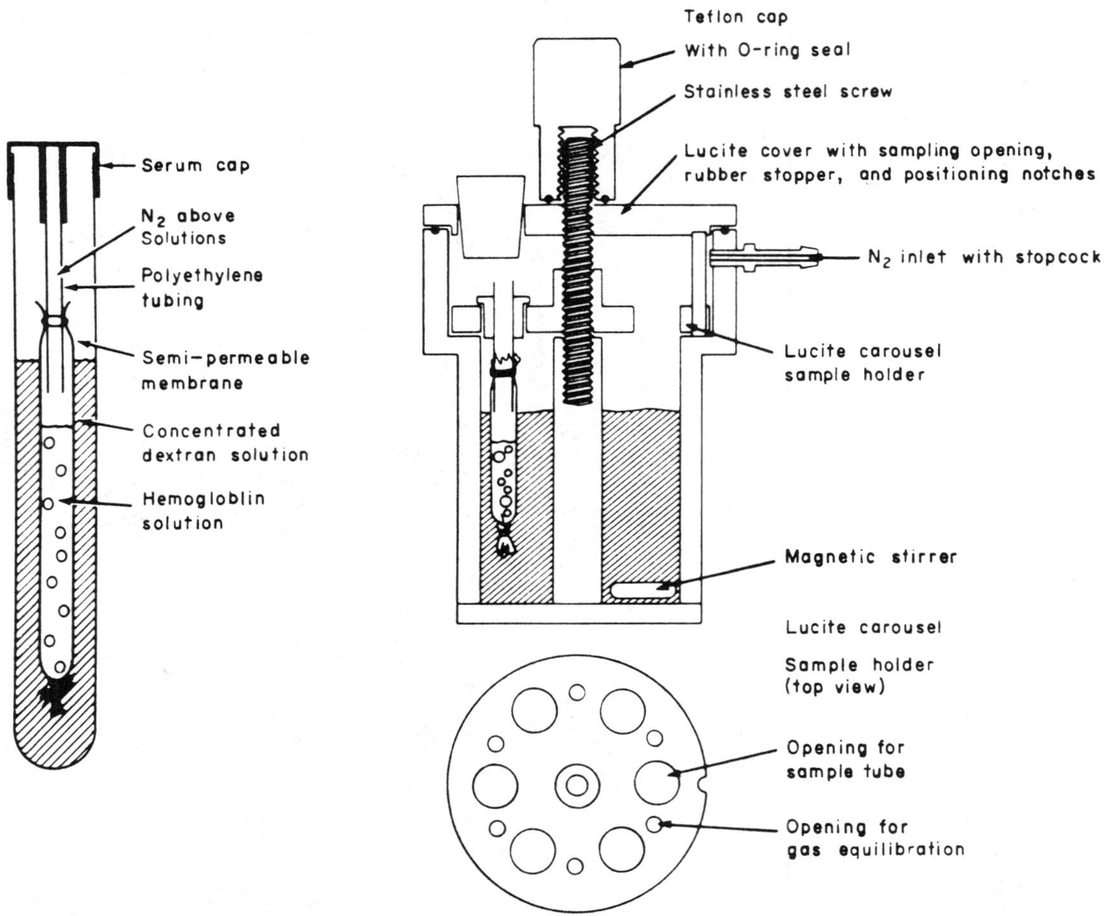

Figure 18-1. An example of experimental application of osmotic stress to study protein assembly. The figure shows the method for equilibrating one to six samples. (For details, see Prouty et al., 1985, from which this

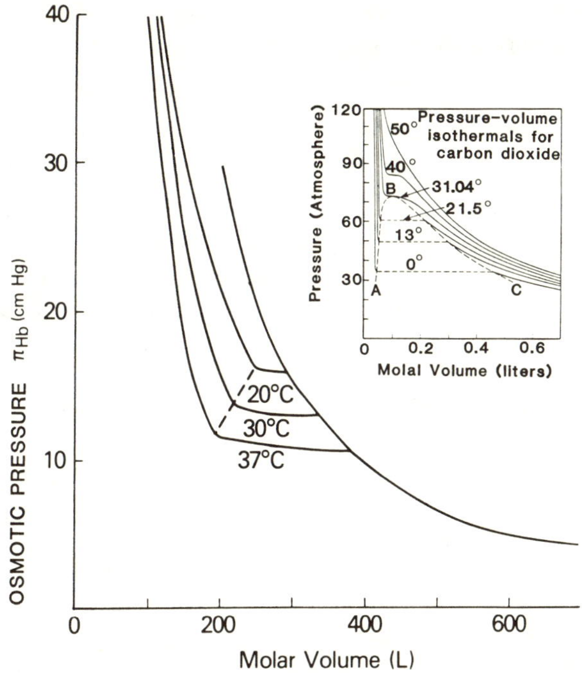

Figure 18-2. Pressure vs. molar volume of sickle cell hemoglobin to illustrate the resemblance to classical phase transitions. The inset shows the analogous pressure-volume isotherms for carbon dioxide condensation. (From Prouty et al., 1985, courtesy of *Journal of Molecular Biology*.)

the condensation of gases into liquids. For example, one can use a variant of the Clausius-Clapeyron Equation to extract the entropy and enthalpy of the transition. In applying stress, one has actually done physical work on the protein and has made possible measurement of its chemical potential change. We measure the rate of change of transition pressure with transition temperature and the volume change in transition and introduce these into

$$d\pi/dT = \Delta S/\Delta V$$

and

$$\Delta H = T\,\Delta S + \pi\,\Delta V$$

to create the plots of ΔS and ΔH versus temperature. A large positive entropy of polymerization is consistent with the stimulation of the condensation by warming and is what one ordinarily expects in "hydrophobic bonding," or what one usually associates with a process involving loss of ordered solvent from the vicinity of the polymerizing species.

DNA

When exposed to a polymer such as polyethyleneglycol or polyvinylpyrolidone, DNA double helices show a structurally informative condensation reaction. The DNA separates from the added polymer and precipitates into an array of parallel molecules so well ordered as to permit measurement of molecular separation by x-ray diffraction (Lerman, 1971; Maniatis et al., 1974). Since the lattice spacing decreases with increasing osmotic pressure of the outside polymer, one can thermodynamically follow a condensation similar to that just described for sickle cell hemoglobin, but one can also know the packing characteristics of the assembled molecules. One can in fact measure the actual force exerted by neighboring molecules as they are pushed together (Rau et al., 1984), and one learns that they bear little resemblance to the electrostatic double layer forces expected between these molecules. Rather they seem to entail the work of removing water from between molecular surfaces even when those surfaces are 10 to 15 Å apart, almost regardless of the ionic strength of the medium and of its ionic composition. The repulsive force is an exponential with a decay constant of about 3 Å (Figure 18-3).

DNA will also form ordered precipitates in the presence of particular cations such as spermine and related polyamines, cobalt hexamine, and even Mn or Cd. To us, the most striking feature of this precipitation is that the parallel helices remain some 7-10 Å apart even though the condensing spe-

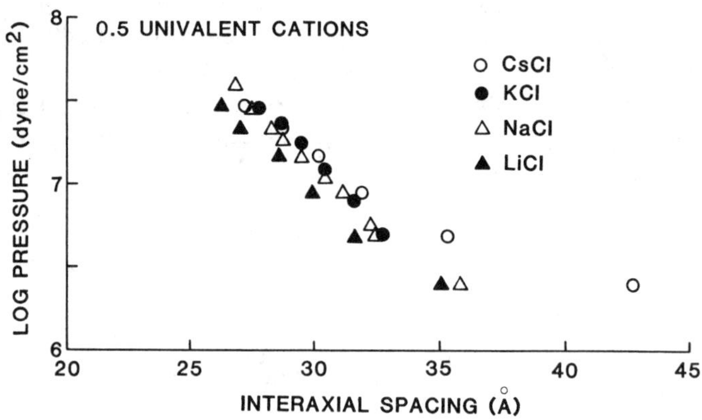

Figure 18-3. Pressure vs. interaxial separation in hexagonal arrays of parallel double-stranded helices of DNA in 0.5 M univalent electrolyte solutions. (For details, see Rau et al., 1984.)

cies is known to sit within the double helical structure. The attractive force of condensation must be acting at distances greater than 10 Å. Precipitation is stimulated by raising the temperature but suppressed by the application of hydrostatic pressure. When the molecules are pushed together from their spontaneous minimum-energy position, the difference between attractive and repulsive forces goes over to an exponential with an approximate 1.5 Å decay length constant (Figure 18-4).

When inadequate amounts of the precipitating cation are available and DNA is subjected to polymer osmotic stress, there are clear transitions between the two kinds of force curves, with a well-defined relation between the pressure of transition and the concentration of limiting cation. Similarly, transitions that cannot occur at low temperature will be effected by the further application of osmotic stress. Combining the pressure, temperature, ionic activity, and helix spacing changes observed, one is able to extract energies and entropies of these transitions. As with the sickle hemoglobin, the transitions to higher macromolecular density reveal positive entropy and enthalpy of a magnitude that suggest attractive energies some 50 times greater than would be expected from van der Waals attraction. We are led to conclude that long-range solvent-driven interactions mediated by bound ions underlie the spontaneous assembly of DNA. What is further noteworthy is that these solvation forces, again "hydrophobic" in appearance, are set up by highly polar surfaces that do not come into direct molecular contact. The actual pressures over which transitions are induced range up to 15 atm or roughly

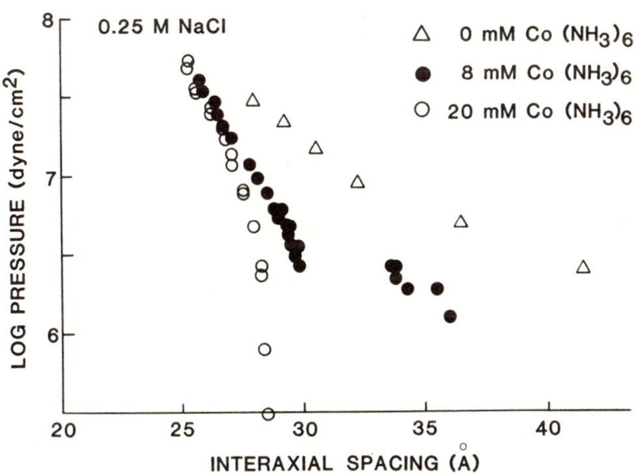

Figure 18-4. Collapse of DNA double helices under the combined influence of cobalt hexamine and osmotic stress.

1/2 osmole, rather small on the scale of perturbation encountered under anhydrous conditions but well within the precision with which cells control or respond to osmotic stress.

Lipids

It has been known for decades that lipid hydrocarbon chains of bilayers in multilamellar arrays will freeze upon desiccation. What has been only relatively recently recognized is that the interlamellar forces between bilayers as they move together when water is removed are primarily a dehydration or a desolvation of the bilayer surface rather than an interaction in which the intervening solvent is treated as a structureless medium. Systematic characterization of these "hydration forces" shows that they can extend over distances several times the size of a water molecule (for review see Rand et al., 1981; Rand and Parsegian, 1986) and closely resemble the analogous forces seen between DNA double helixes.

Combining osmotic stress and x-ray diffraction, one determines force versus bilayer separations and also finds the critical osmotic pressures at which multilayer dehydration creates an acyl-chain structural transition. In principle, one can apply a Clausius-Clapeyron equation to this transition, relating the critical osmotic stress, the transition temperature, and the water volume change to extract the entropy and enthalpy of transition all along the solid chain (gel) and melted chain (liquid crystalline) coexistence curve. Preliminary experiments (Gershfeld and Parsegian, unpublished) show a clean upward shift in the calorimetrically detected chain transition temperature of dipalmitoylphosphatidylcholine kept in contact with a reservoir containing dextran polymer.

Alternatively, one may use the π-volume relation from osmotic stress to compute the free energy needed to increase a phase transition temperature by a given amount. For a molecular cross section A, and bilayer separation d_w with value d_w^o at the limit of multilayer swelling, where there is a balance of attractive and repulsive forces, one may integrate

$$\Delta G = -\int_{d_w^o}^{d_w} \pi(d_w)\, A(d_w)\, (1 + \frac{d \ln A}{d \ln d_w})\, d(d_w)\,,$$

where π has a repulsive (hydration) part

$$\pi_{hyd} = P_o e^{-d_w/\lambda}$$

and an attractive (van der Waals) part

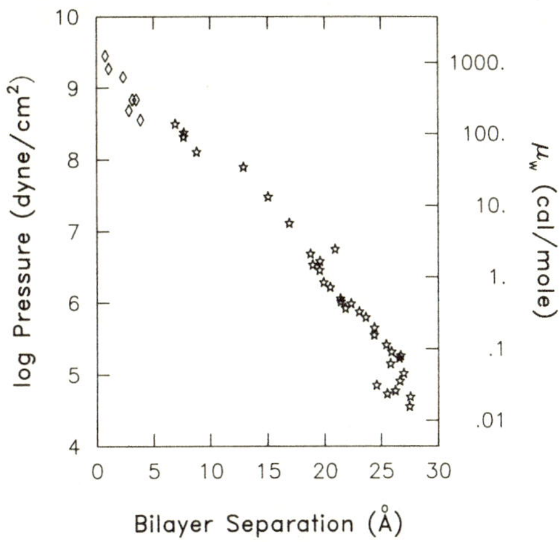

Figure 18-5. The pressure between bilayers of dilauroylphosphatidyl-choline as a function of bilayer separation. Note melting transition at higher pressure. Chemical potential of water relative to bulk water is shown on right. (Reproduced from the *Biophysical Journal*, 1982, vol. 37, pp. 657-666, by copyright permission of the Biophysical Society.)

$$\pi_{vdw} = \frac{H}{6\pi}\left[\frac{1}{d_w{}^3} - \frac{2}{(d_w + t)^3} + \frac{1}{(d_w + t)^3}\right],$$

where t is bilayer thickness and H is the Hamaker coefficient.

Published data for dilauroylphosphatidylcholine at 25°C are $d_w^o = 29$ Å, $P_o \times A = 10^{-4.57}$ dynes, $\lambda = 2.7$ Å, $t = 30$ Å, and $H = 4.9 \times 10^{-14}$ ergs. The integral is dominated by hydration forces (Rand et al., 1981; Lis et al., 1982). In order to push bilayers to $d_w = 7$ Å, about 5.4×10^{-14} ergs, or 1 kT, of work per polar head must be done. This osmotic work of about 1 kT suffices to raise the melting temperature of DLPC from −2°C to 25°C. The osmotic stress at which this occurs, about 10^8 dyne/cm^2 or 100 atm (see Figure 18-5) corresponds to a relative humidity of 93%. Again, as with sickle cell hemoglobin and DNA, small stresses create major structural changes.

Subjectivity of Probes of Cell Water

Figure 18-6 compares the physical stress scale with the relative humidity p/p_o. The action of small perturbations leads us to ask what indeed does constitute bound or nonnormal water influencing cellular organization.

Few controversies in science are as strong and as emotional as those over the condition of water inside living cells. The range of opinions is extreme, including those who hold that all cell water is in some important way different from normal "bulk" water (cf. Clegg and Drost-Hansen, 1979; Ling,

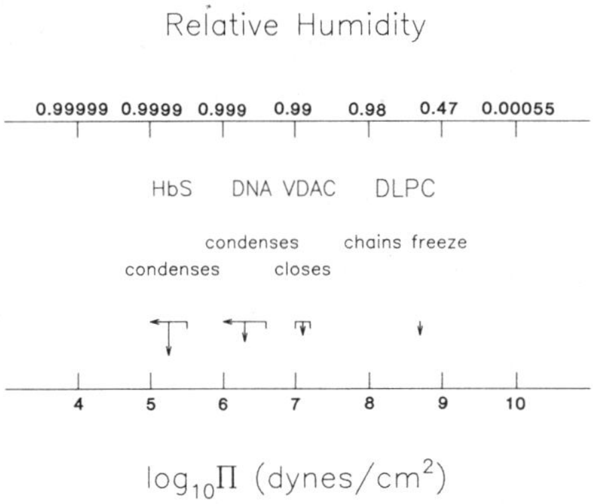

Figure 18-6. Comparison of osmotic stress scales with regions of typical structural transitions. The $\log_{10}\pi$ scale appears to be natural for the exponentially decaying forces shown in Figures 18-3 and 18-5.

1984) and the more conventional position that only a monolayer or less of water on the macromolecular surface is noticeably perturbed by comparison with bulk water. Where does one look for a gauge of water activity relevant to the changes that will occur upon desiccation? We take the position that most of the controversy is needless, that water in a cellular or macromolecular preparation will show different properties, depending on the way that one chooses to describe it, and that the cellular or macromolecular response to changes in water activity depends on the amount of water that is lost or transferred.

In our application of osmotic stress we have found that very small diminution of water activity, at least very small perturbations on the scale of lowered vapor pressures, will qualitatively influence membrane or molecular organization. The reason is that large amounts of slightly perturbed water are transferred during reorganization. In terms of relative humidities p/p_o, the osmotic pressure is

$$\pi = (kT/v_w)\ln(p_o/p),$$

where kT has its usual meaning and v_w is the volume of a water molecule, or we can say the chemical potential of water μ_w is

$$\mu_w = \pi v_w \ln(p_o/p).$$

Pressures on the order of 10^6 dynes/cm^2 used to condense sickle cell hemoglobin (Figure 18-2) are seen to be virtually negligible on the scale of relative humidities. Specifically, $10^6 = (4 \times 10^{-14}/30 \times 10^{-24}) \ln(p_o/p)$; $p/p_o = 0.99925$. Even the much higher pressures encountered in DNA assembly or lipid chain transitions, 10^8 dynes/cm^2, correspond to relative humidities $>90\%$, much higher than those normally used in studying molecular hydration by vapor sorption. Thinking in terms, then, of energies governing molecular organization, one must be aware that the energies pumped into a cell when it is dehydrated are very large even though the actual energy is small per water molecule.

For example, ask what number η of water molecules must be transferred if a dehydrating step is to involve an energy change kT. Since $\Delta W = \pi \Delta V$ under stress π, and π is given by $(kT/v_w)\ln(p_o/p)$, if we say that $\Delta W = kT$ we may write

$$\eta = 1/\ln(p_o/p).$$

Even at 90% relative humidity, one needs only 9 to 10 waters transferred; at 50%, 1 to 2.

One can begin to think of many cellular systems capable of large changes in associated water even in normal operation. Ionic channels in closing might lose much or all of their internal space, as is suggested from models (Unwin and Zampighi, 1980) and from osmotic stress determination of the internal volume change. The voltage-dependent anion channel (VDAC) from mitochondria appears to change aqueous content by some 2 to 4 $\times$ 10^4 $Å^3$ or 1,000 water molecules (Zimmerberg and Parsegian, 1984). A closing energy of kT can be exerted by osmotic stress corresponding to relative humidities on the order only of 99.9%!

And what, then, of the controversies about "bound" or "perturbed" water? Rather than review a large and frustrating literature, it is probably best for us simply to note that the criteria of perturbation have been almost subjective. It depends what specific technique is used to interrogate a particular system. Diffusion or NMR or the dissolution of small particles involves relatively few water molecules. The tiny perturbations that show up in macromolecular interactions or transitions may not add up here to noticeable quantities. Yet it would be a mistake to assume from such measurements that weakly perturbed water is in all ways "normal."

It is indeed sad to see the persistence of the conclusion that only a monolayer or less of water hydrates a protein or membrane surface, a conclusion based on single-molecule probes of water. It should be clear that the reorganization required to achieve the "anhydrous" states encounters weak forces of solvent perturbation that sum to significance over the many water molecules near a macromolecular surface. To ignore these long-range weak perturbations is to miss an essential feature of molecular organization.

Conclusions and Prospects

Given the near inseparability of water and living systems, it should not be surprising that biological materials are so sensitive to small changes in the activity of water. What is of course more surprising is the resilience of many kinds of cells deprived of water. It seems to us that the observation of structural transitions in well-defined macromolecular systems provides a practical physical chemical adjunct to the splendid cell biological work that has preceded it and comprises the major part of this book.

Another major body of literature might well bear on questions of organization under stress. Entropy-driven, or temperature-sensitive, hydrostatic-pressure-sensitive assemblies have long been recognized in cellular organelles (Zimmerman, 1970; Lauffer, 1975) and in protein extracts (e.g. Harrington and Kegeles, 1973). The striking qualitative similarities between the factors governing those systems and the parameters important to DNA and sickle cell hemoglobin assembly suggest to us that the forces being measured in the latter systems are representative of those in many other cellular processes.

The precision with which cells try to control aqueous activity and the alacrity with which many cells respond to osmotic changes suggest an exquisite sensitivity to osmotic stress. Two clear examples of this sensitivity are the protrusion of actin filaments (Tilney and Inoue, 1985) and blebbing in epithelial cells (Harris, 1973), noted by Trinkhaus (1985) in a recent discussion on the forces causing cellular protrusion.

From reports on observations of seed formation under desiccation with subsequent vulnerability to drying after the onset of germination, one might speculate that normal seed development involves a natural condensation of components tailor-made to fit together. The constraints on such a design probably pose severe limitations on the molecular population, a limitation that is necessarily violated with the metabolism of new species during germination. The pieces no longer fit. Rather than driving compact assembly, desiccation can then destroy. The example of sickle cell hemoglobin shows how small a genetic modification can be to create qualitative changes in the ability to respond to osmotic stress.

The varied but precise response of molecular assemblies to osmotic stress might indicate that materials designed to survive the more vigorous perturbation of desiccation seen in "anhydrous biology" purposefully assume successive stable assemblies at the various stages of dehydration. There may in fact have been some evolutionary development to optimize or even to use the "anhydrous" state beyond the challenge of basic survival.

REFERENCES

Bernal, J.D., and Fankuchen, I. 1941. X-ray and crystallographic studies of plant virus preparations. J. Gen Physiol. 25:111-165.

Best, R. 1940. The action of electrolytes on solutions of tobacco mosaic virus nucleoprotein. Aust. J. Exp. Biol. Med. Sci. 18:307-312.

Churaev, N.V., and Derjaguin, B.V. 1985. Inclusion of structural forces in the theory of stability of colloids and films. J. Coll. Interf. Sci. 103: 542-553.

Clegg, I., and Drost-Hansen, W. 1979. Cell Associated Water. Academic Press, New York.

Harrington, W.F. and Kegeles, G. 1973. Pressure effects in ultracentrifugation of interacting systems. S.P. Colowick and N.O. Kaplan, eds. Meth. Enzymol. 27: 306-345. Academic Press, New York.

Harris, A. 1973. Cell surface movements related to cell locomotion in locomotion of tissue cells. Ciba Fdn. Symp. 14: 3-26. Elsevier/North-Holland, Amsterdam.

Kauzmann, W. 1959. Some factors in the interpretation of protein denaturation. Adv. Protein Chemistry 14: 1-63.

Lauffer, M.A. 1975. Entropy-Driven Processes in Biology. Springer-Verlag, New York.

Lerman, L.S. 1971. A transition to a compact form of DNA in polymer solutions. Proc. Natl. Acad. Sci. (USA) 68: 1886-1890.

Ling, G.N. 1984. In Search of the Physical Basis of Life. Plenum Press, New York.

Lis, L.J., McAlister, M., Fuller, N., Rand, R.P. and Pargesian, A. 1982. Interactions between neutral phospholipid bilayer membranes. Biophys. J. 37: 657-666.

Maniatis, T., Venable, J.W. Jr., and Lerman, L.S. 1974. The Structure of ψ-DNA. J. Mol. Biol. 84: 37-64.

Noguchi, C.T., and Schechter, A.N. 1985. The intracellular polymerization of sickle hemoglobin. Ann. Rev. Biophys. and Bioeng. 14: 239-263.

Prouty, M.S., Schechter, A.N., and Parsegian, V.A. 1985. Chemical potential measurements of deoxyhemoglobin S polymerization. J. Mol. Biol. 184: 517-528.

Rand, R.P., and Parsegian, V.A. 1986. Mimicry and mechanism in phospholipid models of membrane fusion. Ann. Rev. Physiol. 48: 201-212.

Rand, R.P., Reese, T.S., and Miller, R.G. 1981. Phospholipid bilayer deformations associated with interbilayer contact and fusion. Nature 293: 237-238.

Rau, D.C., Lee, B., and Parsegian, V.A. 1984. Measurement of the repulsive force between polyelectrolyte molecules in ionic solution: Hydration forces between parallel DNA double helices. Proc. Natl. Acad. Sci. 81: 2621-2625.

Tilney, L.G., and Inoue, S. 1985. Acrosomal reaction of the thyme sperm. III. The relationship between actin assembly and water influx during the extension of the acrosomal process. J. Cell Biol. 100: 1273-1283.

Trinkhaus, J.P. 1985. Protrusive activity of the cell surface and the initiation of cell movement during morphogenesis. Exptl. Biol. and Med. 10: 130-173.

Unwin, H.N., and Zampighi, G. 1980. Structure of the junction between communicating cells. Nature 283: 545-549.

Zimmerberg, J., and Parsegian, V.A. 1984. Polymer-inaccessible space as a measure of the volume change during channel opening and closing: VDAC channels. Biophys. J. 45: 59a.

Zimmerman, A.M. 1970. High Pressure Effects on Cellular Processes. Cell Biol. Series. Academic Press, New York.

19 Stabilization of Small Unilamellar Phospholipid Vesicles by Sucrose during Freezing and Dehydration

George Strauss
Helmut Hauser

Phospholipid bilayers constitute the fundamental structural element of biological membranes. As such they have been widely used as model membranes to study the intrinsic bilayer properties but also to study the interaction of bilayers with ions, small molecules, and macromolecules, particularly proteins.

Bilayers have recently become important in several research areas for which the term *membrane-mimetic chemistry* has been coined. It comprises membrane-mediated processes such as membrane fusion, interfacial catalysis, energy conduction and conversion, drug delivery and targeting. In this more recent field of application there is a need for simple and quick methods for the production of closed unilamellar lipid vesicles. The methodology must be versatile, yielding bilayer vesicles of appropriate size and size distribution, chemical and physical stability, and bilayer properties such as surface charge density, bilayer fluidity, and permeability.

Before one can think of applying unilamellar lipid vesicles to drug delivery or targeting, the problem of long-term storage of these vesicles must be solved. Storage of lipid vesicles as aqueous suspensions is unsatisfactory, because aqueous lipid suspensions are neither chemically nor physically stable. They undergo chemical degradation, and consequently the lipid bilayer is disrupted, and drugs encapsulated in the vesicle cavity may leak out. Small unilamellar vesicles (SUV) have also been shown to be thermodynamically unstable and to aggregate and/or fuse with time.

Desiccation and rehydration at room temperature or freezing of cells, cell components, or plasma membranes have been reported to cause disruption of the membrane structure, resulting in the loss of membrane functions. In desiccation-tolerant organisms, membrane structure and functions appear to be retained in the presence of certain sugars (Chapters 11 and 12). This observation prompted a systematic study of the stabilizing effect of sugars on phospholipid bilayers during freeze-thawing and dehydration-rehydration. The work presented here is addressed to the question of whether or not SUV made of phosphatidylcholine can be stabilized in the presence of sugar with full retention of their structural and permeability properties during freeze-thawing and freeze-drying. If this was indeed the case, it would offer the important advantage of permitting storage of unilamellar phospholipid vesicles in the frozen or dehydrated state.

The work described here was carried out with SUV made by ultrasonication of egg phosphotidylcholine (EPC) dispersions in aqueous media with or without sucrose. SUV are a useful system for testing the effect of freeze-thawing and freeze-drying on the size and bilayer integrity in the presence and absence of sugars. The physical methods used here to monitor vesicle size and bilayer integrity comprise gel filtration on Sepharose 4B, dynamic light scattering, ^{1}H high-resolution NMR, freeze-fracture electron microscopy, and ESR spin labeling. Gel filtration patterns on Sepharose 4B of sonicated EPC dispersions in the presence and absence of sucrose and before and after subjecting the EPC dispersion to freeze-thawing or freeze-drying are shown in Figure 19-1. The elution profile of sonicated EPC dispersions in buffer consisted of two peaks (Figure 19-1a); a minor one eluted at the column void volume V_o and a major broad peak at an elution volume V_e = 22 ± 1 ml, corresponding to a Stokes radius of about 10 nm. Upon freeze-thawing or freeze-drying, sonicated EPC dispersions in the absence of sucrose showed aggregation and/or fusion of SUV. This is evident from the fact that, after the dispersion has been subjected to freeze-thawing, 95% of the phospholipid was pelleted by centrifugation at 12,000 × g for 10 min. When the phospholipid remaining in the supernatant was applied to Sepharose 4B, nearly all was eluted in the void volume V_o.

The elution pattern of sonicated EPC in the same buffer but containing 5% (0.15 M) sucrose is shown in Figure 19-1b. Again two peaks were obtained, a minor one in the void volume and a major one at V_e = 19 ml, corresponding to a Stokes radius of about 12 nm. The effect of freeze-thawing or freeze-drying on EPC-SUV in the presence of 5% (0.15 M) sucrose is different from that shown in Figure 19-1a. No change in vesicle size and size distribution was observed. The two elution profiles obtained with EPC dispersions before (Figure 19-1b) and after freeze-thawing (Figure 19-1c) were superimposable. The difference of the two patterns gave points randomly distributed about the base line (Figure 19-1d), indicating that in the presence of 5% (0.15 M) sucrose the SUV survive freeze-thawing without a

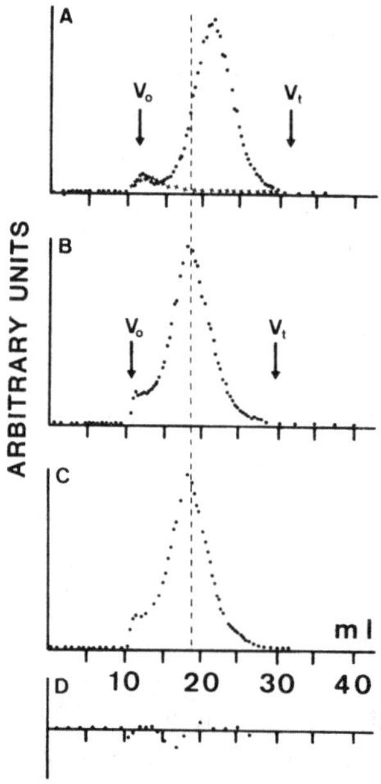

Figure 19-1. Gel filtration of sonicated EPC dispersions on Sepharose 4B in the presence and absence of sucrose. (a) Sonicated EPC (0.5 ml at 10 mg/ml = 0.0133 M) dispersed in buffer without sucrose (0.12 M NaCl, 0.01 M Hepes pH 7, 0.05% NaN_3) was applied to the Sepharose 4B column (46 x 0.9 cm) preequilibrated with the same buffer. The column was eluted with buffer at a flow rate of 5 ml/hr, and the phospholipid concentration was determined in each of the 0.3 ml fractions collected (●). The same sonicated EPC dispersion as chromatographed in A was subjected to freeze-thawing. After centrifugation of the dispersion at 12, 000 × g for 5 min, which pelleted 95% of the phospholipid, 0.5 ml of the supernatant was applied to the column. The elution profile of this sample is included in A(x). (b-d) A second Sepharose 4B column of similar dimension as in (a) was equilibrated in buffer (0.14 M NaCl, 0.01 M Hepes pH 7, 0.05% NaN_3, 0.15 M sucrose); 0.5 ml of sonicated EPC dispersions in the same buffer (10 mg/ml = 0.0133 M) were chromatographed before (b) and after freeze-thawing (c). The difference between the elution profiles in (b) and (c) is shown in (d). Both columns were presaturated with EPC and handled in the same way. Phospholipid recovery was about 90%. The void and total volume, V_o and V_t, respectively, are indicated by arrows.

significant change in the average size and size distribution. Similar results were obtained with EPC-SUV after freeze-drying in the presence of sucrose. The gel filtration data were confirmed by dynamic light scattering.

The fraction of lipid remaining as SUV after freeze-thawing or freeze-drying can be conveniently determined by ^1H high-resolution NMR. This assay of SUV is based on the observation that lipid particles with a diameter smaller than about 100 nm give rise to a high-resolution spectrum, while particles with larger diameters such as large unilamellar vesicles and multilamellar liposomes give broad-line NMR spectra. The signal intensity derived from the ^1H NMR spectrum by integration is a measure of the amount of lipid present as SUV. The results of the NMR assay used here to determine the amount of SUV were in good agreement with those of gel filtration, dynamic light scattering, and freeze-fracture electron microscopy.

The top spectrum in Figure 19-2 is the ^1H high-resolution NMR spectrum of a sonicated EPC dispersion in D_2O recorded at 360 MHz. The assignment of the prominent signals is included. With CH_3COONa as an internal integration standard it can be shown that all EPC molecules ($\approx 100 \pm 10\%$) contribute to the high-resolution NMR spectrum. After the EPC dispersions had been subjected to freeze-thawing, no high-resolution spectrum was obtained; the lines were broadened to the extent that they were lost in the base line

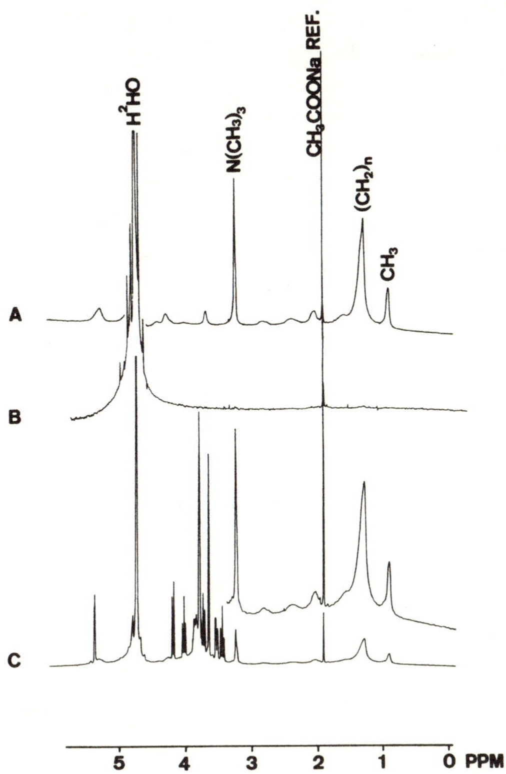

Figure 19-2. ^1H NMR spectra (360 MHz) of sonicated EPC dispersions (10 mg/ml = 13.3 mM) in ^2H_2O. (a) in ^2H_2O at room temperature. (b) in ^2H_2O, after freeze-thawing. (c) in 5% (≈0.15 M) sucrose in ^2H_2O, after freeze-thawing. A vertically expanded version of (c), with amplification as in (a) is also shown. To all samples an appropriate amount of CH_3COONa was added as an internal integration standard.

(Figure 19-2b), indicating that no SUV were present after freeze-thawing. Sonicated EPC dispersions in D_2O containing 10% (≈0.3 M) sucrose gave ^1H NMR spectra similar to that shown in Figure 19-2a. The ^1H NMR spectrum of such a dispersion after freeze-thawing is shown in Figure 19-2c. In this case the high-resolution spectrum is essentially retained. It is almost identical to that shown in Figure 19-2a except for the sucrose resonances between 3.4 and 4.3 ppm. Integration of the spectrum indicated complete recovery of SUV after freeze-thawing. Similar results were obtained when sonicated EPC dispersions containing 10% sucrose were subjected to freeze-drying.

More direct evidence that SUV present in sonicated EPC dispersions containing 10% sucrose are preserved during freeze-thawing and freeze-drying can be obtained from freeze-fracture electron microscopy. Freeze-thawing or freeze-drying of these dispersions had little effect on the average vesicle size and size distribution as derived from electron micrographs of freeze-fractured preparations (data not shown).

We have used an NMR method and ESR spin labeling to demonstrate that the integrity of the EPC bilayer is maintained during a freeze-thaw or a dehydration-rehydration cycle. Figure 19-3a shows the ^1H high-resolution

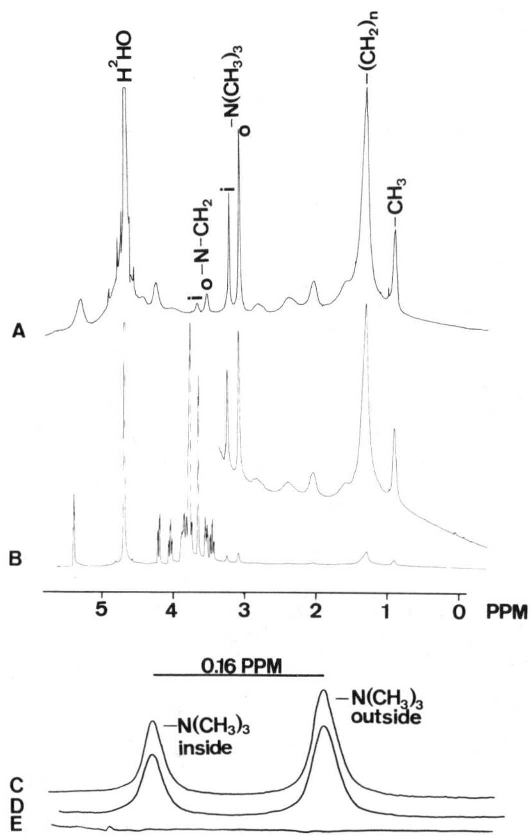

Figure 19-3. ^1H NMR spectra (360 MHz) of sonicated EPC dispersions (10 mg/ml = 13.3 mM) in ^2H_2O to which $K_3Fe(CN)_6$ (final concentration 0.1 M) had been added after sonication. (a) in ^2H_2O at room temperature. Outside (shifted) and inside (non-shifted) signal of the polar groups are labeled o and i, respectively. (b) like (a), but in 30% (0.88 M) sucrose solution in ^2H_2O. A vertically expanded version of this spectrum, with amplification as in (a), is also shown. (c) horizontally expanded region of (b) showing outside and inside —$N(CH_3)_3$ signals. (d) like (c) but after freeze-thawing. (e) difference spectrum of (c) and (d).

NMR spectrum of a sonicated EPC dispersion in D_2O containing 0.1 M $K_3Fe(CN)_6$. The polar group signals from EPC molecules on the external vesicle surface exposed to the paramagnetic $Fe(CN)_6^{3-}$ were shifted upfield. This was also true for sonicated EPC dispersions containing sucrose. When freeze-thawing or freeze-drying of sonicated EPC dispersions occurred in the absence of sugar, the high-resolution spectrum was lost (cf. Figure 19-2b). In the resulting broad-line spectrum the —$N(CH_3)_3$ choline signal was just detectable. It was shifted upfield and appeared in the same position as the external —$N(CH_3)_3$ signal in Figure 19-3a (labeled o), suggesting that the EPC bilayer became leaky during freeze-thawing or freeze-drying and the phospholipid polar groups inside the vesicle cavity became accessible to the $Fe(CN)_6^{3-}$ anion. In contrast, in the presence of 30% sucrose (0.88 M) freeze-thawing or freeze-drying affected neither the upfield shift induced by $K_3Fe(CN)_6$ nor the relative intensity of the signals from the outer and inner polar groups (Figure 19-3b). The difference spectrum of the —$N(CH_3)_3$ choline region of the sample before and after freeze-thawing was a flat base line (Figure 19-3e), proving that EPC-SUV had stayed intact during freeze-thawing in the presence of sucrose.

The integrity of EPC bilayers is confirmed by ESR spin labeling. For this purpose, about 1 mol % of the spin probe CAT 16 shown in Figure 19-4 was incorporated in the EPC bilayer of SUV made by ultrasonication. The free radical of this spin probe will be present in the polar group region of the phospholipid bilayer. The ESR signal arising from the label in the outer monolayer of the vesicle bilayer was quenched by adding 50 mM ascorbate. The remaining signal was therefore due to the spin label present on the inner monolayer; this amounted to 25-35% of the total signal intensity. This remaining signal intensity was measured as a function of temperature (Figure 19-4). When such an EPC dispersion was cooled down, the inner ESR signal was suddenly and almost completely quenched at about −13°C, the freezing point of H_2O in the ESR sample tube. The signal loss was irreversible; no signal recovery was observed on rewarming. The freezing point was recognized by a sudden change in the microwave absorption by the sample. Sonicated EPC dispersed in 30% (0.88 M) sucrose showed a different behavior. Essentially no loss of the ESR signal intensity occurred during cooling to −33°C and rewarming. This indicates that the EPC vesicles remained tight; the bilayer did not become permeable to ascorbate when subjected to freeze-thawing in the presence of sucrose. This result is consistent with the NMR experiment using impermeable shift reagents.

A question of interest is whether the protective effect of sucrose is a specific or a general effect of sugars. The stabilization of EPC-SUV in the presence of different sugars is shown in Figure 19-5. The ^1H-signal intensity

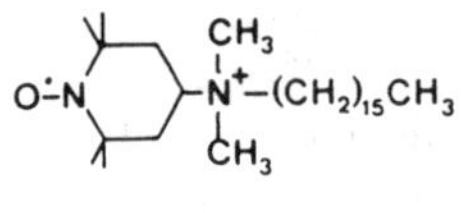

CAT 16

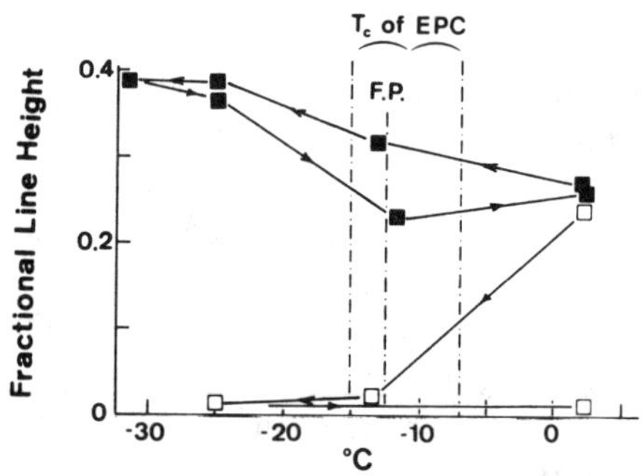

Figure 19-4. Detection of the onset of bilayer disruption. The ESR signal intensity of sonicated EPC vesicles containing the spin probe CAT 16 after addition of 50 mM sodium ascorbate to the external medium is plotted as a function of temperature. The chemical formula of the spin probe is also shown. The internal ESR signal intensity is expressed as the fraction of the total ESR signal intensity as measured in the absence of ascorbate at the same temperature. In all experiments the lipid concentration was 67 mM and that of the spin probe was 1 mM; the sucrose concentration (where used) was 30% (0.88 M). The transition temperature T_c or the transition temperature range and the freezing point of the solution, FP, are marked. The connecting lines show the sequence of the measurement only and do not represent linear changes. Sonicated EPC in H_2O with (■) and without sucrose (□).

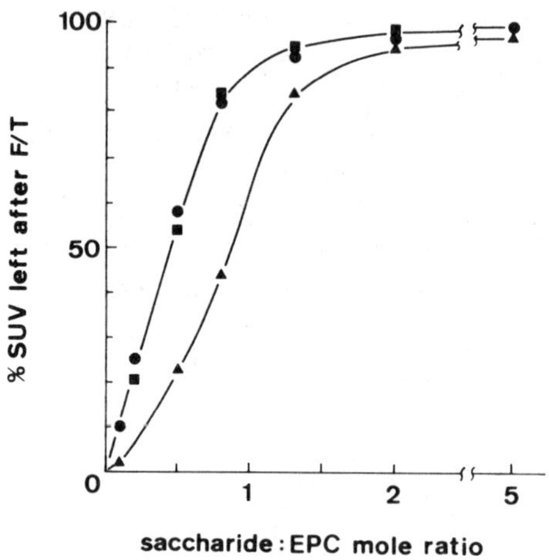

Figure 19-5. Comparative study of the protective effect of different sugars on SUV of EPC during freeze-thawing. The sonicated EPC dispersion (5 mg/ml = 6.67 mM) in ^2H_2O was subjected to freeze-thawing (F/T) and the ^1H NMR signal intensity remaining after freeze-thawing (expressed as % of total phospholipid) was determined as a function of the saccharide/EPC mole ratio. (●) sucrose, (■) trehalose, (▲) glucose.

remaining after sonicated EPC had been subjected to freeze-thawing in the presence of different mono- and disaccharides is plotted as a function of the saccharide/EPC mole ratio. Apparently sucrose was as effective as trehalose, and both these disaccharides were more effective than glucose. These experiments suggest that the protection of SUV is a general effect of sugars; sugars may, however, differ in their efficiency.

The next experiments are relevant to the question of the mechanism by which sugars stabilize phospholipid SUV. Figure 19-6a shows a plot of the ^1H NMR signal intensity remaining after freeze-thawing as a function of EPC concentration. The sucrose concentration was kept constant at 2.7 mM in this series of experiments. The effect of sucrose apparently depends on the sucrose/EPC mole ratio and not on the absolute sucrose concentration. The fact that the remaining signal intensity (= the fraction of SUV remaining after freeze-thawing) passes through a maximum is interpreted as indicating that the interaction between sucrose and phospholipid is governed by the mass action law.

Further evidence that the sucrose is interacting with the phospholipid polar group, probably the phosphodiester group, is presented in Figure 19-6b. Increasing concentrations of Eu^{3+} gradually inhibited the protective effect of sucrose: under the conditions used, 5 mM $Eu(NO_3)_3$ produced about a 50% inhibition of the protection observed in the presence of 2.2% (0.067 M) sucrose. Eu^{3+} has been shown by ^{31}P NMR to interact with the phosphodiester group of EPC, giving rise to shift changes of the ^{31}P resonance. This change in chemical shift induced in the presence of Eu^{3+} was shown to contain a contact contribution, indicating that Eu^{3+} ions must be in close proximity to the ^{31}P nucleus (Hauser et al., 1976). The Eu^{3+} experiments are

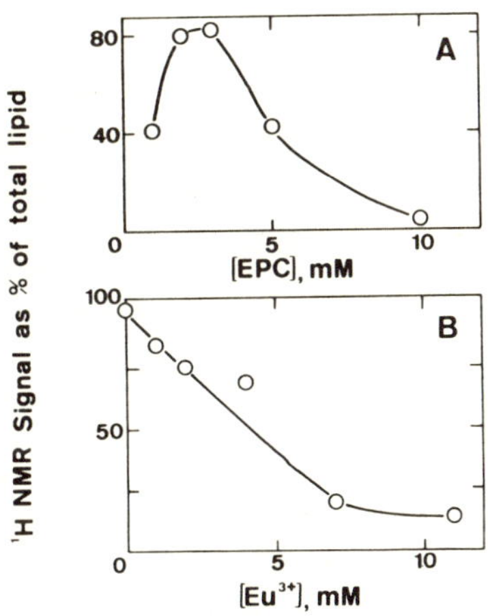

Figure 19-6. ^1H NMR signal intensity of the hydrocarbon chain protons of EPC as a function of the EPC concentration. EPC vesicles dispersed in ^2H_2O were prepared by ultrasonication. (a) [EPC] varied from 1.33 to 10 mM, while the sucrose concentration was kept constant at 2.7 mM. (b) suppression of the protective effect of sucrose by adding $Eu(NO_3)_3$·[EPC]=13.3 mM, [sucrose] = 66.7 mM, [Eu^{3+}] = 0 to 11 mM. Sonicated EPC dispersions containing both sucrose and $Eu(NO_3)_3$ were subjected to freeze-thawing.

interpreted as indicating that the metal ion is competing with sucrose for a common binding site, which is probably the phosphate group.

The experiments reported here allow several conclusions to be formulated.

In the presence of relatively small quantities of sucrose, small unilamellar EPC vesicles dispersed in aqueous media are preserved during freeze-thawing and freeze-drying. In the absence of sucrose these procedures cause aggregation and fusion of SUV to large multilamellar phospholipid particles.

In the presence of 5-10% sucrose the bilayer integrity is maintained during freeze-thawing and freeze-drying.

The protecting, stabilizing effect of sucrose on SUV of EPC is not specific for sucrose; other mono- and disaccharides give similar effects, although there might be quantitative differences.

The mechanism of stabilization involves the interaction of the sugar with the phospholipid polar group, probably hydrogen bonding to the phosphodiester group of the phospholipid.

REFERENCES

Crowe, J.H. and Crowe, L.M. 1982. Induction of anhydrobiosis: membrane changes during drying. Cryobiology 19: 317-328.

Crowe, J.H., Crowe, L.M. and Mondrian, R. 1983. Stabilization of biological membranes at low water activities. Cryobiology 20: 346-356.

Hauser, H., Phillips, M.C., Levine, B.A. and Williams, R.J.P. 1976. Conformation of the lecithin polar group in charged vesicles. Nature 261: 390-394.

Strauss, G. and Hauser, H. In press. Proc. Natl. Acad. Sci. USA.

20 Dehydration Problems Associated with the Preservation of Seed and Plant Germplasm

Phillip C. Stanwood

Plant genetic resources provide the foundation upon which modern agriculture is based. Only recently have the agricultural and scientific communities come to a widespread realization of the importance of understanding, collecting, evaluating, and preserving these resources (Frankel and Hawkes, 1975). Plant germplasm can be defined as populations which represent the genetic diversity of a plant species or plant type. This view theoretically implies that there is a "gene pool" of an entire group of plants of particular interest (Bradshaw, 1975; Oldfield, 1984). In reality, when one attempts to conserve plant germplasm, only a fraction of this total pool can be expected to be collected. Consequently, the best possible preservation techniques should be used to ensure the continued availability of these collected samples.

The greatest risks associated with preservation efforts are that physiological deterioration may occur in storage (Withers, 1980) or that susceptibility to environmental stresses may depress performance when the stored seeds are planted in the field. A preservation scheme that can attenuate, minimize, or overcome these risk factors is a major goal of plant preservation research.

The capability of a given plant material to tolerate desiccation is an important factor in preservation, especially at subzero and liquid nitrogen (−196°C) temperatures.

Plant materials can be classed as orthodox (i.e. capable of being desiccated without damage, or desiccation tolerant), or as recalcitrant (i.e. damaged by desiccation below about 30% fresh weight basis [43% dry weight basis] Roberts, 1973). Differences in desiccation tolerance can be characteristics of species or genera and are most notable in seed, pollen, and bud tissue. For instance, most wheat (*Triticum vulgare*) and other small-grain cereals have orthodox seeds, but their pollen appears to be recalcitrant. Hop (*Humulus lupulus*) seed and pollen, on the other hand, are both orthodox in behavior (Haunold and Stanwood, 1985). *Anthurium* seed and pollen are both thought to be recalcitrant (Stanwood, 1985). Microplantlets, apical meristem, calli, and cell suspensions are generally sensitive to desiccation and therefore would be classified as recalcitrant.

Cryopreservation

Cryopreservation using liquid nitrogen as a storage medium provides a controlled environment that can ensure long-term preservation (Bajaj, 1983; Stanwood and Bass, 1981; Withers, 1980). The tissue water content appears to be a major factor limiting the use of liquid nitrogen storage for most plant materials. Freezable water results in ice crystal formation, which in turn may disrupt membranes or cells, causing irreversible damage to the tissue. Cryoprotectants and precisely controlled protocols are used to avoid this freezing damage (Karatha, 1985; Li and Sakai, 1978, 1982).

Selection of Plant Materials for Preservation

Selection of the most appropriate material is crucial to any plant germplasm preservation process and is predicated on many factors, of which storability is one. The most common form of preservation is the storage of seeds at low or subfreezing temperatures and at extremely low moisture contents. Most of the common agronomic and horticultural species are preserved in this manner. Cryopreservation of the orthodox class of seeds has been extremely successful; over 200 species are capable of being stored in liquid nitrogen without damage (Stanwood and Bass, 1981; Stanwood, 1985).

Seeds of many of the tropical and subtropical flower species and temperate-zone tree species fall into the class of recalcitrance (Chin and Roberts, 1980). Seeds from these species have defied routine long-term (lasting more than 1 year) preservation attempts at any storage temperature.

Some plant species do not produce seed, or if they do, it is in very small quantities or is genetically unstable. Preservation of these species is usually accomplished by continually growing the plants under laboratory, greenhouse, and/or field conditions (Withers, 1980). Many of the fruit and nut crop species are conserved in this manner (Brooks and Barton, 1977). From a preservation standpoint this technique is difficult and expensive because of the tremendous amount of resources needed to keep the plants under cultivation or optimal growing conditions.

The preservation of pollen is often desirable (Towill, 1985), especially for intercrossing species that require a long time to mature or flower and are difficult to bring to flower at the same time. The main drawbacks in using pollen are (1) the difficulties of obtaining enough pollen from the host plant in the first instance and (2) being able to preserve that pollen for sufficiently long periods of time, so that the mother plant or genetically equal plants can be regrown to produce additional pollen when needed. Pollen preservation must therefore be coupled with other preservation techniques which would allow for the reproduction of the host or mother plant. The application of cryopreservation techniques to pollen would help satisfy the long-term storage requirement. Pollen, like seeds, is either orthodox or recalcitrant and displays storage problems and characteristics similar to those of seeds (see Chapter 6).

The use of microplantlets, dormant buds, meristems, calli, and cell suspensions in a plant preservation scheme is attractive from a physiological point of view because the storage environment can be controlled, and thus the risk of loss can be reduced. Depending on the system used, these techniques may require fewer resources for maintenance than field-grown material would demand. The disadvantage of preserving these materials is that the lower the order of tissue development (e.g. microplantlets to buds to calli to cell suspensions), the greater the difficulty in reproducing the mature plant and the more genetic variation seen in the reproduced plant. The maintenance of genetic integrity in the cultured materials also becomes a major concern (Bajaj, 1983; Withers, 1980). Currently, except in a few rare instances, easily propagated tissues, namely microplantlets, buds, and apical meristem, are the most likely candidates for routine preservation of cloned plant germplasm.

Cryopreservation of vegetative materials is hampered by their high moisture content. Except for desiccation-tolerant dormant buds (Becwar et al., 1981) these materials are all recalcitrant and present freezing problems when stored at subzero temperatures. In spite of these freezing problems, there have been some encouraging results in the cryopreservation of some vegetative materials (Sakai, 1985).

Preservation Technology

The central problem of preservation technology concerns the relationship of moisture content to the storage of seed materials. Since many of the problems associated with moisture content are exemplified in seeds, they become a good tool for studying the relationships between moisture, temperature, and preservation. Several concepts of seed preservation have developed over the years. The most appropriate concepts fall into three groups: (1) moisture and temperature in storage; (2) the use of high-moisture freezing limits; and (3) application of colligative water substitution techniques for recalcitrant seeds.

Differential Thermal Analysis (DTA)

Differential thermal analysis (DTA) can be used for detecting freezable water in biological materials as a function of cooling rate and temperature. DTA monitors the release of energy in tissue as water changes phase from a liquid to a solid or vice versa. This exothermic reaction is evaluated against a nonbiological, water-free reference and is plotted as a differential temperature (Burke et al., 1976).

DTA is particularly useful in plant preservation research; not only can freezable water be detected, but the temperature at which the freezing event occurred is also determined (Becwar et al., 1982). For example a seed at 20% moisture content[1] may contain freezable water, which, if frozen, would kill the seed. However, the temperature at which the freezing occurred could be as low as −40°C (Steponkus, 1980). So even though the seed has freezable water, it can sometimes be safely stored at temperatures approaching −40°C. DTA would easily detect this characteristic and would allow a proper storage temperature to be prescribed.

Interaction Between Seed Moisture and Storage Temperatures

Moisture content is the primary factor influencing seed longevity in storage, especially at temperatures above freezing. Longevity decreases exponentially as the storage temperature increases (Roberts and Ellis, 1982). Figure 20-1 demonstrates this interaction. It can be clearly seen that, at a storage temperature of 32°C and a seed moisture content of 10%, all germination is lost after 16 years of storage. A moderate reduction of moisture content to 7% increased survival of the seed to approximately 40% germination. However, at a lower storage temperature of 12°C the 10% moisture content seed had a

1. Moisture contents in this paper are reported on a fresh weight basis unless otherwise noted; i.e., moisture % = (fresh weight − dry weight)/fresh weight · 100.

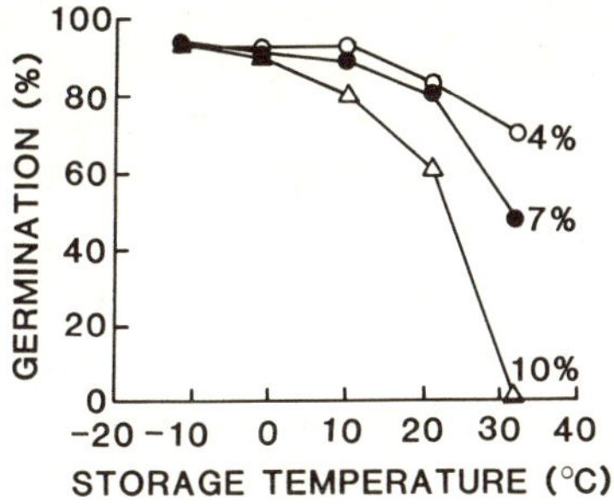

Figure 20-1. Average germination of 'RS 610' hybrid *Sorghum vulgare* seed after 16 years of sealed storage as affected by seed moisture (4, 7, and 10%) and storage temperature (-12°C to 32°C). (Adapted from *Crop Science*, Vol. 18, No. 4, July-August, 1978, p. 576, by permission of the Crop Science Society of America, Inc.)

survival of 70% germination, and a reduction in moisture content to 7% increased survival to 90% germination. This type of interaction occurs in all seed materials, but the absolute relationship between moisture level and storage temperature varies considerably among species and selections within a species.

There have been a number of attempts to characterize the moisture and temperature interaction, ranging from very simple rules of thumb to complex models. Harrington (1963) suggested that storage life is doubled for each 1% reduction in moisture content or for each 5°C lowering of storage temperature. The range over which his rule applies is very limited, and the rule is too general to be useful for all species.

Roberts and Ellis (1982) have developed more comprehensive models, attempting to quantify temperature-seed moisture content relationships on a species by species basis. In their model the following equation was developed:

$$v = (K_i - p)/[10^{K_e - C_w \log m - C_h t - C_q t^2}]$$

in which v is the percentage viability as a probit value, m is seed moisture content (% fresh weight basis), t is storage temperature (°C) and K_e, C_w, C_h, C_{qr}, and K_i are empirically derived constants. What is particularly useful with this approach is that variables due to seed type or condition can be factored into the equation, and a wider range of temperature and moisture conditions can be considered.

From a practical stand-point, seed moisture content can be reduced only to a certain level (usually 4 to 6%), and further extension of storage life depends upon lowering storage temperatures. There were suggestions at the conference that, if all water is removed from a tissue, degrading biological reactions are restricted, and therefore storage life could possibly be extended indefinitely. Two problems or unknowns arise: (1) there are some suggestions that important biological degradation reactions occur in the absence of water (Chapter 5), and (2) it is not known whether all the water from biological material can practically be removed. These points will be left for others to debate. It is clear from practical experience that removal of all water cannot be assured; thus, reduced seed moisture is only part of the answer for long-term preservation of seeds, and lowered temperatures offer further options to increase longevity.

High Moisture Freezing Limit

The interaction between seed moisture content and freezing survival is critical for germplasm preservation. In seeds and other desiccation-tolerant plant materials, there is a distinct tissue moisture content below which freezable water cannot be detected (Becwar et al., 1981; Ishikawa and Sakai, 1978; Junttila and Stashnoff, 1977; Roos and Stanwood, 1981; and Chapter 1). If seed moisture is adjusted to a given moisture content, exposed to a freezing environment, rewarmed, and evaluated for survival, a graph of the survival can be plotted (Figure 20-2). Sesame seed, for example, showed complete freezing survival up to a moisture content of approximately 12%. Above this level, survival decreased very rapidly, with no survival above 14% moisture content. We call this threshold the high-moisture freezing limit (HMFL; Roos and Stanwood, 1981). The HMFL is generally very narrow (2-3%)

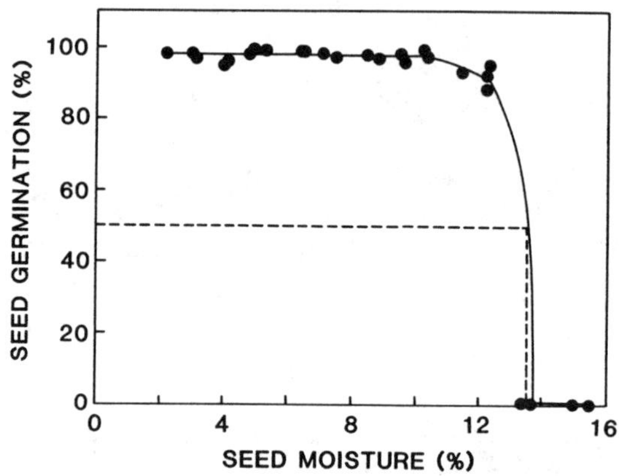

Figure 20-2. High moisture freezing limit (HMFL) of *Sesamum indicum* seed. Seed was adjusted to varying moisture levels, subjected to a freezing environment, rewarmed and germinated. Curve represents a 3rd order polynomial fit. HMFL estimated at 50% germination.

within a species but varies considerably among species. Table 20-1 shows the HMFL for a range of species, with *Phaseolus* (bean) being the highest, approximately 27%, and *Sesamum* (sesame) the lowest, at 9.3% moisture content (Stanwood, 1984).

The temperature at which freezing occurs is also very important. DTA provides us with that information. As can be seen in Figure 20-3, the temperature at which freezing occurs varies among species. In the case where two or more exotherms are observed in seeds (Figure 20-4), the interpretation is that the warmer exotherm represents intercellular ice formation and generally is not lethal. The exotherm which occurs at a lower temperature is considered to represent intracellular ice formation and is lethal (Burke et al., 1976). In cases where only one exotherm is observed, both intra- and intercellular ice formation are thought to be occurring.

There have been some reports of exotherms as low as −30°C in hydrated lettuce seed (Junttila and Stushnoff, 1977) and −35°C in dry (less than 6% moisture content) filbert seed (Stanwood, 1985). More research is needed to clarify these extremely low exotherms and to determine their site in the

Table 20-1
High moisture freezing limit and lethal freezing temperatures for a variety of crop seed species

			Germination after exposure to liquid N_2		
	HMFL		Seed moisture below HMFL	Seed moisture above HMFL	
Crop	Wet weight basis	Dry weight basis	Germ. (%)	Lethal freezing temperature (°C)	Germ. (%)
Barley	20.8 (1.2)	28.0	98	−12; −13	18
Bean	27.2 (1.2)	37.3	99	−25	84
Cabbage	13.8 (0.3)	16.1	90	−28	0
Carrot	21.7 (1.6)	27.8	83	−25	0
Cauliflower	14.2 (1.1)	16.5	97	−25	0
Clover	25.6 (0.5)	34.6	95	−15	2
Cucumber	16.4 (0.9)	19.7	98	−23; −26	1
Fescue	23.0 (3.8)	29.8	98	−25	2
Onion	24.7 (0.8)	32.7	70	−18; −22	0
Pepper	18.6 (1.2)	22.9	99	−22; −25	0
Radish	16.8 (0.9)	20.3	99	−25	4
Sesame	9.3 (1.6)	10.2	97	−18; −26	0
Tomato	18.5 (1.6)	22.7	93	−20; −25	0
Wheat	26.8 (4.7)	36.6	96	−7	25

Note: Parenthetical numbers are standard deviations. Lethal freezing temperatures were determined by differential thermal analysis (from Stanwood, 1984).

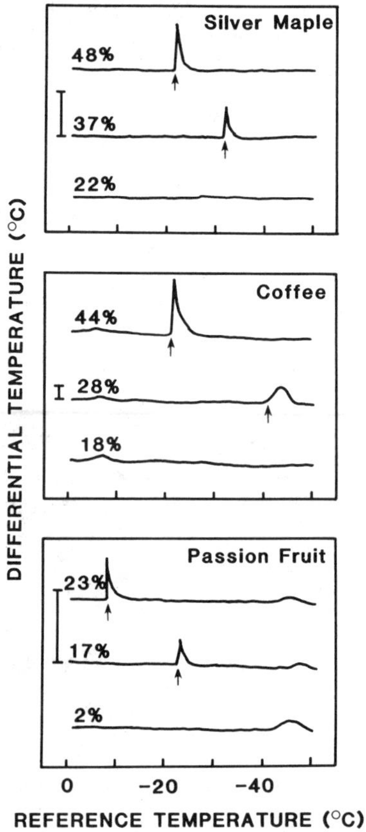

Figure 20-3. Differential thermal analysis cooling profiles of seed tissues at various moisture contents. Exotherms indicate phase transitions from a liquid to solid state. Exotherms with arrows denote freezing of water in the seed tissues. The scale bar at the left of the DTA profiles represents 1°C temperature difference. (From Becwar et al., 1983)

tissue, their compositional source (i.e. water, oil, etc.), and their biological impact.

The point at which HMFL occurs in plant materials should be known if subfreezing conditions are to be used. Freezing temperatures must be avoided in high-moisture materials, or the moisture content must be reduced below the HMFL to assure safe storage.

The source of high-moisture freezing damage is not clear. A considerable amount of research being conducted on hydrated plant materials attempts to identify the mechanism of freezing damage (Levitt, 1980; Steponkus, 1984). Membrane damage is clearly involved, yet no specific mechanism has been identified. Some research suggests a physical type of damage; other work suggests that chemical alterations are the source of the damage. Concentration of solutes in and around the membrane at very slow cooling rates may be another source of damage (Franks, 1985).

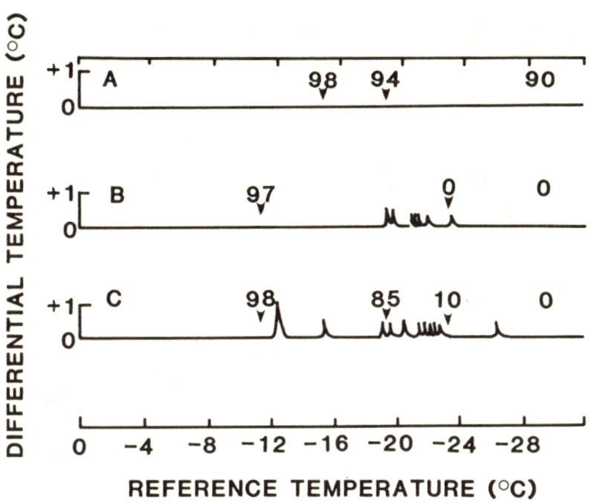

Figure 20-4. DTA cooling profiles and germination of lettuce seed at various moisture contents. Arrows indicate percent seed germination after exposure to the indicated temperatures. Moisture contents were: (a) 5.7-16%; (b) 26%; (c) 32%. Ten seeds in each sample. Freezing rate, 20°C per h. Various moisture levels were obtained by imbibing the seeds in PEG 6000 at 30°C for various time periods. (Redrawn by permission from *Nature*, vol. 269, pp. 325-327. Copyright © 1977 Macmillan Journals Limited.)

Colligative Cryoprotection

Recalcitrant seeds are not currently adaptable to storage because of their requirement for high moisture contents. They are usually sensitive to cool storage conditions and are killed by freezing temperatures (Chin and Roberts, 1980). To extend the storage life of recalcitrant seeds, one must reduce the metabolic rate of the material. High storage temperatures and chemical inhibitors have been tried but with only limited success (Goldbach, 1979; Villers, 1975). Another option is to lower the storage temperature enough to bring about a significant reduction in the metabolic rate within the seed. Therein is the dilemma.

Researchers working with animal cells and tissue, most notably blood, remove water osmotically from the cell and replace it with a nontoxic compound that helps the cell alleviate freezing damage by means of "colligative cryoprotection" (Meryman and Williams, 1980). This approach assumes that, in part, desiccation damage results from shrinking the cells below a level of elasticity such that upon rehydration cell lysis occurs (Chapter 17). Altered spatial relations in cell membranes, concentration and precipitation of solutes, and changes in pH are other possible damaging factors that may be associated with dehydration/freezing injury (Steponkus, 1980). The cryoprotectant may also interact with these aqueous factors and provide "antifreeze protection."

A large number of cryoprotectants have been used on plant materials. Table 20-2 is a list of compounds of related effectiveness used on cabbage cells. Polyethylene glycol, glucose, DMSO, sorbitol, mannitol, proline, hydroxyproline, and a few other amino acids have recurred in the literature (Finkle et al., 1985; Withers, 1980). Infiltration studies on *Anthurium* suggest that L-proline readily infiltrated the seeds in relatively high concentration without any apparent toxicity (Coniff, 1983). The use of these cryoprotectants has varying degrees of success, depending on procedural protocols and the type of plant materials being cryopreserved. Trehalose, discussed elsewhere in this symposium, has not been extensively studied as a cryoprotectant in plant systems.

Anthurium seeds, a recalcitrant type, can be used to demonstrate colliga-

Table 20-2

Effectiveness of various compounds used as cryoprotectants for cabbage cells

Slight protection (−7°C)	Medial protection (−10°C)	Medial protection (−15°C)	High protection (−20°C)	Highest protection (−25°C)
Raffinose	D-Glucuronic acid	Ascorbic acid	Sucrose	Glucose
Mannitol	Hydroxyproline	Acetoamide	Sorbitol	β-Methyl-D-glycoside
Glucosamine	Glutamic acid	Methyl acetoamide	Lactate	Xylose
Galacturonic acid	Methyl formamide	Dimethyl formamide	Panthothenic acid	Rhamnose
Dimethyl acetoamide	Formamide	Betaine aldehyde	Methanol	Erythritol
Acetyl glycine	Dimethyl urea	(Cabbage sap)	Ethylene glycol	Xylitol
α-Aminobutyrate	N-methyl urea	(*Pseudoacacia:* bark sap)	Glyceraldehyde	Glycerol
α-Glycerophosphate	Dimethyl thetine		Dimethyl sulfoxide	Betaine
β-Glycerophosphate	Acetyl choline		1,3-Dimethyl urea	Trimethyl-amine N-oxide
Water (N.C.): −5°C	N-Methyl pyrrolidone		Methyl glycine	
	Triethylene glycol		Dimethyl glycine	
			Methionine sulfoxide	
			Lactoamide	

Note: Table shows nominal concentrations of 0.5 M aqueous solutions (from Sakai and Yoshida, 1968, courtesy of *Cryobiology*).

tive cryoprotection of plant materials. Freshly harvested seed has approximately 40-50% moisture content. If the seeds are slowly dried at room temperature under low humidity (20-30% RH), they begin to lose viability at between 30-35% moisture content and lose almost all viability by the time they reach 20% moisture content (25% dry weight basis). If stored at moderate temperatures (15-25°C) at their original moisture level, they will lose all viability within 25-30 days because of natural aging processes (Becwar et al., 1983).

L-proline, a natural plant amino acid, can be used in high molar concentrations to remove water from the seed. Seeds placed in a 5.5 M solution will become dehydrated within 72 hr to a moisture level between 15 and 20%. Movement of the proline into the seed is occurring at the same time. Mass/volume experiments suggest that for every milligram of water removed, approximately 0.5 mg of proline moves into the seed. Survival of the seed under these dehydrated conditions is roughly 50% of the original germination of the seed lot.

The moisture level obtained by this method is low enough to limit freezing damage. DTA data suggest that freezable water is not present below 18-20% moisture content. Once the seed has been chemically dehydrated, it is placed in the vapor phase of liquid nitrogen for storage. The cooling rate of the sample in the vapor phase is 20-30°C/min. A sample is retrieved by taking it out of the liquid nitrogen vapor phase storage and placing it in a 38°C water bath for rapid rewarming (100°C/min or faster). Once rewarmed, the proline is rinsed out of the seed, generally through one or two baths of decreasing concentrations of proline. Once the proline has been removed from the seed, the sample can be handled or germinated in a normal fashion. Survival of storage in liquid nitrogen is 84% of the dehydrated control (Table 20-3).

Table 20-3
Survival of *Anthurium* seed (percentage germination) after dehydration with L-proline for 72 hr

	Proline concentration (M)		
Treatment	0.0	3.5	5.5
Proline + dehydration	84	42	48
Proline + dehydration + LN2 exposure	0	0	40

Note: Approximate moistures for varying treatments were 45% (0.0 M), 25% (3.5 M) and 17% (5.5 M). After dehydration seed were also exposed to liquid nitrogen for 24 hrs, rewarmed and germinated.

More research is needed in this area to improve the survival of the seed during the dehydration process. A slower extraction of the water from the seed using a stepwise exposure of proline concentrations and use of combinations of cryoprotectants are two possible areas of practical research. Membrane/hydration interactions in orthodox and recalcitrant seeds need to be defined to better our understanding of the fundamental problems associated with dehydration and survival. As knowledge increases in this area, more refined and effective dehydration schemes can be developed, resulting, we may hope, in the possibility of preserving a wider variety of plant materials, including buds, meristems, calli, and cell suspensions as well as recalcitrant seeds.

REFERENCES

Bajaj, Y.P.S. 1983. Cryopreservation and international exchange of germplasm. *In* Plant Cell Culture in Crop Improvement. S.K. Sen and K.L. Giles (eds.). Plenum. New York. 19-41.

Bass, L.N. and Stanwood, P.C. 1978. Long-term preservation of sorghum seed as affected by seed moisture, temperature, and atmospheric environment. Crop Sci. 18:575-577.

Becwar, M.J., Stanwood, P.C. and Leonhardt, K.W. 1983. Dehydration effects on freezing characteristics and survival in liquid nitrogen of desiccation-tolerant and desiccation-sensitive seeds. J. Amer. Soc. Hort. Sci. 108: 613-618.

Becwar, M.R., Rajashekar, C., Hansen Bristow, K.J. and Burke, M.J. 1981. Deep undercooling of tissue water and winter hardiness limitations in timberline flora. Plant Physiol. 68: 111-114.

Becwar, M.R., Stanwood, P.C. and Roos, E.E. 1982. Dehydration effects on imbibitional leakage from desiccation-sensitive seeds. Plant Physiol. 69: 1132-1135.

Bradshaw, A.D. 1975. Population structure and the effects of isolation and selection. *In* Crop Genetic Resources for Today and Tomorrow. O.H. Frankel and J.G. Hawkes (eds.). Cambridge Univ. Press. London. 37-52.

Brooks, H.J. and Barton, D.W. 1977. A plan for national fruit and nut germplasm repositories. Hort. Sci. 12: 298-300.

Burke, M.J., Gusta, L.V., Quamme, H.A., Weiser, C.J. and Li, P.H. 1976. Freezing and injury in plants. Ann. Rev. Pl. Physiology. 27: 507-528.

Chin, H.F. and Roberts, E.H. 1980. Recalcitrant Crop Seeds. Tropical Press SDN. BHD. Kuala Lumpur, Malaysia. 151 pp.

Coniff, K. 1983. Uptake of label cryoprotectants by *Anthurium schezerianum* L.: A desiccation sensitive seed species. M.S. thesis. Colorado State University, Fort Collins, Colorado. 88 pp.

Finkle, B.J., Zavala, M.E. and Ulrich, J.M. 1985. Cryoprotective compounds in the viable freezing of plant tissues. *In* Cryopreservation of Plant Cells and Organs. K.K. Kartha (ed.). CRC Press, Boca Raton, Florida. 75-114.

Frankel, O.H. and Hawkes, J.G. 1975. Genetic resources: The past ten years and the next. *In* Crop Genetic Resources for Today and Tomorrow. O.H. Frankel and J.G. Hawkes (eds.) Cambridge Univ. Press. London. 1-11.

Franks, F. 1985. Biophysics and Biochemistry at Low Temperatures. Cambridge Univ. Press. London.

Goldbach, H. 1979. Imbibed storage of *Melicoccus bijugatus* and *Eugenis brasiliensis* (E. *dombeyi*) using abscisic acid as a germination inhibitor. Seed Sci. & Technol. 7: 403-406.

Harrington, J.F. 1963. Practical instructions and advice on seed storage. Proc. Intern. Seed Testing Assoc. 28: 989-994.

Haunold, A. and Stanwood, P.C. 1985. Long-term preservation of hop pollen in liquid nitrogen. Crop Sci. 25: 194-196.

Ishikawa, M. and Sakai, A. 1978. Freezing avoidance in rice and wheat seeds in relation to water content. Low Temp. Sci. Ser. B 36: 39-49.

Junttila, O. and Stushnoff, C. 1977. Freezing avoidance by deep supercooling in hydrated lettuce seeds. Nature 269: 325-327.

Kartha, K.K. 1985. Cryopreservation of Plant Cells and Organs. CRC Press. Boca Raton, Florida. 276 pp.

Levitt, J. 1980. Responses of Plants to Environmental Stresses: Chilling, Freezing, and High Temperature Stresses. Academic Press. New York. 497 pp.

Li, P.H. and Sakai, A. 1978. Plant Cold Hardiness and Freezing Stress: Mechanisms and Crop Implications. Academic Press. New York. 416 pp.

Li, P.H. and Sakai, A. 1982. Plant Cold Hardiness and Freezing Stress: Mechanisms and Crop Implications. Vol. 2. Academic Press. New York. 694 pp.

Meryman, H.T. and Williams, R.J. 1980. Mechanisms of freezing injury and natural tolerance and the principles of artificial cryoprotection. *In* Crop Genetic Resources: The Conservation of Difficult Material. L.A. Withers and J.T. Williams (eds.). IBPGR, FAO, Rome, Italy. 5-37.

Oldfield, M.L. 1984. The value of conserving genetic resources. U.S. Dept. of the Interior, National Park Service Publication, Washington D.C. 360 pp.

Roberts, E.H. 1973. Predicting the storage life of seeds. Seed Sci. and Technol. 1: 499-514.

Roberts, E.H. and Ellis, R.H. 1982. Physiological, ultrastructural and metabolic aspects of seed viability. *In* The Physiology and Biochemistry of Seed Development, Dormancy and Germination. A.A. Khan (ed.). Elsevier Biomedical Press. 465-485.

Roos, E.E. and Stanwood, P.C. 1981. Effects of low temperature, cooling rate, and moisture content of seed germination of lettuce. J. Amer. Soc. Hort. Sci. 106: 30-34.

Sakai, A. 1985. Cryopreservation of shoot-tips of fruit trees and herbaceous plants. *In* Cryopreservation of Plant Cells and Organs. K.K Kartha (ed.). CRC Press, Boca Raton, Florida. 135-158.

Sakai, A. and Yoshida, S. 1968. The role of sugar and related compounds in variations of freezing resistance. Cryobiology 5: 160.

Senaratna, T., McKersie, B.D. and Stinson, R.H. 1985. Simulation of dehydration injury to membranes from soybean axes by free radicals. Plant Physiol. 77: 472-474.

Stanwood, P.C. 1984. Cryopreservation of seed germplasm: A preliminary guide to the practical preservation of seed germplasm in liquid nitrogen (LN2, -196°C). IBPGR Advisory Committee on Seed Storage, Report of the Second Meeting. IBPGR, FAO, Rome, Italy. 8-27.

Stanwood, P.C. 1985. Cryopreservation of seed germplasm for genetic conservation.

In Plant Cryopreservation. K. Kartha (ed.). CRC Press. Boca Raton, Florida. 199-225.

Stanwood, P.C. and Bass, L.N. 1981. Seed germplasm preservation using liquid nitrogen. Seed Sci. & Technol. 9: 423-437.

Steponkus, P.L. 1980. Dehydration and Desiccation Injury: Membrane Damage in Crop Genetic Resources. The Conservation of Difficult Material. L.A. Withers and J.T. Williams (eds.). IBPGR, Food and Agriculture Organization, Rome, Italy. 101-103.

Steponkus, P.L. 1984. Role of the plasma membrane in freezing injury and cold acclimation. Ann. Rev. Plant Physiol. 35:543-584.

Towill, L.E. 1985. Low Temperature and Freeze-/Vacuum-Drying Preservation of Pollen. K. Kartha (ed.). CRC Press. Boca Raton, Florida. 171-198.

Villers, T.A. 1975. Genetic maintenance of seeds in imbibed storage. *In* Crop Genetic Resources for Today and Tomorrow. O.H. Frankel and J.G. Hawkes (eds.). Cambridge Univ. Press. London. 297-316.

Withers, L.A. 1980. Tissue Culture Storage for Genetic Conservation IBPGR Report. IBPGR, Food and Agriculture Organization, Rome, Italy. 91 pp.

APPENDICES

Methods of Analysis of Membranes and Water

Appendix A

Detection and Measurement of Membrane Fusion and Leakage

George Strauss

Perturbation of the bilayer structure of cells and organelles resulting from dehydration or freezing generally leads to the aggregation and fusion of adjacent bilayer membranes, accompanied by leakage of solutes across the bilayers. When studying these processes in model systems represented by unilamellar phospholipid vesicles, we must distinguish between (1) clustering of intact vesicles; (2) fusion of small unilamellar vesicles (SUV) and mixing of their entrapped solutes to form larger uni- or oligolamellar vesicles; and (3) loss of entrapped solutes by leakage into the external aqueous phase during the fusion of SUV. Several techniques are effective for these purposes.

Energy transfer. The transfer of electronic excitation energy (i.e. the induction of fluorescence in an acceptor by selective excitation of an energy donor) is a sensitive measure of proximity, since the energy transfer efficiency is proportional to $1/R^6$, R being the donor-acceptor distance. Clustering or fusion of vesicles can be determined quantitatively by incorporating donor and acceptor probes into separate populations of vesicles which are then mixed and subjected to the perturbation under test (Mehreteab and Strauss, 1978). Energy transfer will then be proportional to the fraction of vesicles that have aggregated or fused.

In order to avoid energy transfer in the absence of vesicle interaction, the probes must be structured so as to remain firmly attached to intact bilayers. This requirement is met by compounds such as amphiphilic cyanine dyes, which, like phospholipids, have two long acyl chains and a polar head group. A suitable pair of probes is dioctadecyloxacarbocyanine as donor and dioctadecylindocarbocyanine as acceptor, as illustrated in Figure A-1. Energy

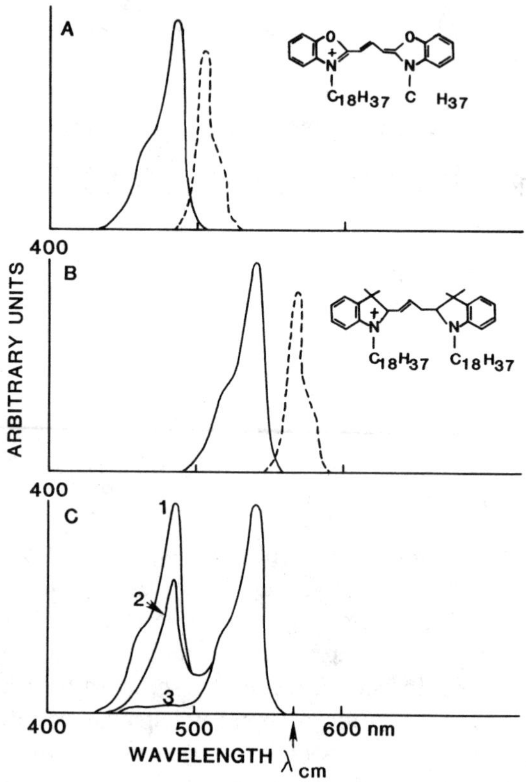

Figure A-1. Absorption spectra (———) and fluorescence emission spectra (-----) of dioctadecyloxacarbocyanine (a) used as donor and of dioctadecylindocarbocyanine (b) used as acceptor.
Excitation spectra (c) obtained with lipid vesicles labeled with the dyes noted above. Curve 1, vesicles labeled with both dyes, giving "100%" energy transfer. Curve 2, a mixture of two vesicle populations labeled with donor and acceptor probes, respectively, and subjected to a perturbation, giving partial energy transfer. Curve 3, the mixture as for Curve 2 but not perturbed. The peak height at 540 nm (excitation maximum of the acceptor) has been normalized for the three curves. The emission of the acceptor was read at 568 nm.

transfer is recognized and measured either by loss of the fluorescence of the donor in the presence of the acceptor or, preferably, by exciting the donor and observing fluorescence emission by the acceptor.

To quantify the extent of vesicle interaction, energy transfer in the system under test is measured relative to the limiting cases representing zero and 100% vesicle interaction, respectively. Fluorescence excitation spectra are recorded by plotting the emission intensity of the acceptor as a function of excitation wavelength. Such a spectrum shows a peak at the absorption maximum of the donor, representing sensitized fluorescence due to energy transfer, and a second peak at the absorption maximum of the acceptor, representing direct excitation of the latter. The ratio of these peak heights (or more precisely, the ratio of the integrated peak areas) is determined for the test system and for the two reference systems for an assay of vesicle interaction. "Zero transfer" refers to a mixture of labeled vesicle populations left unperturbed; "100%" refers to vesicles doubly labeled with both probes. The intrinsic fluorescence quantum yield of these probes could be affected by the physical state of the bilayer membrane (i.e. could change as the membrane is dehydrated). Therefore, the reference systems must be subjected to the same perturbation processes as the test system.

Mixing of aqueous vesicle cores. The energy transfer method, being a proximity test, does not distinguish between vesicle clustering and actual fusion. Occurrence of the latter can be evidenced only by observing a direct mixing of the interior aqueous compartments of different vesicles. Such a merging of cores can be recognized and measured by the method of Wilschut and Papahadjopoulos (1979). Separate vesicle populations are prepared which in their inner aqueous compartments contain (1) Tb^{3+} (preferably, as a weak complex with citrate), and (2) dipicolinic acid (DPA), both of which are essentially nonfluorescent. When vesicle cores containing these labels unite, a strongly fluorescent Tb·DPA complex is formed. In order to distinguish a mixing of cores from leakage, disodium-EDTA is added to the external solution. This converts any Tb^{3+} or Tb·DPA leaking out into a nonfluorescent Tb·EDTA complex. The fluorescence intensity of the Tb·DPA complex resulting from a given membrane perturbation process, when compared with "zero" and "100%" reference systems, gives the fraction of vesicles that fused without leakage. To be meaningful, this result must be combined with a separate determination of the extent of leakage.

Leakage of solutes. In an adaptation of the above method vesicles are labeled with the preformed Tb·DPA complex, and disodium-EDTA is added to the external solution. Any loss of fluorescence intensity resulting from a perturbation can be interpreted as a loss of entrapped aqueous vesicle contents. The "100%" condition is obtained by repeating the experiment in the absence of external EDTA.

The fluorescence of the Tb·DPA complex is measured by exciting the sample at 276 nm and reading the emission at 540 nm. These wavelengths do not interfere with the excitation and emission wavelengths used in the energy transfer method when using the specific dyes shown in Figure A-1. It is possible, therefore, to use a quadruply labeled system in which both vesicle aggregation/fusion and leakage can be measured (Gibson and Strauss, 1984). One vesicle population contains the donor cyanine dye within the bilayer vesicle wall and Tb^{3+} in the aqueous core; the other population contains the acceptor cyanine dye and DPA in the corresponding locations.

REFERENCES

Gibson, S. M. and Strauss, G. 1984. Reaction characteristics and mechanisms of lipid bilayer vesicle fusion. Biochim. Biophys. Acta 769: 531-542.

Mehreteab, A. and Strauss, G. 1978. Energy transfer and energy losses in bilayer membrane vesicles (liposomes). Photochem. Photobiol. 28: 369-375.

Wilschut, J., and Papahadjopoulos, D. 1979. Ca^{++}-induced fusion of phospholipid vesicles monitored by mixing of aqueous contents. Nature 281: 690-692.

Appendix B

Comments on ^{31}P NMR

Helmut Hauser

The nonspherical distribution of electrons about the ^{31}P nucleus manifests itself in the anisotropy of the chemical shielding properties: the chemical shielding of this nucleus is orientation dependent and represented by a second-rank tensor. If a single crystal of a phosphate-containing compound is brought into the magnetic field of an NMR spectrometer, a single ^{31}P signal is observed. The magnetic field, or chemical shift, at which the ^{31}P signal occurs depends on the orientation of the single-crystal axes with respect to the magnetic field. The orientation of the principal axes of the ^{31}P chemical shielding tensor in phospholipids is unknown. The reason is that suitable single crystals of phospholipids required for the determination of the orientation of these axes have not been produced as yet.

The ^{31}P chemical shielding tensor of barium diethylphosphate determined by Griffin and his coworkers has been used as a first approximation in the interpretation of ^{31}P NMR spectra of phospholipid bilayers and biological membranes. Let the three principal axes of the ^{31}P chemical shielding tensor be represented by the Cartesian coordinates, X, Y, and Z. If the single crystal is oriented such that the magnetic field H_o is either parallel to X or Y or Z, the corresponding ^{31}P resonances are observed at positions marked σ_{11}, σ_{22}, and σ_{33}, respectively, as shown in Figure B-1a.

In a crystal powder all orientations are equally probable. The resulting powder spectrum (Figure B-1b) is the sum of all spectra originating from the randomly oriented crystallites. The spectral range of the powder spectrum is defined by σ_{11} and σ_{33}.

When the phospholipid crystal is heated above a certain temperature, the Krafft point, the hydrocarbon chains melt. The resulting liquid crystalline

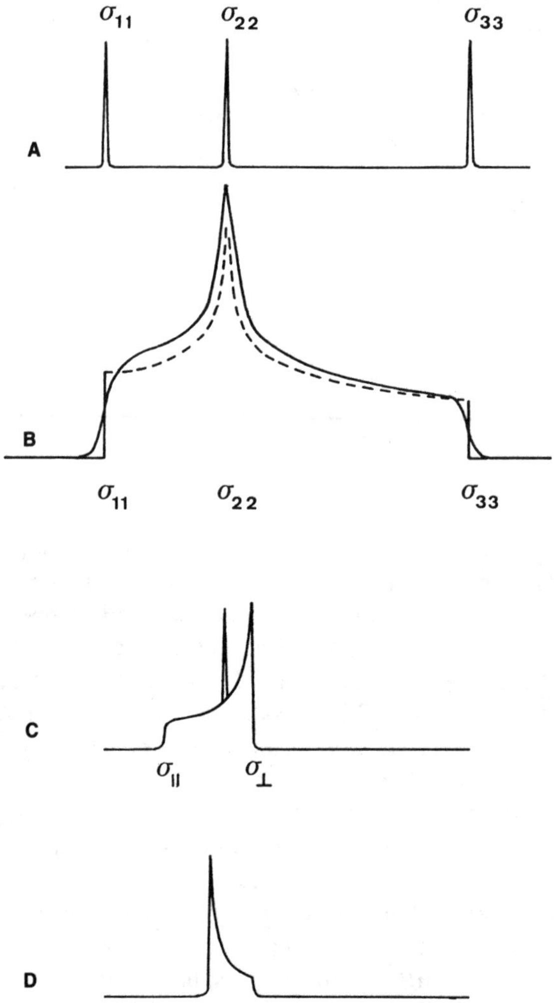

Figure B-1. Diagram of ³¹P NMR spectra, obtained with phospholipids differing in their structural arrangement (phase). (a) Spectra obtained from a single crystal oriented at three different directions. The peaks labeled σ_{11}, σ_{22}, and σ_{33} originate from the single crystal oriented along the three axes x, y, and z of the ³¹P chemical shielding tensor, respectively. (b) A powder spectrum obtained from ground crystals of the same phospholipid. (c) A spectrum from a smectic (lamellar) liquid-crystalline phase of a phospholipid; the sharp resonance on top of the powder spectrum is from aggregates undergoing isotropic motion—e.g. from micelles or small unilamellar vesicles. (d) A spectrum from a phospholipid forming a hexagonal phase.

phase is a highly dynamic one. Various molecular and segmental motions contribute to the averaging of the ^{31}P chemical shielding tensor. While the powder spectrum of a crystalline phospholipid is characteristic of an asymmetric chemical shielding tensor (Figure B-1b), the motional averaging in the liquid crystal gives rise to a spectral line shape which is characteristic of a symmetrical shielding tensor (Figure B-1c). Such a line shape results from the phospholipid molecule as a whole, or the phosphate group as a segment, undergoing rotation about the bilayer normal. The correlation time of this rotation must be fast on the NMR time scale. The edges of the ^{31}P NMR spectrum defined by $\sigma_{\parallel}$ and $\sigma_{\perp}$ correspond to orientations of the magnetic field parallel and perpendicular to the bilayer normal, respectively. If there is a phase change from the smectic to the hexagonal phase, the lipid molecule may experience additional motional averaging owing to lateral diffusion of the molecule about the cylinder axis of the hexagonal phase. Such a phase change would express itself in a significant change in the line shape of the ^{31}P NMR spectrum (Figure B-1d). Compared with the "bilayer" spectrum in Figure B-1c, the polarity of the spectrum is reversed and the spectral range defined by the edges of the spectrum is halved. This spectral range, or breadth, referred to as chemical shielding anisotropy, is a measure of the anisotropy of motion. If the motion of the phospholipid molecule becomes isotropic, the powder spectrum collapses to a singlet as indicated by the narrow signal in Figure B-1c. Isotropic signals are observed with cubic and micellar phases, and sometimes with smectic phases, provided the particles are small enough so that Brownian tumbling contributes to the motional averaging. ^{31}P NMR may therefore be used as a diagnostic tool to differentiate between different phospholipid phases.

Appendix C

X-Ray Diffraction as a Technique for Studying the Mesomorphic Phase Properties of Lipids

Martin Caffrey

I shall briefly describe the use of x-ray diffraction for lipid mesomorphic phase identification and phase transition temperature determination. Some sample diffraction patterns, characteristic of the more commonly encountered lipid phases, such as the lamellar, ripple, hexagonal, cubic, and fluid isotropic phases, are included for reference purposes.

X-ray diffraction has long been used as the definitive method for lipid-phase identification and phase transition temperature determination (Luzzati, 1968; Levine, 1973; Shipley, 1973). By its very nature, however, the diffraction method requires the phase to be well ordered and extensive. Where these conditions are not satisfied, the method becomes ambiguous, and other techniques such as electron microscopy may prove more useful (an excellent discussion of this point is presented in Gruner et al., 1985).

While the theory of x-ray diffraction is complex, in practice it provides a relatively straightforward method of lipid phase identification. All that is needed is to record the diffraction pattern under well-defined conditions and to make some simple distance measurements and trigonometric calculations. The experimental arrangement for making the diffraction measurement is depicted schematically in Figure C-1. A collimated x-ray beam ($\bar{S}_o$) of known wavelength (λ) strikes the sample. Diffracted x-rays ($\bar{S}$) intercept a detector, most commonly x-ray-sensitive film, upon which the diffraction

This work was supported in part by a grant from the National Institutes of Health, U.S. Public Health Service (HL-18255), to G.W. Feigenson and by a grant-in-aid of research from Sigma Xi, The Scientific Research Society, and a grant (no. 1R23AM36849-01) from the National Institutes of Health to the author.

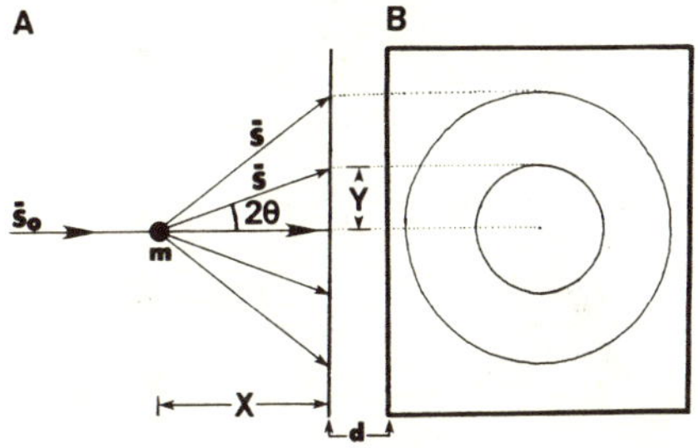

Figure C-1. Schematic representation of the x-ray diffraction experiment and the geometries involved. A side view is presented in (a), and (b) shows a powder diffraction pattern as it might appear on a two-dimensional detector (d) such as x-ray sensitive film. $\bar{S}_o$, incident x-ray beam; $\bar{S}$, diffracted x-ray beam; 2θ, scattering angle; *m*, sample; *x*, sample-to-detector distance; *y*, radius of the first powder diffraction ring.

pattern is recorded. The sample-to-detector distance (*x*) is accurately known or can easily be measured.

Unoriented lipid samples usually give rise to powder diffraction patterns which consist of a series of concentric rings (Figures C-1 and C-2). The radius of a given diffraction ring (*y* in Figure C-1) is measured from the film. This corresponds to a given scattering angle (2θ), which is the angle between the incident and diffracted beams. Thus, θ can be calculated from the relationship:

$$\tan 2\theta = y/x,\ \theta = \tfrac{1}{2}\tan^{-1}(y/x).$$

In order to calculate the *d*-spacing (*d*) of the lattice associated with this reflection, the Bragg equation is used:

$$n\lambda = 2d \sin\theta,\ d = n\lambda/2\sin\theta.$$

Values for λ and θ are known, and the integer *n* depends on the order of the reflection. Units of *d* are those of λ, typically ångströms or nanometers.

One point that occasionally needs clarification is the use of the symbol *s* in plots of diffracted intensity *versus* distance in reciprocal space. Here *s* has units of $(\text{Å})^{-1}$ or $(\text{nm})^{-1}$ and is usually defined as

$$s = 2(\sin\theta)/\lambda = 1/d.$$

The advantage of using *s* is that it normalizes for λ, which can vary depending on the x-ray source.

It is usual to distinguish two regions in lipid diffraction patterns, namely, low angle and wide angle. As its name implies, low-angle diffraction arises from scattering close to the direct beam at low 2θ values. Scattering in this region derives from order in the sample, which is long range at ≥5 Å, as, for example, in the lamellar period of the lamellar phase (*d* in Figure 14-2a). The

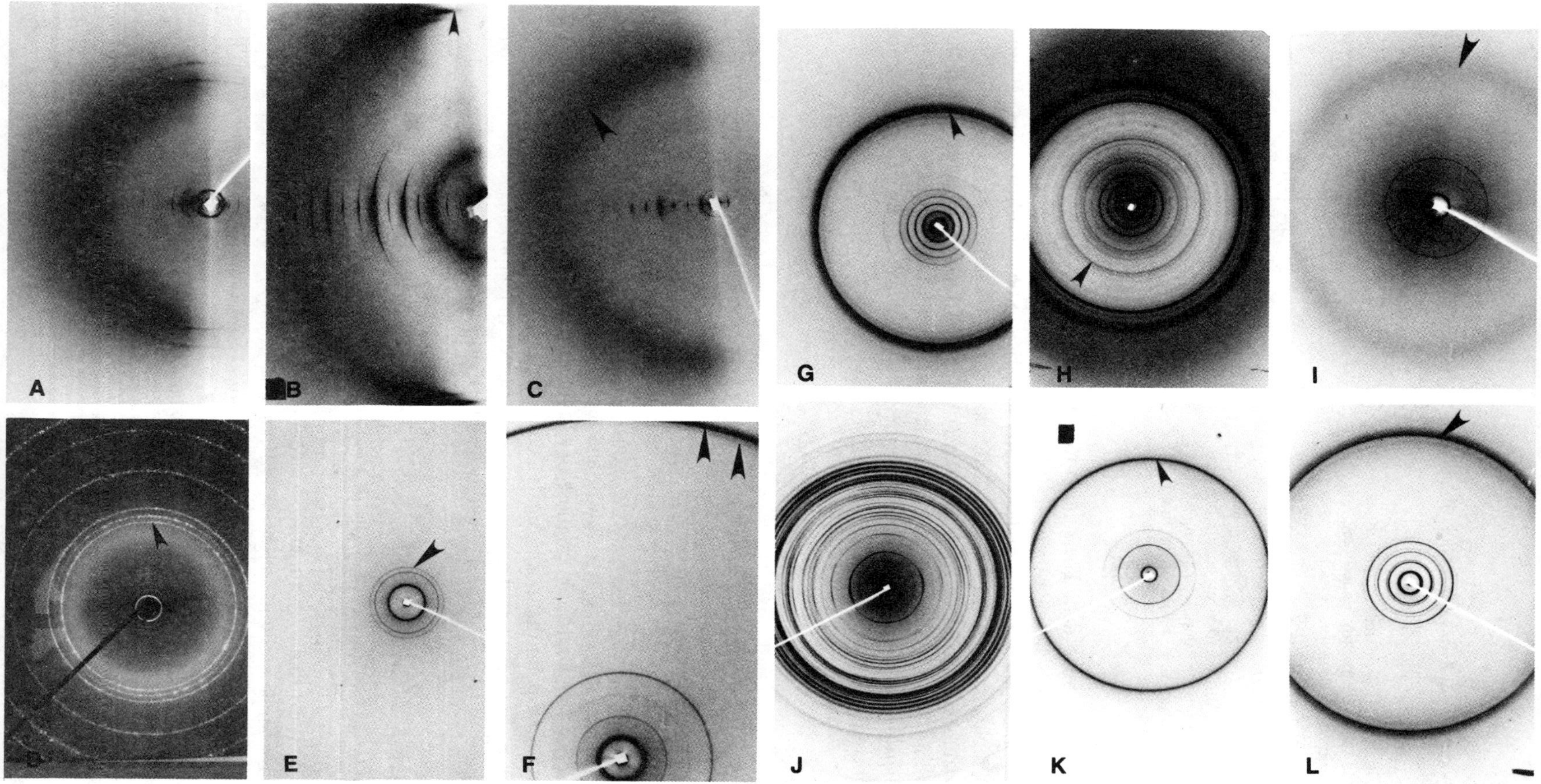
A
B
C
G
H
I
E
F
J
K
L

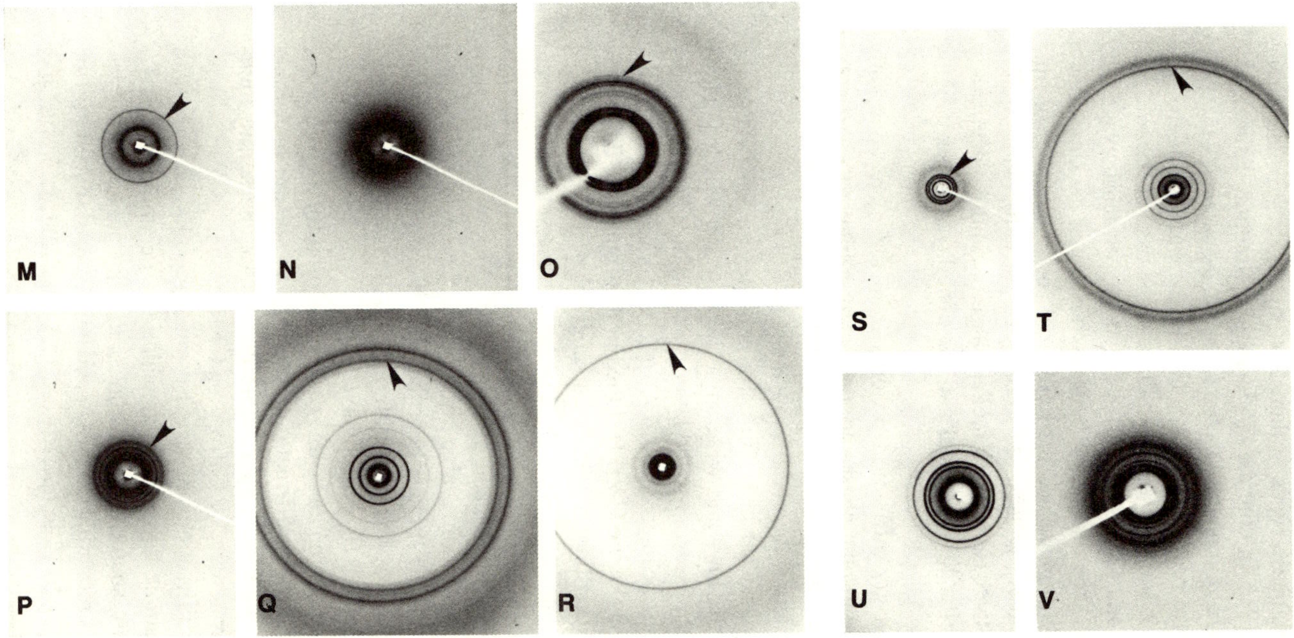

Figure C-2. X-ray diffraction patterns of the various polymorphic and mesomorphic lipid phases. A schematic view of these phases is shown in Figure 14-2. (a) Egg phosphatidylethanolamine, oriented (lamellar/inverted hexagonal phase coexistence). (b) Dipalmitoylphosphatidylcholine, oriented (lamellar gel phase, L'_β). (c) Egg phosphatidylcholine, dry-oriented (lamellar liquid crystal phase, L_α). (d) Nitroxide-labeled (7,6) phosphatidylcholine, −12.5°C (lamellar liquid crystal phase plus ice). (e) Monoelaidin, 3.5 *M* NaCl, 77°C (inverted hexagonal phase). (f) Monoelaidin, hydrated, 15°C (lamellar gel phase). (g) Dipalmitoylphosphatidylcholine, hydrated 26°C (lamellar gel phase, L'_β). (h) Dipalmitoylphosphatidylcholine, hydrated 5°C (lamellar subgel phase, L'_c). (i) Dihexadecylphosphatidylethanolamine, hydrated 75°C (lamellar liquid crystal phase, L_α). (j) Dipalmitoylphosphatidylcholine, dry, 4°C (lamellar crystalline phase, L_c). (k) Dihexadecylphosphatidylethanolamine, hydrated, 21°C (lamellar gel phase, L_β). (l) Dipalmitoylphosphatidylcholine, hydrated (ripple phase, P'_β). (m) Monoelaidin, hydrated, 30°C (lamellar liquid crystal phase, L_α). (n) Monoelaidin, hydrated, 115°C (fluid isotropic phase). (o) Monoolein, hydrated, 22°C (body-centered cubic phase, space group number 12, Ia3d). (p) Monoelaidin, 3.5 *M* NaCl, 76.5°C (inverted hexagonal, H_{II}, phase). (q) Ca^{2+}-phosphatidylserine, 26°C (lamellar gel phase). (r) Phosphatidylserine, hydrated, −5°C (extensively swollen lamellar gel phase). (s) Monoelaidin, hydrated, 52°C (body-centered cubic phase, space group number 8). (t) Distearoylphosphatidylcholine, hydrated, 21°C (lamellar gel phase, L'_β). (u) Dioleoylphosphatidylcholine/gramicidin (10/1, mole/mole), hydrated, 22°C (inverted hexagonal phase, H_{II}). (v) Monoolein, hydrated, 22°C (cubic primitive phase, space group number 4).

low-angle region identifies lattice type and symmetry and enables us to distinguish the various phase types as lamellar, ripple, hexagonal, cubic, fluid isotropic, and so on. In contrast, wide-angle diffraction derives from short-range order in the sample with characteristic *d*-spacings of ≤5 Å. Scattering in this region provides information on the packing of the acyl chains, which can be crystalline, semicrystalline in the case of the so-called subgel phase (Ruocco and Shipley, 1982), hexagonally or quasihexagonally packed (one should note the distinction between hexagonal packing of acyl chains within the lamellar phase and the hexagonal packing of rodlike lipid aggregates in the H_I and H_{II} phases; see Figures 14-2c, f, h), with tilted or untilted acyl chains in the case of the lamellar phase or fluidlike acyl chains in the case of the lamellar liquid crystal and the nonbilayer phases (Figure 14-2).[1]

The characteristic x-ray diffraction patterns from the more commonly encountered lipid mesomorphic phases are presented in Figure C-2. A complete description of the conditions used in generating these patterns is not given, since the intent here is merely to show the nature of typical patterns as they might appear on x-ray-sensitive film, for example. In the case of the lamellar phases, both the low- and wide-angle regions are included. However, for the nonbilayer phases, only the latter is shown, because here wide-angle scattering is diffuse, centered at 4.6 Å, just as is seen for the lamellar liquid crystal (L_α) phase (Figure C-2). Included in this figure is a diffraction pattern for a lamellar crystalline (L_c), lamellar subgel phase (L_c'), lamellar gel (with tilted [L_β'] and untilted chains [L_β]), lamellar liquid crystal (L_α), ripple (P_β'), inverted hexagonal (H_{II}; the normal hexagonal phase [H_I] has a diffraction pattern similar to that of H_{II}, and the two can be distinguished by their differential response to hydration), cubic primitive (C_p), body-centered cubic (C_i), and the fluid isotropic (FI) phase. For the sake of completeness, examples are also given of patterns obtained with oriented preparations of the L_β, L_α, and H_{II} phases and of L_α/H_{II} phase coexistence.

An x-ray diffraction pattern can be used as a fingerprint for a given phase type. Therefore, recording the diffraction pattern as a function of sample temperature provides us with a very reliable method for measuring lipid-phase transition temperatures (T_t; Caffrey and Feigenson, 1984; Caffrey, 1985). A change in the diffraction pattern corresponds to a phase change, and the determination of sample temperature at which this occurs identifies T_t. With conventional x-ray sources (sealed tubes, rotating anodes), the procedure usually involves recording patterns sequentially in so-called still or static mode at a series of fixed temperatures. This approach, however, has

1. The notation of Luzzati (1968) is used in describing the various lipid phase types.

the disadvantage of possibly missing a phase or phase coexistence region that prevails over a narrow temperature range. Continuous methods have been developed which overcome this problem. With conventional sources the procedure involves a slow scan in temperature, with continuous recording of a portion (usually a thin strip along the diameter of the pattern) of the diffraction pattern on a length of film (Khan, 1985; Klug and Alexander, 1984; Lindblom et al., 1979). Orders-of-magnitude-higher intensities are available at synchrotron x-ray sources, and when these are used in conjunction with suitable x-ray imaging devices, a very rapid and continuous two-dimensional record on videotape of the diffraction pattern can be made as sample temperature is scanned. The advantages of this method, called time-resolved x-ray diffraction, include (1) that data can be collected at prodigious rates, (2) that all temperatures in the range of interest are explored, and (3) that in addition to establishing T_t, the identity of the phases on either side of T_t and the relative amounts of coexisting phases in mixed-phase regions can, in principle, be determined. This method and its uses have been detailed elsewhere (Caffrey and Bilderback, 1983, 1984; Caffrey and Feigenson, 1984; Caffrey, 1984, 1985).

The situation is somewhat more complex when using the x-ray diffraction method to study lipid phase states and phase changes in intact or isolated biological systems. In the first place, the heterogeneity that exists in lipid composition in these natural systems means that T_t is broad. Second, in the case of cellular membranes at least, unless these are derived from naturally stacked sytems such as chloroplasts or the myelin sheath or are specifically treated and/or oriented, little useful information can be gleaned from the low-angle diffraction data without significant effort. Nonetheless, one useful determination can be made quite simply and with high sensitivity. This is the high-temperature limit (T_{th}) of hydrocarbon chain melting (i.e. the highest temperature at which gel phase lipid exists in these native systems). As seen in Figure C-2, the gel phase is characterized by a sharp diffraction line in the wide-angle region at $(4.2\ \text{Å})^{-1}$. Thus, the temperature at which the gel [$(4.2\ \text{Å})^{-1}$ line] phase appears (disappears) upon cooling (heating) identifies T_{th}. Here x-ray diffraction is particularly suitable, since it is sensitive to ~2% of the total lipid in the gel or solid phase. For a more complete description of the method and its application, consult McKersie and Thompson (1977), Mitsui (1978), Caffrey and Feigenson (1984), Caffrey et al. (in press), Caffrey and Lew (in press), and the references cited in them.

I have attempted to outline in the simplest possible terms how x-ray diffraction can be used to investigate the mesomorphic phase properties of lipids. My hope is that the method, so presented, will prove less intimidating to the uninitiated and will stimulate an interest in and enhanced use of this most powerful technique. Finally, a word of caution is in order. When investigating phenomena of the type discussed above, it is of paramount importance that we bring as many complementary techniques as are available to bear on the problem. Total reliance on a single method, such as x-ray diffraction, can easily spell disaster. On the other hand, a result reached by means of a number of independent and complementary measurements has superior reliability.

REFERENCES

Caffrey, M. 1984. X-radiation damage of hydrated lecithin membranes detected by real-time x-ray diffraction using wiggler-enhanced synchrotron radiation as the ionizing radiation source. Nucl. Instr. Meth. 222: 329-338.

Caffrey, M. 1985. Kinetics and mechanism of the lamellar gel/lamellar liquid crystal and lamellar/inverted hexagonal phase transition in phosphatidylethanolamine: a real-time x-ray diffraction study using synchrotron radiation. Biochemistry 24: 4826-4844.

Caffrey, M. and Bilderback, D.H. 1983. Real-time x-ray diffraction using synchrotron radiation: system characterization and applications. Nucl. Instr. Meth. 208: 495-510.

Caffrey, M. and Bilderback, D.H. 1984. Kinetics of the main phase transition of hydrated lecithin monitored by real-time x-ray diffraction. Biophys. J. 45: 627-631.

Caffrey, M. and Feigenson, G.W. 1984. Influence of metal ions on the phase properties of phosphatidic acid in combination with natural and synthetic phosphatidylcholines: an x-ray diffraction study using synchrotron radiation. Biochemistry 23: 323-331.

Caffrey, M. and Lew, R.R. The Ca^{2+}-induced phase transition in soybean root microsomal membranes in a consequence of phospholipase D activation. Submitted for publication.

Caffrey, M., Werner, B.G. and Priestley, D.A. A crystalline lipid phase in a dry biological system: evidence from x-ray diffraction analysis of *Typha latifolia* pollen. Submitted for publication.

Gruner, S.M., Cullis, P.R., Hope, M.J. and Tilcock, C.P.S. 1985. Lipid polymorphism: the molecular basis of nonbilayer phases. Ann. Rev. Biophys. Chem. 14: 211-238.

Khan, Y. 1985. A microcomputer-controlled Guinier-Lennē-type dynamic-temperature x-ray diffraction camera. J. Phys. E: Sci. Instrum. 18: 1054-1057.

Klug, H.P. and Alexander, L.E. 1974. X-Ray Diffraction: Procedures for Polycrystalline and Amorphous Materials, 2nd ed. Wiley and Sons, New York, pp. 245-254.

Levine, Y.K. 1973. X-ray diffraction studies of membranes. Progr. Surf. Sci. 3: 279-352.

Lindblom, G., Larsson, K., Johansson, L., Fontell, K. and Forsēn, S. 1979. The cubic phase of monoglyceride-water systems. J. Am. Chem. Soc. 101: 5465-5470.

Luzzati, V. 1968. X-ray diffraction studies of lipid-water systems. *In* Biological Membranes, vol. 1, ed. D. Chapman. Academic Press, New York, pp. 71-123.

McKersie, B.D. and Thompson, J.E. 1977. Lipid crystallization in senescent membranes from cotyledons. Plant Physiol. 59: 803-807.

Mitsui, T. 1978. X-ray diffraction studies of membranes. Adv. Biophys. 10: 97-135.

Rilfors, L., Lindblom, G., Wieslander, A. and Christiansson, A. 1984. Lipid bilayer stability in biological membranes. Biomembranes 12: 205-245.

Ruocco, M.J. and Shipley, G.G. 1982. Characterization of the sub-transition of hydrated dipalmitoylphosphatidylcholine bilayers. Biochim. Biophys. Acta 691: 309-320.

Shipley, G.G. 1973. Diffraction studies of biological membranes and membrane components. *In* Biological Membranes, vol. 2, eds. D. Chapman and D.F.H. Wallach. Academic Press, New York, pp. 1-89.

Appendix D

The Glassy State and Survival of Anhydrous Biological Systems

Michael J. Burke

The subject of glass formation has been reviewed extensively by Franks (1982, 1985), and the interested reader is directed to those references for a more complete discussion. This appendix is intended to direct interest to a property of concentrated solutions that may be relevant to anhydrous biology. Glasses are metastable, they are viscous, and they are either supercooled or supersaturated (Kauzmann, 1948). They have several properties that may be relevant to near anhydrous organisms.

Glasses can be formed from cell solutes like sugars that are known to provide protection from denaturation of large molecules and molecular aggregates. The interface between glass and macromolecule might be quite similar to the interface of a macromolecule and a more dilute solution.

Glasses may fill space in a tissue, and during dehydration may serve to stop additional increases in tissue collapse, solute concentration, and pH alteration. Operationally, glasses are expected to have lower water vapor pressures than the corresponding crystalline solid and thereby add resistance to further dehydration. It will be seen that relevant glasses are stable for temperatures below +90°C, and so they also prevent freezing of water.

Glasses are exceedingly viscous and should stop all chemical reactions that require molecular diffusion. In so doing, they may assure dormancy and stability over time.

Two well-characterized glasses are the PVP (polyvinyl pyrrolidone)/water glass (Figure D-1) and the sucrose/water glass (Figure D-2; Mac-

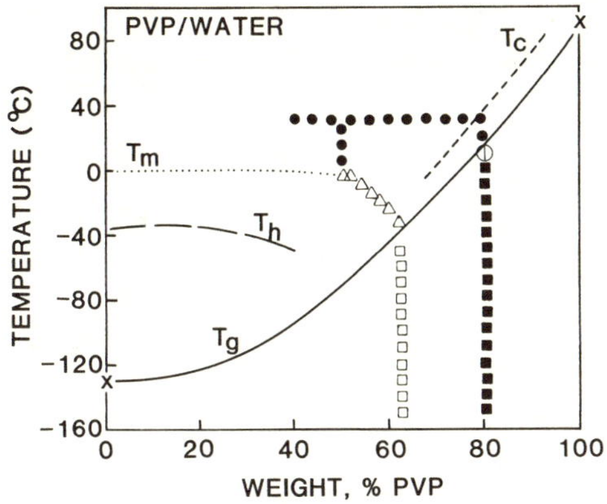

Figure D-1. The supplemented phase diagram of PVP/water adapted from MacKenzie and Rasmussen (1972, courtesy of Plenum Press). T_m is the equilibrium melting curve, T_g the glass transition curve, T_h the homogeneous nucleation curve, and T_c the collapse temperature curve above which freeze-drying fails. Two dehydration and cooling paths are indicated, both starting with ●. In the upper path, the solution is dehydrated at 30°C until the solution is at or below the collapse temperature, T_c, at 80 percent PVP. The sample is then cooled and reaches the glass transition near 0°C (⦶). The only degree of freedom below the glass transition is temperature, indicated by ■. In the lower path, the sample is dehydrated at 30°C until the solution is 50 percent in PVP and then cooled until the freezing point is reached just below 0°C (△). Additional slow cooling follows the T_m curve with ice forming and the solution becoming more concentrated in PVP. Such a sample would reach the glass transition near −30°C. Further cooling below the glass transition in indicated by □.

Kenzie, 1977), for which phase diagrams are shown in the two figures. Glasses of interest in anhydrous biology would most likely be supersaturated solutions that have high solute concentrations (cf. the right sides of Figures D-1 and D-2). If one concentrates a solution and then cools it, the following may occur in order:

1. The solution may become supersaturated as it drops below the crystallization temperature of the solute (T_s in Figure D-2).
2. If crystallization does not occur, and the solution is cooled further, the solution becomes progressively more viscous.
3. The glass transition temperature is reached (T_g in Figures D-1 and D-2) and a phase transition occurs. In a one-component system (i.e. pure sucrose, PVP, or water) a second-order "phase transition" occurs at T_g (67°C, sucrose; 87°C to 175°C, PVP; and −137°C, water; Franks, 1985; MacKenzie, 1977; MacKenzie and Rasmussen, 1972; Shen and Eisenberg, 1966). A second-order phase transition is a discontinuity in the second derivative of the free energy (G) with reference to tempera-

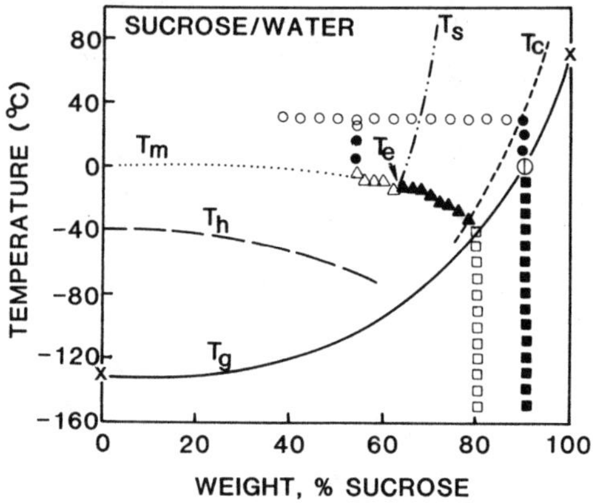

Figure D-2. The supplemented phase diagram of sucrose/water adapted from MacKenzie (1977, courtesy of *Phil. Trans. R. Soc. London*). Labels are defined as in Figure D-1 with the addition of T_s, the crystallization temperature of sucrose, and T_e, the eutectic point for sucrose/water solutions. Again two dehydration and cooling paths are indicated; both start with ○. In the upper path, the solution is dehydrated at 30°C until the solution is at or below the collapse temperature, T_c, at 90 percent sucrose (●). The solution is then cooled and reaches the glass transition near 5°C (⦶). In the lower path, the sample is dehydrated at 30°C until the solution is 50 percent sucrose and then cooled until the freezing point is reached below 0°C (●). Additional slow cooling follows the T_m curve with ice forming and the solution becoming more concentrated in sucrose (△). If crystallization does not occur at T_e, this freeze concentration continues to near the glass transition, T_g (▲). Such a sample would reach the glass transition near −20°C. Further cooling below the glass transition is indicated (□).

ture (T). Therefore, on cooling, a sudden drop in heat capacity (C_p) is observed at T_g (Equation D-1).

$$\delta^2 G/\delta T^2 = -\delta S/\delta T = -\delta C_p/2T \qquad \text{(D-1)}$$

A second-order phase transition can be detected using differential scanning colorimetry as an abrupt endothermic shift of the base line (Figure D-3). Like first-order "phase separations," this base-line shift is more diffuse for two and more component liquids as various solution domains are expected. For example, with the PVP/water solution, the PVP-rich domain will glass at a warmer temperature than the water-rich domain.

4. At and below T_g, the shear viscosity is above 10^{14} N sec/m², with molecular diffusion periods above 10^5 sec for rotation or translation through one molecular distance (i.e. 10^6 sec/nm or 300,000 yrs/cm; Franks, 1982, 1985). The glass is an amorphous solid because of the extreme viscosity, and as such it is much like a crystalline solid. How-

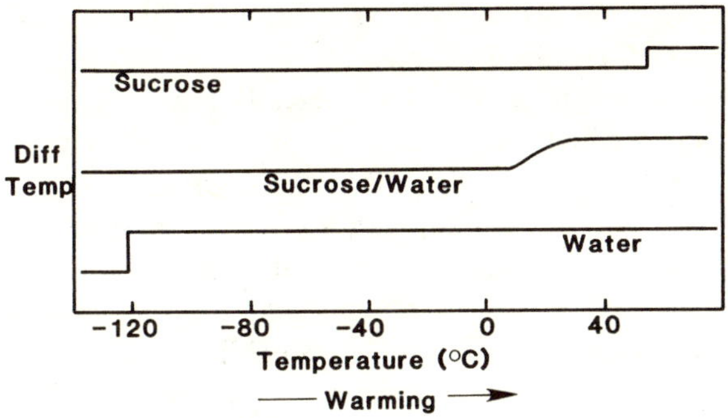

Figure D-3. Hypothetical and idealized differential scanning calorimetry showing glass transitions in pure sucrose, a sucrose/water solution, and water. Exothermic is toward the top.

ever, unlike a crystalline solid, it has surface characteristics that lead to much lower water vapor pressures.

A second method of glass formation is possible for solutions that contain freezable water and in which slow cooling occurs (see Figure D-2).

5. The solution reaches its melting temperature, and a phase separation occurs (i.e. ice separates from solution). As more ice forms, the solution becomes progressively more concentrated.
6. At the eutectic temperature, T_e (Figure D-2), the solution becomes supersaturated as it drops below the crystallization temperature of solute.
7. If crystallization does not occur, then cooling follows (2), (3) and (4) above.

There are large gaps in our knowledge of aqueous glasses. The published literature leaves some questions relating to the vapor pressure of the glass (e.g. see figure 4 in MacKenzie and Rasmussen, 1972) and the nature of second-order phase transitions (separations) in liquids with two and more components.

Desorption of Water from Aqueous Glasses and the Glass Transition

The low vapor pressure of an aqueous glass deserves some discussion. MacKenzie and Rasmussen (1972) show very clear desorption of most of the water from what should be PVP/water glasses. They also show the T_g for pure PVP at 87°C (Figure D-2, possibly a low estimate; see Shen and Eisenberg, 1966). Could their pure PVP actually have been a 10 or 20% water/PVP solution which glassed at 87°C? If so, the PVP/water glass contains more water at zero water activity than the desorption suggests (MacKenzie and Rasmussen, 1972). In addition, glass formation might occur over a wider range of temperatures than previously thought. For example, successful freeze drying must be at temperatures below or above what is called the collapse temperature (T_c, Figures D-1 and D-2; Tsourouflis et al., 1976). At and just below T_c in the freeze-drying experiment, PVP- or sucrose-rich domains remain as a glass skeleton from which the interstitial water is removed. This PVP- or sucrose-rich domain does not collapse on freeze drying because of its amorphous solid characteristics. The interstitial water is more easily removed because it is not yet glassy. All of this suggests that, unlike pure substances that glass at T_g, glass formation in two-component solutions occurs over a broader temperature range that might begin at T_c and progress through T_g and not be complete until the temperature was somewhat colder than T_g.

Glasses and Protection from Severe Dehydration

Some sugars that form high-temperature aqueous glasses can replace "bound" water and protect macromolecules, and membranes from denaturation (Chapter 11). These include trehalose, sorbital, fructose, sucrose, and glucose, all of which form high-temperature aqueous glasses, some stable up to 90°C. As such they may stabilize hydrophobic bonding, prevent salting out, and attenuate adverse effects of pH changes. In addition, because of the low vapor pressure, water in an aqueous glass escapes, coming to equilibrium with the outside environment so that considerable water is retained even at very low water activity. Cell and tissue spaces might remain filled with the aqueous glass if there was sufficient solute (glass forming sugar) present. Complete dehydration could then be avoided at temperatures below the melting point of the glass (possibly below 70°C). In addition, reduced diffusion of even small molecules in a glass at warm temperatures (50°C) must stop essentially all chemical reactions and ensure complete dormancy. Finally, glasses have no thermal transitions below T_g and thereby could provide protection from temperatures between 0°K and +90°C.

Do Aqueous Glasses Occur in Anhydrous Organisms?

Hirsh et al. (1985) have provided good evidence for the occurrence of aqueous glasses below −28°C in some of the plants most resistant to freezing. Their evidence was based on the absence of ice crystals in freeze-fractured samples and on the direct observation of second-order transitions in the differential scanning calorimeter. They describe the glasses as a "novel" method of cryoprotection; however, the method is probably quite common among the freezing-resistant plants.

For anhydrous organisms, one might begin to look for glass transitions by differential scanning calorimetry at moderately high temperatures (50°C). T_g should depend on water content, with lower T_g as water content increases. One might also look for the large quantities of solute necessary for an aqueous glass.

REFERENCES

Franks, F. 1982. Water in biological systems at low temperatures. *In* Water: A Comprehensive Treatise (F. Franks, ed.). Academic Press, New York, pp. 215-338.

Franks, F. 1985. Biophysics and Biochemistry at Low Temperatures. Cambridge Press, Cambridge.

Hirsch, A.G., Williams, R.J. and Meryman, H.J. 1985. A novel method of natural cryoprotection. Plant Physiol. 79: 41-56.

Kauzmann, W. 1948. The nature of the glassy state and the behavior of liquids at low temperatures. Chem. Rev. 43: 219-256.

MacKenzie, A.P. 1977. Non-equilibrium freezing behavior of aqueous systems. Phil. Trans. R. Soc. London B. 278: 167-189.

MacKenzie, A.P. and Rasmussen, D.H. 1972. Interactions in the water-polyvinylpyrrolidone system at low temperature. *In* Water Structure at the Water-Polymer Interface (H.H.G. Jellinek, ed.). Plenum Press, New York, pp. 147-171.

Shen, M.C. and Eisenberg, A. 1966. Glass transition in polymers. Prog. Solid State Chem. 3: 407-480.

Tsourouflis, S., Flink, J.M. and Karel, M. 1976. Loss of structure in freeze-dried carbohydrates solutions: Effect of temperature, moisture content and composition. J. Sci. Food and Ag. 27: 509-519.

Index

Page numbers in italic type indicate illustrations. Those followed by "t" indicate tabular material. "Pl." refers to the color plates, found between pp. 236 and 237.